AGRICULTURAL ECONOMICS AND MANAGEMENT

Kenneth L. Casavant
Washington State University

Craig L. Infanger
University of Kentucky

Deborah E. Bridges
University of Nebraska

Prentice Hall
Upper Saddle River, New Jersey

Library of Congress Cataloging-in-Publication Data
Casavant, Ken.
    Agricultural economics and management / Kenneth L. Casavant, Craig
  L. Infanger, Deborah E. Bridges.
      p.   cm.
    Includes index.
    ISBN 0-13-660184-7
    1. Agriculture—Economic aspects.   2. Farm management.
  I. Infanger, Craig L.   II. Bridges, Deborah E.   III. Title.
  HD1433.C368   1999
  630'.68—dc21   99-16481
                                                             CIP

Acquisitions Editor: Charles Stewart
Assistant Editor: Kate Linsner
Production Editor: Lori Harvey, Carlisle Publishers Services
Production Liaison: Eileen M. O'Sullivan
Director of Manufacturing & Production: Bruce Johnson
Managing Editor: Mary Carnis
Marketing Manager: Melissa Bruner
Production Manager: Marc Bove
Interior Design/Compositor: Carlisle Communications, Ltd.

Cover Designer: Jayne Conte

© 1999 by Prentice-Hall, Inc.
A Pearson Education Company
Upper Saddle River, New Jersey 07458

Printed in the United States of America

ISBN 0-13-660184-7

Prentice-Hall International (UK) Limited, London
Prentice-Hall of Australia Pty. Limited, Sydney
Prentice-Hall Canada Inc., Toronto
Prentice-Hall Hispanoamericana, S.A., Mexico
Prentice-Hall of India Private Limited, New Delhi
Prentice-Hall of Japan, Inc., Tokyo
Pearson Education Asia Pte. Ltd., Singapore
Editoria Prentice-Hall do Brasil, Ltda., Rio De Janeiro

CONTENTS

CHAPTER 3   COST RELATIONSHIPS IN PRODUCTION   52

CHAPTER 4   MANAGEMENT DECISIONS: HOW MUCH TO PRODUCE?   72

Part II

MANAGERIAL TOOLS   257

CHAPTER 10 COMPETITIVE DECISION MAKING   259

CHAPTER 11 NONCOMPETITIVE DECISION MAKING   274

PREFACE

This book provides an introduction to economic analysis in a management environment. Although arising from the discipline of agricultural economics, the approach of the book is to emphasize management problem-solving techniques under various situations, both agricultural and nonagricultural in nature. The objective of the book is to make the student aware of the need for and usefulness of economic decision making and managerial tools in many problem areas.

This book is designed both as an introductory text for students of agricultural economics and management and as a text for those students who, in the course of their academic learning, have occasion to take a course in agricultural economics. Exposure to the business side of agriculture is being required in more and more curricula as the usefulness of the problem-solving nature of agricultural economics is being realized by other academic majors. The writing style is designed to introduce the student to the basic terminology and concepts in a relaxed fashion in the early chapters, with a more technical writing style used in the later chapters after basic concepts have been mastered.

The material in this book draws heavily from the principles of basic microeconomic theory in identifying the problem-solving methods in management. The basic decision-making procedures embodied in management theory are then strengthened by the addition of various pragmatic, managerial concepts. This practical, applied approach to problem solving serves throughout the book to make economic analysis a proven tool rather than a theoretical exercise.

This theme of practical decision making is carried throughout the book, especially as the environment surrounding the manager's decisions is examined. The latter section of the book emphasizes the changing environment—business, political, social—that makes the objectives of good management more difficult to attain and makes improvement in management expertise even more critical.

The book is divided into three parts, allowing a systematic presentation of concepts and tools, applications, and the new management environment.

PART I

This part is concerned primarily with presentation of the traditional theory of the firm, using selected commodity and nonagricultural examples. Prior to the presentation of basic economic principles, the raison d'être of the manager is examined.

This "need" for a manager and managerial tools is emphasized throughout the book. The latter part of this section introduces the individual consumer's decision-making process, supply-and-demand concepts, and the price determination process in the capitalistic economy. Real-world economic adjustments experienced in our economy serve as teaching devices, for example, Russian wheat sales, food prices, scarcity of consumer products, and so forth. The presentation is couched in terms of both producer and consumer experiences because, at this stage of academic learning, most of the students' experiences have been those of consumers.

PART II

This part of the book relaxes some of the assumptions specified in the previous presentation of marginal analysis. Here the pragmatic side of economic analysis is reinforced. Problems of competition, market power, time, risk, and uncertainty are presented and analyzed from a management perspective. Available managerial tools and approaches are used in applied problem situations.

PART III

The decision-making environment of the manager and how this environment affects managerial decisions is the theme of this part. Alternative managerial options under government agricultural policy, under environmental pollution controls, and within agricultural marketing are the focal points of this part. The linkages between the macroeconomy and agriculture and the importance of international trade to agriculture are also presented in this section.

REMEMBRANCES

To my wife, Dorothy, and my daughters, Michele and Colette; we all continue to learn together, forever.   *Ken Casavant*

To my colleagues and students who make me a better teacher and to my wife who makes me a better person.   *Craig Infanger*

To my husband, Tom, whose faith and confidence in my abilities continues to sustain me; our journey continues.   *Deborah Bridges*

A special thanks to Pat Beintema, whose typing assistance came at a critical time and is much appreciated.

The authors would like to thank our reviewers, William M. Park, University of Tennessee; John Fox, Kansas State University; Timothy A. Woods, University of Kentucky; and James G. Beierlein, The Pennsylvania State University. We have attempted to incorporate their helpful suggestions and to heed their insights.

PART I

TOOLS FOR DECISION MAKING

CHAPTER 1

Management, Economics, and Decision Making: An Introduction

TO BEGIN

The idea of management is not new to most of us. In our own experiences we either consider ourselves managers, know a manager, or feel we understand something about management. But there is a logical sequence underlying management that most of us have not examined in any specific fashion. Consider this general statement: **Management** consists of making decisions, decisions based on the ideas of choice between alternatives; and the mechanism by which choices are made is often, if not always, based on economic analysis and evaluation. The manager, then, is the decision maker who chooses among alternatives based on some form of economic reasoning (typically by comparing the benefits to the costs). The manager need not have a titled position in a business firm. A manager can be anyone who participates in and affects the decision-making process, regardless of how informal and unstructured that process might be. Managers include a nine-year-old trying to decide which candy bar to purchase on a Saturday afternoon, a high school senior evaluating alternative universities or jobs, a baseball coach deciding who the starting pitcher will be, a farm manager developing a cropping pattern for the coming year, or an instructor in an agricultural economics class deciding on final grades. All of these people are managers, all are engaged in a decision involving choice, and all are utilizing some form of economic evaluation.

Because we are continually making decisions as managers it becomes useful and maybe even necessary to study management and economic analysis. Everyone uses economic reasoning to some degree when making choices. Some of us abuse that same economic analysis by using inappropriate objectives, inaccurate information, or simply bad judgment. Some decision makers are better at management than others, and their salary level, job responsibility, and career path often reflect that superiority. Even when the decisions are not concerned with monetary rewards, questions such as "Where shall we live?" or "Where should we go for a vacation this year?" or "Should our community consolidate schools?" require the weighing of alternatives and choosing among them. Our nation also relies on management rules or tools when making decisions about pollution controls, rate of growth, agricultural trade, housing standards, wilderness protection, and related issues through the democratic political system.

The role of management and decision making in agricultural economic analysis is one of directing resources at both the firm and societal levels. In agribusiness, good management allows the objectives of the firm to be met in as complete a fashion as financial and managerial capabilities will allow. The discussion of management in this book is oriented toward helping the individual decision maker improve his or her performance. In the following sections of this chapter we indicate the interrelationships among management, economics, and agricultural economics, and look specifically at the decision-making environment affecting managerial activities.

WHAT IS ECONOMICS?

Economics has been defined in several different ways, with the definitions ranging from the very simple to the verbose. This is the definition of economics that we like to use: **Economics** *is the allocation of scarce resources between competing ends for the maximization of those ends over time.* Other definitions state that while this allocation is taking place provision must be made for maintaining and modifying the system of choice. In this book, because we are looking specifically at the management aspect of economic analysis, we will ignore the additional qualifiers from other definitions.

Before this definition can be of any educational use, we need to understand what it is really telling us. From this formal definition of economics we can pull five elements that constitute the underpinnings for the discipline of economics:

1. **Allocation:** Making decisions about how to use our resources or capabilities.
2. **Scarcity:** If we did not have scarce resources in a firm, family, or society we would have no need for allocation (or managers).
3. **Unlimited wants:** The most basic assumption in economic analysis is that each individual has a desire for more, more, and more. The human being is truly insatiable. This principle gives rise to the economic problem of scarcity.
4. **Goals or objectives:** Each economic decision must point toward a specific objective. These objectives vary by person, firm, and economy. Two basic objectives of the firm are profit maximization and/or cost minimization, and consumers want to maximize well-being.

5. **Time:** Many writers include time as an ingredient in economics. Time allows differing courses of action to occur and goals to vary during the decision process, reflecting the dynamic system surrounding the decision maker and his or her shifting value system.

Resources are mentioned several times in the preceding discussion. But what is a resource? A definition we like to use is this: *A* **resource** *is an input provided by nature and modified by humans using technology to produce the goods and services that satisfy human wants and desires.* Resources are also called *inputs* or *factors of production.* Resources have three important characteristics:

1. Resources have economic value: Producers generally must pay to use resources.
2. Their supply is limited: Because the supply of resources is limited, the goods and services produced from those resources are also limited.
3. Resources have alternative uses: Because resources have alternative uses, trade-offs must be made.

As mentioned earlier, unlimited wants give rise to economic problems of scarcity since production of goods and services is unable to fulfill all of the wants and desires of consumers. Scarcity also brings up the economic issue of the *distribution of goods and services.* That is, how should goods and services for consumption be distributed among various persons and groups in society? How should the returns (from selling those goods and services) be distributed among the factors of production (the old labor versus management argument)? Economists spend a great deal of time and effort trying to answer these questions for society.

Another economic issue is **value.** The value of a good has something to do with the desire to have it. But value is not merely a monetary measure, it is also societal (i.e., designer clothes, cars as status symbols, etc.).

Because resources are limited, the goods and services produced from those resources are also scarce, which means consumers must make choices or trade-offs between different goods. Resources can also have alternative uses, so producers must make choices and trade-offs to decide how a particular resource will be used to produce different products. A measure of these trade-offs is **opportunity costs,** defined as *the cost to an individual or firm of using a good for one purpose that is equal to the value that the good could have earned in another use, its next best alternative*—or "what you are giving up to do what you are doing."

A good or service is scarce when we must give up (sacrifice) some amount of one good to obtain some of another good or service. As we make our choices in the face of scarcity, costs are generated. These costs are called *opportunity costs* because they constitute the value of alternative opportunities foregone or sacrificed (i.e., the wage a firm pays to a worker is equal to what that worker could earn in his or her next best job opportunity). Economic decisions, or choices, are based on opportunity costs. In effect, these choices are management decisions. Thus, economics (management) is essentially a reasoning method that compares the benefits (income or other desired outcome) resulting from an action with the sacrifices (costs) of that action.

The process the individual decision maker follows to attain his or her goals has a corollary at the national level as the economic system seeks to satisfy basic societal needs. Some questions an economic system needs to answer are these:

What to produce? That product most desired in society has a higher price, and hence is an incentive to producers.

How to produce? The more readily available resources or technologies have a lower relative price, causing the producer to increase usage of them for producing the product.

Who should receive income from sales? How should the returns from production be distributed among labor, management, capital, and resource owners, who helped produce it?

When to produce? In providing for maintenance and progress as well as in rationing our scarce resources, price varies, in different time periods, to reflect societal desires regarding savings, investments, and growth.

In the U.S. economic system, price is the "director," seeking to allocate resources in answering these questions. The reactions of individuals to price signals become the reactions of the economy. This role of price is not a characteristic of all societies but the same questions must still be answered. In a "command" society such as Cuba, the role of price as a director is augmented by "orders" or "quotas" as the governing body seeks to achieve its goals. In a "traditional" society such as the rice farming regions of rural Asia, the answers to these questions are dictated by hundreds of years of experience in rice production under almost unchanging production techniques from generation to generation. Although we can question the efficiency, progress, and equity of such different systems, we still must concede that each system has succeeded in answering these same basic economic questions.

Microeconomics Versus Macroeconomics

Economics is divided into two broad categories: microeconomics and macroeconomics. **Macroeconomics** deals with the aggregate results of individual decisions by examining growth, savings, investment, inflation, unemployment, and so forth, at the national level. Macroeconomics encompasses the performance of the national and the international economies; it is the study of what determines the total quantity of goods and services produced and the distribution of income within an economy. On the other hand, **microeconomics** deals with individual decisions of production and consumption; it is the study of economic decisions and behavior at

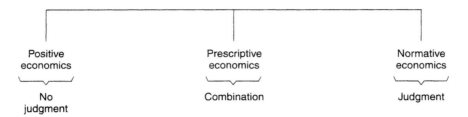

FIGURE 1.1 Classification of economic questions.

the individual producer and consumer level. In microeconomic analysis we usually assume that factors outside the immediate topic of study (such as prices of other goods and income) do not change.

Economists tackle many questions that are important to society. Some questions require value judgments to be made, whereas others deal with estimating a quantifiable relationship between selected variables. Economists have developed a classification scheme, presented in Figure 1.1, for these questions.

Positive economics addresses questions that do *not* involve a judgment (or opinion) in comparing the values that different individuals place on an option. (What is the impact of an increase in the target price on wheat production? What is the effect of an increase in the price of fertilizer on corn production?) These questions typically involve estimating quantifiable relationships between different variables.

Normative economics examines questions that require someone, or some group, to decide what is "good or bad," "fair or unfair," or to develop some standard for determining what is "good or fair." (Should government policy guarantee a "fair market price" for grain production? Is saving the salmon worth the economic impacts on the Pacific Northwest?)

Prescriptive economics addresses questions that deal with ways to achieve a desired result in the most efficient, profitable, or acceptable manner. Prescriptive economics identifies alternative ways to reach a goal and provides methods for choosing among them.

In this text we will study the economy by looking at the firm as an individual economic unit. Further, the questions examined in this text deal with questions that are usually classified as positive economics. This approach emphasizes choices producers should make in order to increase profits and allows us to learn about management tools (such as marginal analysis) necessary to make that profitable decision. We also learn about the national economy by using the micromanagement approach because understanding how managers will react to social and economic changes helps us predict how the industry will react. For example, if wheat allotments become part of the farm program again or if support prices are raised, watching the individual farm manager's response can give us clues as to what will happen to total production.

AGRICULTURE AND ECONOMICS

Many agricultural economics textbooks have detailed the extent and importance of agriculture in an economic system. From the viewpoint of an economic system and the perspective of a manager, agriculture is only one industry within the overall economy, not *the* industry as suggested by some writers. The same economic questions must be answered; the same management decisions must be made. Agriculture does deal with one of the basic human needs, food, and has contributed significantly to the ability of our country to answer the economic questions. These characteristics make agriculture important to any economic system but not to the exclusion of other sectors of the country. This book emphasizes agricultural decisions but includes other applications as well.

Characteristics of Agriculture

Let's take a look at some of the characteristics of agriculture that make it important to the economic system of the United States and other economies.

In a sense agriculture has always been a "special" industry because it allowed the function of labor and management to be combined into one economic entity and one person. This allowed some relief from the class struggle existing in the European countries from which most of the early U.S. settlers came. This combination generated the "goodness" as ascribed by Thomas Jefferson to those who work the soil in a productive manner. Out of this combination of labor and management in one individual and the availability of low-cost, plentiful resources came a tremendous evolution in agriculture, an evolution that can be considered the success story of American agriculture.

At the time of the American Revolution 90 percent of the population was directly active in farming and 60 percent of a consumer's income was spent on food. In 1982 just slightly more than 3 percent of the labor force was active in farming, and only 17 percent of the consumer's income was spent on food. In 1994 2 percent of the labor force was active in farming, and slightly more than 11 percent of the consumer's income was spent on food. This gradual change allowed a release of labor to other growth industries in the U.S. economy and provided a demand for other products as the importance of food in the family budget decreased.

The agricultural sector that has evolved during this time period is extremely diverse. The weather, soil structure, and location have made different products profitable in different parts of our country. The size of farms also differ significantly, so significantly that it is common to refer to a dual agriculture composed of commercial and noncommercial farms. In 1993 those farms with annual sales of more than $250,000 comprised only 6.4 percent of the farms in the United States but represented or accounted for 56 percent of the market value of agricultural products sold.

Agriculture, as a sector, is important to the annual overall output of the economy, the gross domestic product (GDP). In 1994 about 14 percent of the GDP originated from agriculture, but only 2 percent came directly from farming or production agriculture. However, when you look at the jobs created and supported indirectly by agriculture, the role agriculture plays in the national economy becomes much more evident. In 1994 the agricultural services and input industry employed 0.8 million people; the processing and agricultural marketing sector employed 3.3 million people; and wholesale and retail establishments employed 14 million people. Overall, agriculture was responsible for providing approximately 21.6 million jobs, accounting for 15.8 percent of total employment in the United States in 1994.

The success story of agriculture's productivity is also indicated by the fact that in 1994 the 2.1 million farms produced enough food to feed 260 million Americans and still export 20 to 30 percent of production to the world markets. Crop output in the United States has more than doubled since 1940, even though the number of acres increased only 4 percent and labor utilized in crop production decreased almost 80 percent during the same fifty years.

The impacts of this evolution are identifiable, as suggested earlier. The number of people living on farms has shown an overall decrease, from 23.1 million in 1950 to about 8 million in 1982 to around 5 million in 1994. Farms are fewer but larger. Since 1950 farm numbers have dropped from 5.6 million to 2.07 million while average farm size has grown from just over 200 acres per farm to around 470 acres in 1994. The amount of money (capital) required to operate a farm has tremendously increased due to a steady rise in land prices, increased mechanization (tractors, combines, etc.), and an increased reliance on off-farm inputs. For example, since the 1930s fertilizer use increased fifteenfold and machinery (tractors and so forth) increased fivefold. In 1992 fertilizer purchases represented close to 15 percent of annual production expenses. Other significant off-farm purchased inputs include pesticides, herbicides, fuels and energy, credit, and labor.

The agricultural sector of the American economy has many facets; among them a dichotomy exists between production agriculture and the food and fiber industries (referred to as agribusiness agriculture) that serve farming by providing inputs or processing and distributing the products of agriculture. In production agriculture the family farm (where labor and management were so efficiently combined) is participating less and less in the volume of agriculture production. Farm tenure, or how land is held, is changing from full-owners to part-owners (part-owners own part and rent part of the land they farm). Full-owners now account for about 50 percent of the farmers, part-owners about 31 percent, tenants about 11 percent, and farm managers about 1 percent. However, the part-owner has control of more than 55 percent of the land in farms, the full-owner about 31 percent, tenants about 11 percent, and managers about 13 percent.

Nonfarm corporation farming (farms owned by corporations), which often uses managers, is still small in number of farm units and number of acres under its control but is increasing in importance, particularly in enterprises that are very capital intensive, such as irrigated crops and poultry or hog confinement operations. Much concern has been noted in agriculture about corporate farming (some states have attempted to outlaw it) but more and more family farms are turning to incorporation to take advantage of tax laws as a means of intergenerational transfer of land and capital.

Off-farm income is a more noticeable characteristic of farming each year, accounting for almost 50 percent of all farm income. This income comes primarily from off-farm employment but also from rent, dividends, social security, and pensions. Off-farm income to larger farms is usually in the form of dividends and interest.

Agriculture is still an important sector of the U.S. economy. Production agriculture may be decreasing in number of participants but the volume and value of its production have increased. Agribusiness agriculture is certainly a growth industry, important to the economy, and it is becoming an important source of employment and professional opportunity to future managers.

What Is Agricultural Economics?

A definition of agricultural economics is less specific than the formal definition of economics we reviewed earlier. Current definitional notations range from "applying economics to agricultural production" to "applying a social science to seek solutions to economic problems existing in agriculture." Some writers have suggested that agricultural economics should be concerned only with commercial agriculture, leaving other social aspects of rural existence to rural sociologists. The definition we use is based on economics as a discipline but broadens the application of that discipline:

Agricultural economics: Economics applied to agriculture and rural areas.

The rationale for including problems of rural areas in agricultural economics can be seen by taking a brief look at the evolution of the discipline. Farm management was the first area of concern of economists working in agriculture; it was soon followed by the more theoretical field of production economics. As researchers noticed problems in moving and marketing agricultural products a strong interest developed in improving the marketing of these products. Agricultural policy developed as a subset of agricultural economics, as researchers and governmental planners sought to solve the aggregate production–marketing imbalances existing in the 1930s and still present today. As the difference between a public-versus-private viewpoint of agriculture and rural problems was identified, the field of resource economics was developed from land economics. A closely related field, community resource development, examined the problems of rural people and communities as the rural to urban migration continued. Another subset in the agricultural economics discipline that has received a lot of attention over time is econometrics, the combining of economics and statistical-quantification techniques into specific economic models. A more recent area of interest has been agribusiness management, a field developed because the role played by these "off-farm" members of the agricultural sector has become important. So, the discipline of agricultural economics deals with both agriculture and a variety of other related fields.

Why do we focus on agriculture? Because agriculture has such a fundamental importance to society. For example, feeding the world is agriculture's responsibility, a responsibility that must be met using ever-increasing scarce resources. Agricultural production—crops and livestock—is inherently risky given the time lag between production decisions (i.e., planting) and realized production (i.e., harvest) and its dependence on land and water resources. For virtually every agricultural product, the form consumers desire it to be in (i.e., wheat bread, pork chops) is quite different from the product's form at the farm gate (i.e., wheat, hogs). This difference has led to the development of market institutions and mechanisms—cooperatives, food processing firms, etc.—that enable agricultural products to reach the consumers in the desired shape and form.

Agriculture refers to the complex system that begins with natural resources and involves farms, agribusinesses, and governmental organizations in providing products of the land to consumers. Agriculture's economic activities are organized across three broad categories: the farm sector, the agribusiness sector, and the public sector.

The farm sector encompasses farms and ranches, including all the firms that grow crops and livestock, generally for sale. This sector is identified by several unique characteristics. First, crops and livestock are living organisms, thus production is dependent on land and water resources and subject to the vagaries of the natural environment. Second, demand for these products is relatively insensitive to price changes at the retail level. Third, individual producers have little influence over prices. Finally, the consumption of most agricultural commodities does not keep pace with increases in income.

The agribusiness sector includes all the businesses that provide farm services and supplies, as well as the firms that process and market farm commodities. Fertilizer dealers, farm implement dealers, banks that provide credit, meat packing firms, and vegetable processors are just a few examples of agribusiness firms.

The public sector includes agricultural research institutions, primary and secondary education, higher education, extension and information services, and government services. Agricultural colleges at state universities, the U.S. Department of Agriculture (USDA), and agricultural programs in local high schools are just a few examples of the public sector.

A continuing theme of this book is our emphasis on the management aspect of each of these subdisciplines (farm management, marketing, agricultural policy, etc.) and subsectors (farm sector, agribusiness sector, etc.). This allows you, the student, to develop an awareness of the many problem areas or situations in which the economic-managerial tools can be successfully applied. The tools of economic analysis as applied to agriculture and related areas are common to each, and they will serve as a good "classroom" through which you can become managers.

MANAGEMENT: ART OR SCIENCE?

Earlier we said management could be defined as making decisions about choices facing the decision maker, with the decision being based on economic analysis. Other definitions have been developed over time, ranging from "a concern for the organization and the running of daily operations in achieving identifiable goals" to "making decisions affecting the profitability of a business" or "leadership and control of resources." In making these decisions the manager or decision maker draws on much information and knowledge available from various science disciplines.

Consider these disciplines and some examples of how all decision makers might use them.

Philosophy. This discipline includes philosophic value theory, logic, ethics, and beliefs of individuals and groups. The manager uses this discipline, for example, in deciding on objectives and reasoning as to the appropriate decision to be made.

Sociology. This field involves studying social systems, institutions, behavioral norms, and group interrelationships. A manager often considers the cultural and social habits of people he or she works with, buys from, or sells to.

Psychology. Psychology generally deals with attitudes, motivations, and behavior of people. Getting the most out of his or her employees or understanding how and why a customer makes a purchase is important to the decision maker.

History. This field involves reviewing and analyzing significant events that affect individuals, families, communities, institutions, and nations. Past experience, price patterns, and customer response to previous sales provide useful information when making better future decisions.

Political science. This discipline is concerned with the structure and operation of political and governmental institutions. Management must be aware of how government constraints affect their decisions—for example, pollution controls—and must be aware of what services are available from government and how to work with political groups to receive the services desired.

Law. This is the science that deals with the body of legal customs, practices, and laws governing individuals and groups in a society. A decision maker should be aware of the legality of alternative actions, how this legality has been determined by society, and, where possible, how these laws can be modified by citizen initiative.

Economics. This field involves the allocation of scarce resources to satisfy human wants. Managers use economic tools and logic to decide between alternatives—whether monetary or social.

Management uses elements of each of these and other disciplines as it performs assigned tasks. Because each science can be considered a specific self-contained body of facts and theories, management must then be considered an art rather than an individual science, since no one science has been developed that gathers all of the needed qualities of each discipline into a science of management.

In this book we borrow from and use all of these disciplines, but we stress the economics of management. Much of our discussion deals with the "rules of management" or "decision-making criteria" or "economic theory." The theories are simply abstractions from reality or simplified versions of complex situations used as teaching and learning devices. They can be fun and useful, especially when continually applied to real examples of management decisions on-farm and off-farm. These decision rules fit a broader palette than just business-firm decision making so they can be used in many social situations as well. However, we use business-firm (farm and nonfarm) problems here because they are easy to understand and many real examples are available to us.

Management of a firm involves making specific decisions as the production process is undertaken. These producer decisions include the following: What should be produced? How much should be produced? How should it be produced? When should buying and selling occur? Where should buying and selling be done? These questions or decisions provide a useful framework for learning some economic logic and rules in the following chapters. Please note that these individual producer decisions are very similar to the response of any economic system to the larger questions it has to answer, as was discussed earlier in this chapter. When we add individual producer decisions together, we have an economic system attempting to answer those necessary questions.

The decisions made by individual producers are greatly affected by the goals chosen by the decision makers. Whether the goal is profit maximization or highest wheat yield per acre (as you will find out later, these goals are not necessarily the same), a goal can only be met through proper management. In fact, if not for the variation in goals and the risk and uncertainty associated with decision making caused by lack of complete knowledge, there would be no need for the good manager. All managers would make identical decisions.

Managerial Economics

We should mention a "new area" that has been developing, the area of **managerial economics**. Some writers define it as business economics, and others as economics applied to practices of businesses. Still others see it as the use of microeconomic theory, augmented by accounting, finance, and quantitative technique, to improve the performance of a firm. It really does not seem to us that this is anything new to the economics or agricultural economics disciplines, but that this field is rather economic analysis in a management environment! Our approach in this book is to emphasize management problem-solving techniques for both agricultural and nonagricultural situations, thus helping you, the student, become aware of the need and usefulness of economic decision making and managerial tools in many different problem areas. So—maybe we are teaching managerial economics?

DECISION-MAKING ENVIRONMENT

Management consists of making decisions. But how are those decisions made? What comprises a decision? Let's examine the elements comprising a decision and the steps taken in analytical decision making.

Elements in a Decision

Many authors and textbooks have examined the decision-making process and have identified the "elements" comprising a decision. Let's look in a somewhat abstract manner at just what or who comprises this concept called a "decision."

The Decision Maker

The decision maker is normally a manager who chooses among alternative courses of action. The manager profits from good decisions and must accept the problems associated with making bad ones.

Objectives or Goals

To evaluate alternative courses of action, the manager must have a clear idea of what is the objective or goal of the decision. In this book and in most business decisions, profit maximization is assumed to be the intended objective of management decisions unless otherwise stated. You should realize that profits are only a derived goal, since they are intended to be used to pursue satisfaction and happiness. These at times can be conflicting, as in the case of the family farm. Farming is unique to business organizations in that it contains a business firm and a separate household within its structure. The goal of the firm component in the farm is typically profit maximization, whereas the household as a family unit is concerned with happiness as a goal. For example, on a Saturday afternoon when paying attention to livestock might enhance the salability of the animals and, hence, profitability, the farm family might prefer to go fishing and boating at a nearby lake. This causes a conflict of objectives. (We hope this conflict will be resolved in favor of an afternoon at the lake—professors aren't always concerned with profits!)

Conditions Facing the Decision Maker

The choices available to the decision maker depend on what information is available. If a small farm or business is affected by limited financial solvency or if another form of resource control exists, then the manager is obviously constrained by the available options. Most managers also have little direct control over institutions, the law, or the distribution of wealth in the country. Further, in most agricultural commodity markets, the manager has no control and, in some cases, no information about price movements or projections. Individually he or she cannot affect the demand for the products or the supply of inputs needed for the firm. Finally, the decision maker is faced with a continually changing business and social environment under which decisions must be made.

It is the ability of the manager to assess correctly the conditions facing the firm and use them in achieving the identified goals that sets managers apart.

Rewards flow to the manager who, rather than simply bending to or complaining about the conditions, evaluates those conditions and uses them as part of his or her decision framework.

The Measuring Stick

Just as a high jumper needs to measure his or her athletic performance, so the manager needs a device to choose among actions and then evaluate the appropriateness of that action. In economic analysis this measuring stick is **efficiency,** defined as *valuable output divided by valuable input.* Any increase in this ratio caused by one course of action compared to another gives the manager the ability to decide between two actions. In this textbook we assume that, other things being equal (*ceteris paribus* is the formal term), the most efficient course of action will be preferred. We also point out that even when the goal is not profit maximization, better decisions are made if efficiency evaluation is part of the analytical process.

Steps in an Analytical Decision

The decisions made by managers are influenced by goals and objectives. As a decision maker examines alternative courses of action to attain those goals and objectives, the process of management proceeds. This process has been evaluated by many farm management economists and academic researchers with varying numbers of steps included in the process. It has also been evaluated under the titles of "functions of management," "the scientific method," "the process of management," and "the sequence of decision making," among others. Many authors have examined the decision-making process and have suggested the steps used in making a decision. For the most part, the steps involved are similar in approach, and vary essentially in how steps are combined or sequenced. We combined the core of those approaches with a bit of the scientific method of research to arrive at the following steps in an analytical decision:

1. Get ideas and make observations. Formalize the problem and then formulate hypotheses of alternative actions.
2. Analyze observations, including reformulation of problems and ideas concerning their solution, further refining of the problem, and testing of the hypotheses.
3. Make decisions.
4. Take managerial action.
5. Accept responsibility for actions. Evaluate and learn from the outcomes, whether positive or negative.

You should be aware that these steps we have developed are quite interdependent and are commonly combined when faced with a particular decision. Do not memorize the steps, but do analyze them as to the sequence and logic associated with them. Former students of ours have come back from jobs in farming or agribusiness and commented on how often they use this series of steps, no matter what they are called.

BOOK APPROACH

In this text we study the economy by looking at the firm and consumer as individual economic units. This approach emphasizes the choices producers and consumers should make in order to achieve their economic objectives and allows us to learn the management tools, such as marginal analysis, necessary to make appropriate decisions.

Producer Versus Consumer Perspective

Economic analysis is often divided into production economics and consumption economics as reflecting the supply and demand components in the marketplace. They are not separated in this textbook because each individual, producer and consumer, serves as manager in his or her own decision process. We will show that economic tools of analysis are useful to both producers and consumers as they deal with similar budget constraints and objectives. Further, most individuals will find themselves in both roles in the future so dual application of the managerial tools covered in subsequent chapters is totally logical. As we proceed through that discussion we point out the difference in goals, information, and resources available to each of these actors in the marketing system. The bottom line, though, is that each producer and consumer exists only because of the other.

Useful Assumptions

Economics is concerned with overcoming the effects of scarcity by improving the efficiency with which scarce resources are allocated among their many competing uses, so as to satisfy human wants. The study of economics deals with questions such as these: What will be produced from limited resources? How will it be produced? For whom will it be produced? When will it be consumed? However, the real world is extremely complex. To facilitate answering these and other questions, economists generally make several simplifying assumptions in order to understand the economic relationships acting in a complex world. These simplifying assumptions follow:

1. Individuals want to maximize well-being.
2. Firms want to maximize profit.
3. "Perfect knowledge" exists, meaning that the manager knows with certainty what prices, yields, and costs will be. Thus, decisions are made knowing all pertinent facts.

4. "Pure competition" exists, meaning that there are many buyers and sellers of an identical product and no one is big enough to influence price. Because individual firms and consumers have no influence on price, individuals are price takers in the market. (We discuss this assumption in more detail in Chapter 10.)

5. The level of technology or "state of the art" remains the same, meaning no new products or innovations occur during the period in which a decision is being made.

6. Inputs are used efficiently (inputs are used only in Stage II as explained in the next chapter) meaning that efficiency in input use is a prerequisite for profit maximization.

These assumptions are used to allow the economist to focus on the problem at hand. Although some of these assumptions are not always realistic, they have validity in most situations and are essential to the study of economic issues. As we progress through the text, we examine the effects of relaxing these assumptions and what that means for price determination and market behavior. Use of these assumptions allows us to build a firm foundation of economic theory and provides a yardstick by which to measure real economic behavior.

Looking Ahead

In this chapter we discussed the major goal of this textbook—to learn to use economic tools applied to management problems. The close relationship between economics, agricultural economics, and management was developed. We saw how the questions a manager must answer are very similar to the questions an economy must answer. These discussions should have brought you to the point where you are willing, maybe even eager, to learn the management rules and tools that will aid you in making the right managerial decisions.

The next five chapters deal with three of the questions a manager must answer: how much, how, and what to produce. Chapters 2, 3, and 4 are certainly the most important chapters in this text because they introduce you to new economic terminology and lay the groundwork for most of the following chapters. When you have finished Chapter 4 you will know how to maximize profits to a certain extent. Chapters 5 and 6 will show how to maximize profits using alternative products and different ways of producing those products. Chapters 7, 8, and 9 look specifically at consumer demand for products and producer supply of products and also examine how market prices are determined and changed in our economy.

Part II (Chapters 10, 11, and 12) relaxes the assumptions and examines more realistic and complex problems faced by the manager. Competition and market power and problems of time, risk, and uncertainty are presented and analyzed from a management perspective. The last part of the book (Chapters 13 through 17) looks at the decision-making environment of the manager and how this environment has an impact on manager or producer decisions. Methods of extending the manager's capability and productivity are the main subjects of this part. Government programs in agriculture and environmental and institutional concerns are evaluated as to how they affect managerial decisions. The importance of international trade to agriculture and the links between macroeconomic policy and agriculture are also examined.

QUESTIONS

1. In his or her role as a manager, how is an agricultural economics instructor utilizing "economic evaluation" when assigning final grades?
2. Evaluate the following statement: "Air is readily available to everyone free; therefore it isn't scarce, and we need not manage our air."
3. What are the essential elements of economics and economic analysis? Are they different in agricultural economics? If so, how?
4. Give an example of how a farm manager might employ philosophic principles in his or her decision-making process.
5. Suppose you are a farm implement dealer. What conditions at the macroeconomic and microeconomic levels are likely to affect your role as manager?
6. Identify the steps in the decision-making process by examining a recent decision you have made. Use the same decision to identify the elements of a decision.
7. Would you consider management to be an art or a science? Why?
8. In the U.S. economy, price is the director of resources. Identify several countries where this may not be the case and indicate what takes the place of price.

ADDENDUM 1: GRAPHS ARE ONLY A TOOL

A graph is a picture or abstraction of how two "things" relate to each other. In economics we refer to these things as *variables* because they vary (change), and economists are interested in knowing or seeing the relationship between two variables as they change. This makes graphs an important tool in economic analysis. Examples of the relationships of interest to economists include the relationship between the price of apples and the quantity of apples supplied, the price of shrimp and the quantity of shrimp consumers demand, or the relationship between the level of fertilizer use and the yield of wheat per acre.

In this textbook, graphs are used as a teaching tool. It is important that you not only understand how to construct a graph and develop good graphing practices, but also how to interpret the graphs used in this textbook and in other economic analyses. Let's work through an example to illustrate how to construct and interpret a graph.

As mentioned, a graph is used to illustrate the relationship between two variables. Thus, prior to constructing a graph it is necessary to have information on the two variables whose relationship is of interest. This information can be presented either in written form, as a mathematical expression, or, as more commonly seen,

TABLE 1.1   Hours of Study and Score on Exam

Hours of Study	Score on Exam
0	20
5	30
10	40
15	50
20	60
25	70
30	80
35	90
40	100

in tabular form. Consider the relationship between the hours spent studying for an exam and the score received on the exam. Assume that the instructor prepares exams that are fairly difficult so that if a fairly bright student studies zero hours the student will only score twenty points. However, for each additional five hours the student spends studying for the exam, the student's score will increase by ten points. (The information is in written form.) This information about the relationship between hours of study and exam score is summarized in Table 1.1 in tabular form.

This information can also be summarized using a graph to show the relationship between hours of study and exam scores. The graph in Figure 1.2, like most graphs, has horizontal and vertical lines, called *axes*. The vertical axis is customarily labeled the *Y* axis, and the horizontal axis is customarily labeled the *X* axis. (Note that *X* and *Y* are simply labels in this context.) In graphing it is customary to plot the independent variable (the variable whose physical quantity is more readily controlled or which can be measured with the highest precision) on the *X* axis. This means that the dependent variable (the variable whose quantity is dependent or results from using the independent variable) is plotted on the *Y* axis.

In our example, the exam score results from the number of hours studied and the hours studied are under direct control of the student. Therefore, the independent variable is hours of study, so it is plotted on the *X* axis. The dependent variable is the exam score, so it is plotted on the *Y* axis. Each combination of hours of study and associated test score is unique and describes one point on the graph. Plotting these unique combinations yields the graph presented in Figure 1.3.

As shown in Figure 1.3, if the student does not study at all (0 on the horizontal axis) the student receives a score of 20 (20 on the vertical axis). Thus, 0 and 20 (0, 20) describes a point of 0 on the *X* axis and 20 on the *Y* axis. The next point (5, 30) describes a point of 5 on the *X* axis and a point of 30 on the *Y* axis. So each combination of hours of study and corresponding exam score describes a point on the graph in Figure 1.3.

After plotting each unique combination, we can complete the graph by connecting each point with a line. The complete relationship between the two variables, hours of study and test score, is now described in the graph presented in Figure 1.4.

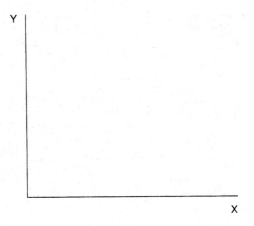

FIGURE 1.2
Graph axes.

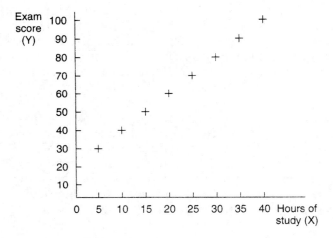

FIGURE 1.3
Relationship between exam
score and hours of study.

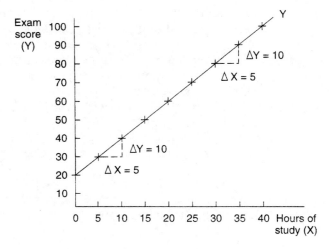

FIGURE 1.4
Relationship between exam
score and hours of study.

Before continuing our discussion on how to interpret the graph, you should note two important points. First, notice that the graph in Figure 1.4 is completely labeled. The axes are identified with the variable names (what is being measured) and the units of measurement. Second, the scale of the axes indicates the quantity (unit of measurement) associated with each distance unit, in this case 10 for the Y axis and 5 for the X axis. If the graph were not labeled and the scale not identified, it would be impossible to interpret the relationship illustrated in the graph. So always completely label your graphs.

Since the graph depicts the complete relationship between the hours of study and the resulting exam score, we can use it to answer any question we may have concerning this relationship (even if the information is not available in tabular form). What if you wanted to know what kind of score you would earn if you only had 20 hours to study? For the answer, simply go to 20 on the X axis (horizontal axis), then go straight up until you touch the line describing the relationship between the two variables, then go straight across to the Y axis and read 60 as the test score. So if you only had 20 hours to study for the exam, you would expect to earn 60 on the exam.

What if you wanted to know how many hours you would have to study to earn 90 points on the exam? For the answer, you would go to 90 on the Y axis (vertical axis), then go straight across until you touch the line describing the relationship between the two variables, then go straight down to the X axis (horizontal axis) and read 35 hours of study. So to earn 90 points on the exam you would have to study 35 hours.

Suppose you wanted to know how the points scored on the exam changed as you changed the number of hours studied. That is, you are interested in the slope of the line representing the relationship between the two variables. The slope of a line is defined as the ratio of change in the Y value to the corresponding change in the X value for any two points on a straight line (slope $=$ rise over run, or $\Delta Y \div \Delta X$). We can determine the slope of the line directly on the graph. If we go from 5 to 10 hours of study ($\Delta X = 5$) our exam score increases from 30 to 40 ($\Delta Y = 10$) so the slope of the line is 2 ($\Delta Y \div \Delta X = 10 \div 5 = 2$). The line representing the relationship between the number of hours studied and the exam score is linear (a straight line). Therefore it has a constant slope, or rate of change, of 2. That is, for every additional hour studied, the exam score will increase 2 points. Because the relationship between the hours studied and exam scores is linear we can represent this functional relationship mathematically using the simple linear equation $Y = a + bX$, where b is the slope of the line and a is the Y intercept term (the point on the Y axis where the line crosses the axis when X is zero). We know that if $X = 0$ then $Y = 20$, and that the slope is 2. So this relationship can be written mathematically as $Y = 20 + 2X$, or the exam score is equal to 20 points plus 2 times the number of hours studied.

The relationship between the exam score received and the hours studied illustrated in Figure 1.4 is positive. That is, the exam score received is positively related to the number of hours studied (the slope of the line is positive). Therefore, the more hours you study, the higher the exam score you receive.

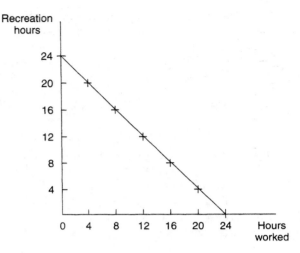

FIGURE 1.5
Relationship between hours worked and recreation.

It is also possible for two variables to be negatively related to each other. In this case the slope of the line would be negative. Consider the relationship between hours spent working and in recreation, as illustrated in Figure 1.5. As shown, the number of hours spent in recreation is negatively related to the hours spent working. If you work more hours, you have fewer hours for recreation. The slope of the line in Figure 1.5 ($\Delta Y \div \Delta X$) is negative (-1) and this relationship can be written mathematically as $Y = 24 - 1X$.

Graphs should not be confusing to you if you take the time to look at the two variables being compared and the values of these variables. They are a good teaching and learning tool and can be a shortcut to new understanding.

ADDENDUM 2: INFLATION AND INDEX NUMBERS

A characteristic of the economic system that we are studying—and, as managers, are working in—is the continuous changing of prices, production, and other information over time. It is difficult to evaluate differing prices, quantities, and the changes of each at different points in time in an attempt to evaluate progress or profitability. This section briefly discusses the source of much of this price movement, inflation, and then introduces the use of index numbers to evaluate price inflation and the indexing of other items of interest to the decision maker.

Inflation can be thought of as a general rise in the price of a substantial number of commodities produced in the economic system. More specifically, it is referred to as a rise in the general price level; as such it results in an overall drop in the real purchasing power of money. We should point out that inflation is an overall increase in the prices in an economy, as contrasted to increases in individual prices such as in fertilizer.

Inflation, until 1982 and early 1983, was a concern of producers and consumers alike as double-digit inflation, even as high as 13 percent, was experienced.

However, remember that inflation has been a consistent American problem for only the last twenty years or so. From 1950 to 1964 the rate of price rise in consumer purchases (defined later as the Consumer Price Index) was only 2 percent per year.

The causes of inflation depend heavily on the perspective of the viewer. A blue collar worker in an automobile factory may feel it is caused by the high profits realized by the automobile firm. The consumer may feel it is caused by the subsidized prices paid to the farmers. The farmer may feel it is the parasitic middleman and his gouging of prices . . . . Whatever the specific perspective, two general explanations exist for the causes of inflation. Cost-push inflation arises when costs of production factors, for example, wages of labor or energy, increase faster than their productivity for a firm; thus the firm is forced to increase the market price of the product being produced. Demand-pull inflation occurs when the quantity demanded from consumers is greater than the available supply of products. In this explanation the fear of future inflation can cause consumers to buy now, thus putting pressure on prices and causing future inflation.

A relevant problem associated with inflation is that it does not affect everyone equally. From the point of view of the agricultural industry, certain impacts can be seen. Even though it pushes up the prices received by farmers, it also pushes up the costs of inputs. These costs (machinery, fertilizer, and so forth) do not directly decrease in response to a later decline in farm prices. Second, demand-pull inflation causes farmers to purchase land and machinery now, before future price rises occur. This also puts pressure on input prices and causes producer demands for higher commodity prices. Inflation gives a competitive advantage in agriculture to those who already own most of their assets. These established farmers, whose land value is increasing, can bid away land from new or beginning farmers. This supports the trend discussed in Chapter 1 toward fewer and larger farms in American agriculture.

As managers and decision makers, we want to evaluate decisions even if costs and revenues are separated or changing over time. To do this, we require the ability to study differing prices and productivity.

Indexing is a mechanical means of allowing a relative comparison between items or groups of items over time. It is useful to students, teachers, analysts, and politicians; in fact, it is useful to all people who evaluate what is happening around them. For example, suppose your family's income has increased by 50 percent over a five-year period or suppose Cargill Grain Company wants to transfer an executive from Minneapolis to New York. In the first case, did the family's purchasing power really increase? In the second, what minimum increase in salary is needed to reflect the different cost of living in New York versus Minneapolis? Both cases require measurement of general price levels, one associated with inflation, the other with locational effects.

In its simplest form, an index number is an indicator; more completely it is a percentage that identifies the relationship between two numbers, using one of the numbers as the base for comparison. As indicated earlier, index numbers can be comparisons over time or between places and can be price, quantity, value, or special items. Indices are most commonly used for price analysis, and a price index can be used to evaluate change for a single commodity, for unlike units such as eggs versus apples, or to compare entire commodity groups.

The simplest index is for one commodity or item over time. It could be for the price per pound of coffee in 1980 ($3.60) versus 1997 ($6.00) or your grade point average (GPA) as a freshman (2.47) versus senior (3.64). Calculation of an index requires a decision about what time should serve as the base year, or reference point. The base year is assigned an index number of 100, and the calculated index numbers are interpreted in relation to the base year.

The simple index numbers for the price of coffee and GPA are illustrated in Table 1.2. This indicates that the price of coffee has increased 67 percent during the seventeen-year period. Also, your performance as a student, measured by the GPA, has improved 47 percent during the four years you have been in school.

The preceding example was for a simple index number; a more useful approach is to use a weighted index number that allows us to combine different items into one relative number for comparison. The weights used depend on the economic question being evaluated. Consider the prices of selected food items in two time periods and the quantity purchased in a typical month during the base period 1985 by a hypothetical consumer as shown in Table 1.3. This index measures the change in the total cost of a fixed list of goods. For example, the 154 figure indicates

TABLE 1.2 Simple Index Number

Year	Coffee Price	Index Number
1980	$3.60	$\frac{6.00}{3.60} \times 100 = 167$
1997	$6.00	

Year in School	GPA	
Senior	3.64	$\frac{3.64}{2.47} \times 100 = 147$
Freshman	2.47	

TABLE 1.3 Weighted Price Index

Food Item	1985	1997	Quantity Purchased in 1985	$P_{85}Q_{85}$	$P_{97}Q_{85}$
Bread	$0.96	$1.40	3	$2.88	$4.20
Bananas	0.44	0.50	1	0.44	0.50
Beer	2.14	3.50	7	14.98	24.50
Hamburger	1.52	1.99	3	4.56	5.97
				$22.86	$35.17

The weighted index is $\frac{\$35.17}{\$22.86} \times 100 = 153.85 = 154$.

that in 1997 it would cost 154 percent of what it cost in 1985 to purchase the same goods in the same amounts.

Some examples of common indices are the Consumer Price Index, wholesale price index, and the index of all prices received by farmers. The *Consumer Price Index* (CPI) is compiled by the Bureau of Labor Statistics and measures changes in prices of goods and services bought by families or wage owners. It does not indicate the change in purchase patterns except when an updated base year is chosen. The *wholesale price index,* also published by the Bureau of Labor Statistics, shows the general rate and direction of the aggregate of price movements at the primary market or where the first major commercial transaction takes place. The last example, the *index of all prices received by farmers,* is a weighted average of an index of value of crops and an index of livestock and livestock products. This index is often used in farm program considerations and evaluations of agricultural posterity.

CHAPTER 2
Physical Production Relationships

INTRODUCTION

Since most human activity is concerned with producing something, whether it is bread, wheat, profit, happiness, or knowledge, the best managers know how to combine goods and services (called *factors of production*) in order to achieve maximum efficiency in resource use. This process occurs at all farm, firm, and even national economy levels. As we saw in Chapter 1, three main questions are at the core of the production process:

1. How much to produce?
2. What to produce?
3. How to produce?

These questions are interrelated. Our discussion of the firm's decision-making process begins by examining the first question in greater detail.

At first glance, the managerial question of how much to produce seems quite simple, with the response being "as much as possible." However, as we will see later, this response is often incorrect, and the decision itself is anything but simple. The information on how best to undertake this decision-making process is extremely important to both decision makers and students of decision making.

As discussed in Chapter 1, the question of how much to produce involves determining the quantity of a product (good or service) that will simultaneously meet the desires or needs of consumers and the objectives of the manager. Underlying the decision of how much to produce is the physical production process that determines what and how much will be produced when factors are combined in different amounts. In other words, a large component of the firm's decision on how much to produce depends on the physical production possibilities of the firm, or what and how much of a product the firm is *capable* of producing given its technology and available inputs. Thus, the first step in answering the question "How much to produce?" is to examine the physical production process.

In This Chapter

The objective of this chapter is to introduce the student to the basics of the physical production process, thereby laying the foundation for the decision rules needed for proper managerial decision making and economic analysis. No prices or values are introduced initially, so this is not an economic question but simply one of understanding the basic production process and the relationships that exist between the factors of production and the resultant output. Prices of inputs and prices of output are introduced in Chapters 3 and 4 to complete the answer to the question "How much to produce?"

CONCEPTS AND DEFINITIONS

The principal decision faced by firm managers and operators is how to combine inputs, or factors of production, in order to achieve some objective. This decision process is important because the amount of production resulting from any activity is dependent on the manner in which factors are combined. A farmer combines fertilizer with water, wheat seed, labor, machinery, and land in order to produce wheat. These factors (or inputs) functionally become the wheat that is produced, and the manner in which the factors are combined determines how much wheat is produced. A student combines hours of study with tuition, room and board, books, and native intelligence in order to achieve a college education. These items consumed by the student, the factors of production, are transformed into that final educational product, with the quality (in effect, quantity) of education produced being determined by how the student combined the factors of production.

Farmers and other business operators manage numerous production processes; hence, numerous decisions about how to allocate resources to produce the goods and services sold in the market must be made. The analytical approach economists use to study this decision-making process, or firm behavior, is based on the following behavioral assumption about producers: *In organizing production, the basic objective of the farmer, or any business operator, is to maximize profits,* where **profits** equal the value of production sold less the cost of producing that output, or profits = revenue − costs. Thus, in studying firm behavior, economists assume that managers strive to maximize their returns (after costs) on production. However, as we will see in Chapter 4, maximizing profit does not necessarily mean maximizing output.

To simplify things, we will assume the firm's behavior is synonymous with the manager's decision-making process. Since the direction of the firm is dictated by the decisions made by the manager, the firm behaves as the manager behaves. This is not an unreasonable assumption for farming or other small business where a single manager is typical.

This analytical approach is also referred to as the *theory of the firm* since it examines the producer's decision process at the firm or farm level (recall that microeconomics deals with individuals). The decision rules derived from the theory of the firm represent the basic production rules that enable producers to make wise choices, and provide the framework, or are "simple handles to hold on to," as we learn how to analyze economic problems. The terminology we will be using often becomes an impediment to learning what is really a very simple, although abstract, way of understanding the productive process and the inner workings of a firm or person.

As we mentioned earlier, profits = revenue − costs, and producers are assumed to be profit maximizers. Thus, to answer the question "How much to produce?" completely, we need to examine the areas of physical production, costs of production, and revenue for the firm. We discuss costs of production in Chapter 3 and the profit maximizing decision in Chapter 4. In this chapter we focus on the physical production process, or what and how much will be produced when inputs are combined in different amounts. Before discussing the exact relationship between inputs and outputs, we need to define several new concepts.

Notations

It is conventional in agricultural economics to use letters to denote factors of production and products. In this book we will use the following words interchangeably:

X	Y
Resource ⎫ Input ⎬ Factor ⎭ →	⎰Production ⎨Output ⎱Product

Resource, input, and factor are all used to denote the goods and services utilized by a firm in the production process. These are usually denoted by the letter X. Production, output, and product are denoted by the letter Y and are the goals or end result of the production process. These terms are used interchangeably, and often any X term is used with any Y term to reflect the production process, for example, input–output or factor–product.

Assumptions Used

Our discussion of firm behavior begins by examining the physical production process in more detail. Note that even for a relatively simple production process, finding the best combination of inputs is a difficult task if all inputs are allowed to vary simultaneously. So to make the problem more manageable, we do what every good economist would do, we make several basic assumptions. These assumptions allow us to isolate the problem at hand in order to examine it in more detail. These are the assumptions we will be using:

1. The objective of the firm is to maximize profits.
2. The firm, or producer, owns, or can obtain, a certain set (or combination) of resources, some of which are variable, some of which are fixed.

As mentioned in Chapter 1, we are assuming that the firm, or producer, makes all decisions under perfect certainty. That is, the producer knows with certainty what prices, costs, and production outcomes will be. The use of these assumptions allows us to examine the physical relationships between the factors of production and the resultant output. As we will see later, these physical relationships are the underpinnings of many economic decisions.

Definitions

We are assuming that the firm has a certain, or given, set of inputs to use in the production process. These inputs are considered either fixed or variable, and the manner in which these inputs are classified is very important.

Fixed factors of production are factors that have to be maintained in the short-run even if production is zero. Land, buildings, and factory equipment are all examples of fixed factors. As the name implies, the level of fixed factors of production (i.e., acreage, building size) remains constant regardless of output level; however, the intensity of their use can change (i.e., double-cropping acreage, double shifts in a factory).

Variable factors of production are factors that vary as output levels change; that is, as the use of variable factors changes, output levels also change. Seed, fertilizer, and number of employees are all examples of variable factors.

The designation of an input as variable or fixed depends on the time horizon of the firm. What do we mean by the time horizon? Economists usually use three categories to classify a time period: immediate short-run, intermediate short-run, and the long-run.

The **immediate short-run** is a time span so short that no resource changes can be made. In the immediate short-run, meaning right now, all factors of production are fixed. For example, at the beginning of a lecture your supplies (pens, paper, etc.) for class are fixed; at harvest time the amount of grain to be harvested is fixed. In the immediate short-run nothing can be done to change the quantities of resources used or the amount of product to be marketed.

The **intermediate short-run** (or **short-run**) is a time span such that some factors are variable and some factors are fixed. In the short-run, use of some resources can be altered while use of others remains fixed. For example, during a single growing season a wheat farmer can vary the amount of fertilizer and irrigation water (variable factors) applied to the planted acreage (fixed factors) to produce wheat. The length of time considered to be the short-run varies depending on the production process (i.e., short-run for a livestock operation is different than the short-run for a wheat farm). Since a manager, or firm operator, seldom has perfect control over all the resources necessary for his or her production process, most economic decisions occur in the short-run. Hence, the short-run is where we will conduct the vast majority of our economic analysis.

The **long-run** is a time span such that no inputs are fixed; that is, all inputs are variable. The long-run allows sufficient time for the owner or manager to change the use of any of the inputs. For example, in the long-run a corn producer can buy more land, a tobacco grower can build a new barn, or a livestock producer can build a new manure handling system. The primary distinction between the long-run and any other shorter time period is the presence of fixed factors of production.

Recall that the question facing the firm or manager is "How much to produce?" So the manager must decide which combination of variable inputs is best to combine with the available fixed inputs in order to maximize profits (the manager's objective). To examine the firm's decision-making process, we utilize the concept of a production function.

THE PRODUCTION FUNCTION

The objective of the producer is to maximize profits. Thus, the manager must decide how to combine the available inputs in the physical production processes of the firm so as to achieve this objective. At the heart of the production process, the "reason to exist" for the firm, is the concept of a **production function,** defined as *the technical relationship between inputs and output indicating the maximum amount of output that can be produced using alternative amounts of variable inputs in combination with one or more fixed inputs under a given state of technology.*

The production function is a mathematical relationship describing the way inputs and outputs are physically correlated. It is based on the idea that a unique and identifiable relationship exists between outputs and inputs, or stated another way, that the amount of output from a production process depends on the amount of input used in that productive process. The production function can be expressed in several ways: tabular, graphical, or mathematical. Each depicts the same information but in different form.

Notationally a common, or general, form for the mathematical relationship is

$$Y = f(X_1, X_2, \ldots, X_n)$$

where Y is the output resulting from the production process, the X's identify the different inputs used in the production process to produce Y, and $f(\ )$ represents the mathematical relationship between the inputs and output, called the production function [where the symbolism $f(\ )$ means "results from" or "is a function of"]. Notationally, a vertical line ($|$) is used to distinguish between the variable and fixed factors of production in the representation of the production function. For example,

$$Y = f(X_1 \mid X_2, \ldots, X_n)$$

where X_1 is the variable input (i.e., fertilizer, number of employees, etc.) and X_2 to X_n are the fixed factors of production (i.e., land, buildings, etc.). This is referred to as a production function with a single variable input and, as we will see later, is very useful for describing how production levels respond to different input levels.

To illustrate the three ways of expressing a production function, consider the following example. A Washington wheat producer knows that if she doesn't apply any fertilizer her wheat yield will be 20 bushels per acre, due to the existing nutrients in the soil, and that for every pound of fertilizer applied wheat yield will increase by 0.4 bushels.

Mathematically, this production function can be expressed as:

$$Y = 20 + 0.4X$$

where X is the pounds of fertilizer applied per acre, and Y is the bushel per acre yield of wheat. This equation states that when no fertilizer is used ($X = 0$) output (Y) will be 20 bushels per acre, and for each additional pound of fertilizer applied output will increase by 0.4 bushels per acre. In this case, fertilizer is the variable input and the intercept term, 20, represents the fixed factors of production.

The mathematical expression of a production function allows us to create a **production schedule,** which is essentially a tabular summary of the input–output relationship. The production schedule for our example ($Y = 20 + 0.4X$) is presented in Table 2.1.

To find the output resulting from each input use level, we simply plug in the input use level (i.e., $X = 30$) into our production function ($Y = 20 + 0.4X$) and solve for Y (when $X = 30$, $Y = 32$). From the production schedule in Table 2.1 we can see that if the farmer uses 0 units of X, wheat yield is 20 bushels; if the farmer uses 50 units of X, wheat yield is 40 bushels.

The information from the production schedule can be plotted to graphically, or visually, display the production function. The graphical illustration of our example ($Y = 20 + 0.4X$) is shown in Figure 2.1, where the functional form of our production function can be described as linear (or a straight line). Recall from your algebra that a linear functional form is expressed as $Y = a + bX$ where a is the Y intercept term (where the line intersects the Y axis) and b is the slope (rise over run) of the line. In our example, $Y = 20 + 0.4X$, a is 20, and b is 0.4.

TABLE 2.1   A Production Schedule

Amount of X used (fertilizer)	Amount of Y produced (bushels of wheat)
0	20
10	24
20	28
30	32
40	36
50	40

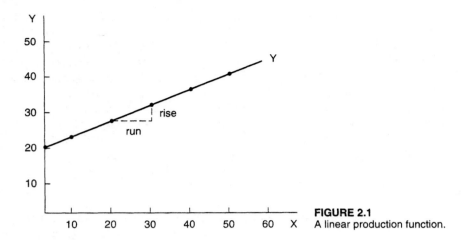

FIGURE 2.1
A linear production function.

As illustrated earlier, there are three ways of depicting the relationship between inputs and output. The tabular presentation is a production schedule, the graphical presentation is a visual picture of the relationship, and the mathematical presentation is a functional form of that production response. You should note that all three contain the same information, but each presents the information in a different manner.

The production process of every firm (including farm firms) can be represented by a specified, or functional, relationship between inputs and output, hence, the name *production function.* For example,

Wheat production = *f*(seed, fertilizer, land, weather)

Milk production = *f*(herd size, breed, feed, time of year)

Grade in ag econ class = *f*(time spent in class, time spent studying, performance on exams, educational background, time spent sleeping)

Can you think of another example of a production function?

Types of Production Responses

The amount of output produced in any physical production process depends on the level of inputs used and the production function. As the level of input use changes, the level of output also changes, with the rate of change in output dependent on the technical relationship between inputs and output. When one factor is varied and other factors are held constant (i.e., production function with single variable input), managers can trace the specific production response that occurs. There are four possible production response relationships, or rates of change, between input use levels and the resultant output, as discussed in the following subsections.

Constant Returns

In this type of response each additional unit of input is as productive as the previous unit. In other words, for each additional unit of input used, output increases at a constant rate (the rate of change in output remains constant). The production function $Y = 2X$ depicts this relationship and is presented tabularly in Table 2.2 and graphically in Figure 2.2.

The increase in production is also depicted by the slope of the production function. When we have **constant returns** the slope, which is "rise over run," or in this case, the change in output (Y) over the change in input (X), is always constant. In this

TABLE 2.2 A Production Schedule with Constant Returns

X	Y	Added Output for Each Added Unit of Input	
0	0		
		>	2
1	2		
		>	2
2	4		
		>	2
3	6		
		>	2
4	8		
		>	2
5	10		

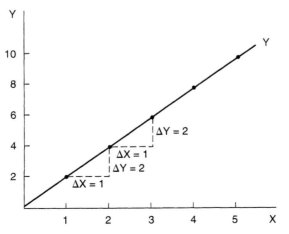

FIGURE 2.2
A production function with constant returns.

example the slope is $\Delta Y \div \Delta X = 2 \div 1 = 2$, where the symbol Δ, called *delta*, means "change in." This indicates that Y increases at a constant rate of 2 units for each additional unit of X that is used, or added, as illustrated in Figure 2.2. For example, when X increases from 1 to 2 units, Y increases from 2 to 4 units (a 2-unit increase); when X increases from 2 to 3 units, Y increases from 4 to 6 units (a 2-unit increase).

Increasing Returns

Increasing returns occur when each additional unit of input added to the production process yields more additional product than the previous unit of input. Table 2.3 and Figure 2.3 show a production response exhibiting increasing returns.

TABLE 2.3 A Production Schedule with Increasing Returns

X	Y	Added Output for Each Added Unit of Input
0	0	
		> 1
1	1	
		> 2
2	3	
		> 3
3	6	
		> 4
4	10	
		> 5
5	15	

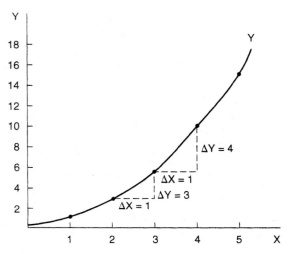

FIGURE 2.3
A production function with increasing returns.

As illustrated in Figure 2.3, when **increasing returns** are observed, the slope of the production function is getting steeper, or increasing, as additional units of input are used. This indicates that output (Y) increases at an increasing rate (the "rise") for each additional unit of input (X) that is used (the "run"). For example, when X increases from 2 to 3, Y increases by 3 (from 3 to 6 units); when X increases from 3 to 4, Y increases by 4 (from 6 to 10 units). Therefore, the fourth unit of input used increases output more than the third unit. From a managerial point of view, this type of response would be extremely desirable because physical efficiency is increasing at each input use level. Unfortunately, this type of return is seldom seen and, if so, usually only when just starting the production process. For example, one person working at a construction site may be able to pour 20 yards of concrete in a day, but when a second person is hired to help, the people work together in carrying and filling so efficiently that 50 yards of concrete are poured. Thus, the added output realized by hiring the second person was 30 yards, more than the 20 yards the first person produced. Unfortunately, in actual situations this increasing returns production response does not continue very long because of other factors of production that become problems and constraints.

Decreasing Returns

The decreasing returns response is seen when each additional unit of input increases the production level, but with a smaller change than the previous input. This relationship is depicted in Table 2.4 and Figure 2.4.

When **decreasing returns** are observed, the level of production continues to increase but at a slower rate, as shown in Figure 2.4. The slope of the production function is getting flatter, that is, decreasing, as the change in increased production divided by the change in input becomes smaller. This indicates that output

TABLE 2.4   A Production Schedule with Decreasing Returns

X	Y	Added Output for Each Added Unit of Input	
0	0		
		>	5
1	5		
		>	4
2	9		
		>	3
3	12		
		>	2
4	14		
		>	1
5	15		

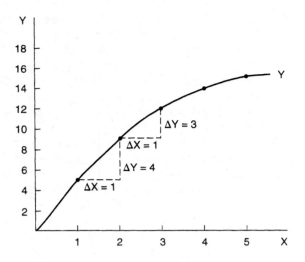

FIGURE 2.4
A production function with decreasing returns.

(*Y*) increases at a decreasing rate for each additional unit of input (*X*) that is used. For example, when *X* increases from 1 to 2, *Y* increases by 4 (from 5 to 9 units); when *X* increases from 2 to 3, *Y* increases by 3 (from 9 to 12 units). Thus, the third unit of input is less productive than the second unit. This result is due to the decrease in additional returns per unit of added input.

The case of decreasing returns, often referred to as *diminishing returns*, is what is most commonly found in both agricultural and nonagricultural decisions. Why? Because fixed resources begin to become constraints on increased production. In other words, as production continues to increase through the use of variable inputs, we begin to reach the maximum capacity of the fixed factors; thus, gains in production will decrease. Decreasing returns illustrates an important economic concept, the **Law of Diminishing Returns** (sometimes referred to as the *Law of Diminishing Marginal Returns*), which states that *as additional units of one input are combined with a fixed amount of other inputs, a point is always reached where the additional product received from the last unit of added input will decline.*

It is this "law" that provides a role for trained managers and decision makers in everyday activities; for without decreasing returns, few, if any, decisions would have to be made as production continues. Theoretically, without decreasing returns we could achieve most of our decision goals by simply increasing the amount of inputs we used in our production activity.

Negative Returns

Negative returns occur when each additional unit of input added to the production process decreases the production level. That is, there is a negative relationship between the additional use of input and the resultant output. This relationship is depicted in Table 2.5 and Figure 2.5.

BOX 2.1   Law of Diminishing Returns

You can trace the notion of decreasing returns in a production function to an English stockbroker, David Ricardo (1772–1823). Descended from Jewish immigrants to England, Ricardo learned high finance in his father's brokerage firm but eventually formed his own firm. By the time he was thirty-five he was a millionaire, allowing him to indulge in mathematics and political economy. He was a contemporary of Adam Smith and carried on correspondence with Thomas Malthus. Together they were prominent members of the classical school of economics. Ricardo's most famous book was *On the Principles of Political Economy and Taxation* (1817).

Ricardo was living at a time when England was undergoing vast social change. The "enclosure" of common grazing areas by the landlords and royalty to facilitate wool production dispossessed thousands of families. Corn laws, which placed tariffs on imported corn, protected agriculture but forced corn prices to all-time highs and further oppressed those at the lowest economic levels. The future looked bleak for rural agriculturalists and for society in general.

Ricardo theorized one unique explanation for the problem of food production which sets him apart from the other economists of his day. A landowner of substantial wealth, Ricardo noted that agricultural land varies in terms of fertility and proximity to markets. The better lands were always brought into production first but as population expanded, farmers pushed out to other lands of poorer quality. Cost of production would necessarily rise (i.e., lower soil fertility, higher transport costs) and returns to labor and capital would decline. It was in this setting that Ricardo first set out the notion which has become the **Law of Diminishing Returns:** *As you applied family labor, modest capital (crude implements and animal power), and biological inputs to food production on marginal land, returns would always decline at some point in time.* The end result would be a future of bare subsistence living as population grew inexorably.

Thus, Ricardo's Law of Diminishing Returns is one reason why economics was dubbed "the dismal science."

When **negative returns** are observed, the level of production declines with each additional unit of input used as shown in Figure 2.5. For example, when X increases from 4 to 5, Y decreases by 3 (from 15 to 12 units); when X increases from 5 to 6, Y decreases by 4 (from 12 to 8 units). Thus, the sixth unit of input used produces fewer units of output than the fifth unit. Negative returns represent a negative relationship between the level of input use and the resultant output level. Output decreases as input use increases; thus, negative returns is a special case of decreasing returns. The important point to remember is that when negative returns

TABLE 2.5 A Production Schedule with Negative Returns

X	Y	Added Output for Each Added Unit of Input
3	17	
		> -2
4	15	
		> -3
5	12	
		> -4
6	8	
		> -5
7	3	

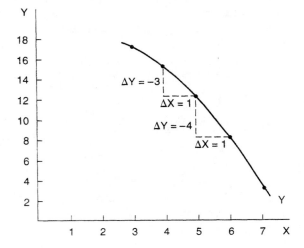

FIGURE 2.5
A production function with negative returns.

are observed, the manager can increase output by decreasing the use of variable input. Negative returns typically occur after the maximum production level is reached and reflect the constraints fixed factors impose on production. A rational manager would not willingly produce in this area of the production schedule because by using fewer units of inputs, and hence lower costs, output could be increased.

It is possible to combine the four types of production responses with the Law of Diminishing Returns to generate what we will call the *typical production function*. This function, with specific responses (and slopes) associated with it, is commonly accepted by research economists and managers as a general function depicting the typical production response when combining one variable input with fixed factors of production. It is used consistently throughout this textbook.

The Typical Production Function

The production function represents the physical correlation between inputs and the resultant output. It tells us the amount of output that can be obtained from a combination of alternative amounts of variable inputs with one or more fixed factors. This implies that by examining the production function we will be able to examine the effects of changing input use levels on output levels. But when one considers how many individual inputs are used to produce a single product, such as wheat or an automobile, examining the production function appears to be a complicated, if not impossible, task. However, we can simplify the task by using the following assumptions:

1. One product is produced, and there is only one means of producing it.
2. One input is variable, all others are fixed.

Assumption 1 states that we are only considering the production of a single product using a single production method. Assumption 2 simply states that we are examining the effect on output from varying the use of a single input, holding all other inputs at a fixed level. Implicit in this assumption is that the supply of the variable input is unlimited. In other words, the manager or firm can use as much of the input as desired without affecting the availability of the input.

At first glance, these assumptions may seem overly simplistic. Yet, in reality it does reflect a typical decision process. For example, if a farmer wanted to quantify the effect of increased fertilizer use on corn yield, she would not simultaneously increase fertilizer use and increase the level of irrigation water. If she increased the use of both inputs at the same time she would be unable to determine if the resulting change in corn yield was due to increased fertilizer or increased water use. Thus, she could not make an economic decision on whether or not using more fertilizer is wise.

Focusing on only one variable input and fixing the quantities of the other inputs allows us to determine the impacts on output from changes in the use of that input. Also focusing on one changing input allows a clearer presentation of the physical and economic relationship between the variable input and output; a relationship that extends to multiple variable inputs.

Product Curves

Three concepts, referred to as *product curves,* are commonly used to study the typical production function. Each product curve gives specific information that is useful in decision making.

The **total physical product** curve, referred to as TPP or TP, illustrates the relationship that exists between output and one variable input, holding all other inputs constant. Total physical product is measured in physical terms and represents the maximum amount of output brought about by each level of input use. Since TPP represents total output, it is often referred to simply as Y.

The **average physical product** curve, referred to as APP or AP, shows how much production, on average, can be obtained per unit of the variable input with a fixed amount of other inputs. It is calculated as the output at each level (TPP) divided by that level of input (X), or:

$$APP = TPP \div X$$

It indicates the average productivity of the inputs being used or, on the average, how productive is each level of input use. For example, APP can tell us how productive, on the average, our ten hours of studying for a test or the four employees we've hired have been. Because Y is synonymous with TPP, we can also write: $APP = Y \div X$.

The **marginal physical product** curve, referred to as MPP or MP, represents the amount of additional (marginal) total physical product obtained from using an additional (marginal) unit of variable input. It is computed as the change in output divided by the change in input use:

$$MPP = \Delta TPP \div \Delta X$$

or, simply, $MPP = \Delta Y \div \Delta X$. Marginal physical product is, by definition, the slope (rise over run) of the total physical product curve (mathematically, MPP is the first derivative of TPP). In our description of the types of responses possible in production we used the MPP to identify the rate of change in output resulting from adding one more unit of input.

We can put these three concepts, which comprise the typical production function, into action using the following example. The manager of Lou's Sandwich Shop, a local restaurant, can use the number of employees to increase the number of sandwiches prepared every hour. In this example, the variable input (X) is the number of employees and the output (Y) is the number of sandwiches made per hour. The production schedule for the sandwich shop is presented in Table 2.6 and plotted in Figure 2.6.

Because MPP represents the additional output received or obtained from using an additional unit of input, it is common to place the values of MPP between the two input levels as was done in Table 2.6 and Figure 2.6. For example, as the number of workers (X) increases from 1 to 2, output (Y) increases from 2 to 5, an increase of 3. Since the additional output (MPP = 3) is due to increasing X from 1 to 2, the value 3 is reported between these two input levels in Table 2.6 and Figure 2.6. Treating MPP in this manner serves to remind us that when we are looking at marginal changes we are looking at the effect on output of using an additional unit of input.

Using our example, we can examine the production function more closely. From the definition of total physical product (TPP) we know that this curve shows

TABLE 2.6   Production Schedule for Lou's Sandwich Shop

X (Input)	TPP (Y)	APP (Y ÷ X)	MPP (ΔY ÷ ΔX)	
0	0	∞		
			>	2.00
1	2	2.00		
			>	3.00
2	5	2.50		
			>	4.00
3	9	3.00		
			>	2.00
4	11	2.75		
			>	1.00
5	12	2.40		
			>	-1.00
6	11	1.83		

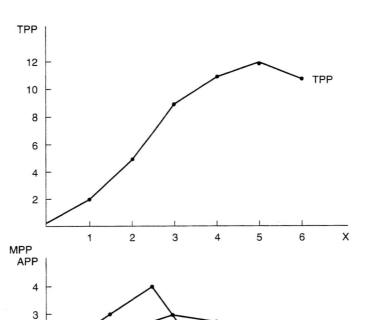

FIGURE 2.6
Production function for Lou's Sandwich Shop.

us the maximum amount of output that can be obtained from a given level of input use. We can see in both Table 2.6 and Figure 2.6 that using 2 employees produces 5 sandwiches per hour and no more; 6 employees produces 11 sandwiches per hour and no more. Average physical product (APP), as defined, indicates the average productivity of each unit of variable input being used. If 4 employees are employed, the average number of sandwiches (or productivity) prepared per hour is 2.75. In other words, an average of 2.75 sandwiches per hour are prepared by each of the 4 employees working. Marginal physical product (MPP) indicates the amount of additional output that can be obtained from using an additional unit of input and, by definition, is the slope of TPP. From Table 2.6 we see that the MPP of using the third unit of input is the additional output obtained from increasing input use from 2 to 3 employees and is equal to 4 sandwiches ($\Delta Y \div \Delta X = 4 \div 1$). Likewise, the MPP of using the sixth unit of input, the additional output obtained from increasing input use from 5 to 6 units, is -1 ($\Delta Y \div \Delta X = -1 \div 1$). (We'll talk later about the negative MPP.)

We can also identify the production responses observed in Figure 2.6 and Table 2.6. Recall that increasing returns occur when each additional unit of input is more productive than the previous unit. The production function in our example exhibits increasing returns through the third unit of input ($X = 3$) as TPP is increasing at an increasing rate (MPP is increasing). Decreasing returns occur when each additional unit of input is less productive than the previous unit. The production function in our example exhibits decreasing returns from the third unit to the fifth unit of input ($X = 5$) as TPP is increasing but at a slower rate (MPP is declining, but still positive). Negative returns, which are defined as a decrease in output for an increase in input, occur past the fifth unit of input (MPP is negative).

Relationship Between Product Curves

It is possible, as we just did, to graph product curves using discrete data, although this results in curves that are discontinuous, meaning straight-line segments connecting points of known information. But for teaching purposes it is useful to use continuous and smoothed curves, which suggest that small fractions of inputs would produce corresponding levels of production. Using smoothed curves allows us to more closely examine the characteristics of each of the three product curves in the typical production function, as well as the relationships between the three curves. These relationships are depicted graphically in Figure 2.7.

As illustrated in Figure 2.7, the shape of the TPP curve shows that as input use increases, the quantity produced increases but the rate of increase changes. The TPP curve exhibits increasing returns up to point (1) in Figure 2.7, meaning output is increasing at an increasing rate. At this point, called the *inflection point* or the *point of diminishing returns*, TPP changes from increasing at an increasing rate to increas-

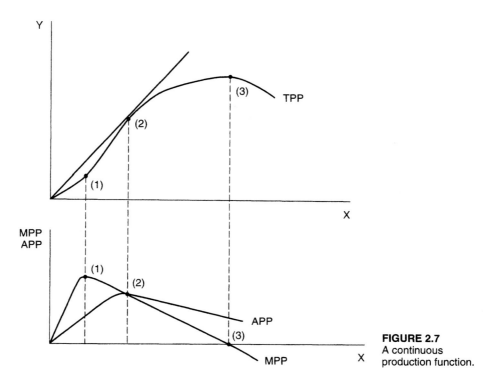

FIGURE 2.7
A continuous
production function.

ing at a decreasing rate. The TPP curve exhibits decreasing returns, where each additional unit of input generates less additional (marginal) product than the previous unit, from point (1) to point (3) where TPP reaches its maximum. Beyond point (3) the total physical product actually begins to decrease with the use of additional input as negative returns set in. If the TPP curve were not curved in this way it would imply that an unlimited amount of product could be produced from a fixed input, which is not possible.

Consider the following example. Suppose you are a farmer with 100 acres of land, machinery, corn seed, fertilizer, and irrigation water. Holding all factors of production constant except irrigation water will enable you to quantify the effect water levels have on yields. Thus, you combine irrigation water (the variable factor) with your fixed factors (in this case, land, seed, machinery, and fertilizer) to produce corn. Without any water, your corn crop would die. But as you add additional water (that level beyond minimum needed for subsistence) your corn yield increases rapidly. As you add more and more water, your corn yield continues to increase, but at a slower rate, eventually reaching some maximum yield per acre. After the maximum yield is reached adding additional water will decrease yields (sprout damage, etc.), and too much additional water will actually kill the plant (flooding, etc.), resulting in nothing to harvest. As described, the production function for corn exhibits increasing, decreasing, and negative returns. This type of production response is seen in other areas as well, for example, milk production, weight gain in livestock, and manufacturing processes.

Average physical product and marginal physical product are derived directly from the input–output relationship of the total physical product curve. Thus, the product curves are related to each other and these relationships are important. We can use Figure 2.7 to examine these relationships in more detail.

The marginal physical product curve describes the slope of the total physical product curve. In Figure 2.7, the marginal physical product curve increases up to point (1), the range in which TPP exhibits increasing returns. At point (1), the point of diminishing returns (where the slope of the TPP changes from increasing at an increasing rate to increasing at a decreasing rate), the MPP reaches its maximum. Marginal physical product is declining, but still positive, from point (1) to point (3) indicating that output is increasing but at a slower rate, which corresponds to the range in which TPP exhibits decreasing returns. Marginal physical product is equal to zero at point (3), which corresponds directly to the maximum of TPP. Marginal physical product becomes negative after point (3), corresponding to the decreasing production as negative returns are observed.

The average physical product curve describes the average productivity of the variable inputs. The average physical product curve in Figure 2.7 initially increases, reaching a maximum at point (2) and then declines. The high point of APP, point (2), is located where a line drawn out of the origin is tangent (barely touching) to the TPP. The slope at each point on this line is the ratio of output (on the vertical axis) to input (on the horizontal axis), a ratio that is the definition of average physical product. Where this line is tangent to the TPP, point (2), identifies the maximum ratio of output per unit of input, which is the definition for the maximum of APP, for this production function. At the point where APP reaches its maximum, point (2), MPP crosses APP, or MPP = APP where APP is maximum. As long as MPP is greater than APP, average physical product continues to increase because the marginal increase in output raises the average for all inputs at a given level. Why? Consider the following example: Suppose you had six calves, and five of those calves each weighed 350 pounds at weaning. The average weaning weight for those five calves would be 350 pounds (350 × 5 = 1,750 total weight; average weight 1,750 ÷ 5 = 350). If the sixth calf weighs only 300 pounds at weaning, the average weaning weight decreases from 350 to 341.7 pounds. Why? Because the additional (or marginal) weaning weight of the sixth calf is lower than the previous average (300 < 350). But, if the sixth calf weighs 400 pounds at weaning the average would increase, from 350 to 358.3 pounds, because the marginal change is greater than the average (400 < 350).

The relationships between the MPP and APP curves illustrated in Figure 2.7 can be summarized as follows:

1. When MPP is greater than APP, APP is increasing.
2. When MPP is equal to APP, APP is maximum.
3. When MPP is less than APP, APP is decreasing.

Production Function Characteristics

The relationships between TPP, APP, and MPP are very specific, so much so that, given complete information about one curve, the other two curves can be derived. For example, if we know that the first, second, and third person we hire have marginal products of 6, 9, and 4, respectively, we have enough information to compute the TPP and APP. In this case it is simply a matter of using what we know about the product curves to find what we don't know. Setting it up in a tabular format, as in Table 2.7, will make this easier to see.

Using the information in Table 2.7, we begin by computing TPP (the computed numbers are in bold). From our earlier discussion we know that marginal physical product is the addition to total output from using an additional unit of input ($MPP = \Delta Y \div \Delta X$). Because the MPP of the first person we hire is 6, it follows that when $X = 1$, $Y = 6$. Why? We know that when $X = 0$, $Y = 0$. We also know that when we increase input use (hire the first person) from $X = 0$ to $X = 1$ output will increase by 6 units ($MPP = 6$); so output increases from $Y = 0$ to $Y = 6$ ($0 + 6 = 6$). Because the second person we hire has a marginal product of 9, when $X = 2$, $Y = 15$. Why? When input use increases from $X = 1$ to $X = 2$, output increases by 9 units, which makes total output equal 15 ($6 + 9 = 15$) when $X = 2$. When X increases to 3, Y increases by 4, so when $X = 3$, $Y = 19$ ($15 + 4 = 19$). After computing the total physical product we can compute the average physical product at each input level (computed values in bold) in the usual manner: When $X = 1$, $APP = 6$; $X = 2$, $APP = 7.5$; and $X = 3$, $APP = 6.33$. On the other hand, if we only knew the input level and the corresponding APP, we could derive TPP by simply multiplying APP by the appropriate units of input, since $APP = TPP \div X$ then $APP \times X = TPP$. Then computing MPP would follow easily. Understanding the definitions of MPP, APP, and TPP, and what the curves represent, will not only enable you to calculate MPP and APP easily, but is crucial to understanding the relationships existing between the curves.

Questions often arise about how "typical" is the typical production function introduced in this section? The answer is that, based on agricultural and business research, most production functions exhibit most of the characteristics developed in this discussion and *all* of them, at some point, exhibit the point of diminishing

TABLE 2.7 Computing TPP and MPP from *X* and MPP

X	MPP			X	Y	MPP	APP
0			$\rightarrow$	0	0		∞
	>	6				> 6	
1			$\rightarrow$	1	6		6.00
	>	9				> 9	
2			$\rightarrow$	2	15		7.50
	>	4				> 4	
3			$\rightarrow$	3	19		6.33

returns. Even if none of them specifically appeared as our graphs have shown them, the physical relationships described are so useful in later economic analysis that a detailed discussion of a "typical" production function is warranted.

For example, working from agronomic data, a research economist used this equation to plot the predicted yield from different lime application levels for corn:

$$\text{Yield} = 30.7 + 119.39\text{LIME}^{0.5} - 31.10\text{LIME}$$

The relationship between corn yield and lime application levels is illustrated in Figure 2.8. As shown, predicted corn yield rises rapidly as you add lime, then becomes less responsive, and finally yield begins to decline as more lime is added. Actual yield, of course, is a function of weather, insects, disease, and management. As you can see, corn yield is very responsive to the first two tons per acre of lime applied to corn fields. However, based on this information a real farm manager would not apply more than three tons of lime to a corn field—*even if the lime were free!* This is an example of an actual production function for the application of lime to corn fields, and is a real-world illustration of decreasing returns in a production function.

Another question that often arises in classroom discussions is "Where does the manager generate the information contained in a production function?" The answer to this question depends on the production process or decision being evaluated. In agriculture an idea of the response of corn to fertilizer application could come from past experience on the farm, the county agent's evaluation of the county cropping patterns, or research results from either fertilizer companies or university experiment stations. In a nonfarm business, similar types of information should be available: other managers, prior experience, or consulting companies, for example. The main point to remember is that the information may not always be available in specific graphical form, but much general information is usually available.

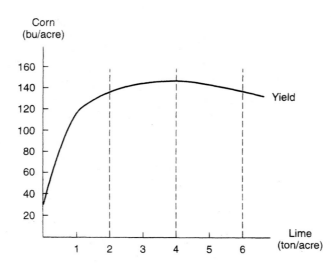

FIGURE 2.8
An estimated production function for lime applied in Midwest corn production. (*Source:* H. Hall, University of Kentucky Staff Paper 349, 1994.)

Production functions exist in farm and nonfarm situations and in both private and public institutions. Although the decisions may at first appear very different, public health agencies, educational institutions, and other public entities are faced with the same economic choices as private firms. The principal difference is not the objective or goal of the decision but how the tools of economic analysis we are learning can be directly applied. We will use some of these examples as the tools of economic analysis are developed further in succeeding chapters.

Stages of Production

Our discussion of the production function and the product curves began with the question "How much to produce?" Given the alternative levels of input and output available in a production function, which one level of input and corresponding output will maximize profits? Although we need information on costs and revenue (price of inputs and price of output) to determine exactly how much to produce, the product curves provide enough information to identify a profitable (sensible) range of production. These regions are referred to as **stages of production** and are classified as *rational* or *irrational*, as illustrated in Figure 2.9. Here we see that Stage I begins where no input is utilized, point (0), and ends where APP is

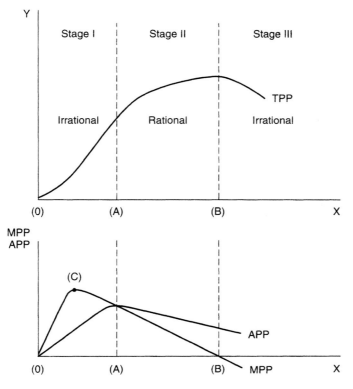

FIGURE 2.9
Stages of production.

at its maximum (APP = MPP), point (A). If we decide to produce at all we will use inputs at least to this point because every succeeding unit of input up to point (A) has a marginal product that raises the average (MPP > APP). Even though we would like to receive the highest marginal product, point (C), for each unit of input, the law of diminishing returns doesn't allow it. So, even though the added output per added unit of input (MPP) is decreasing, we will continue to add inputs at least until APP reaches its maximum. It is similar to test scores where, although we would like to get 100 on the next test, we are still pleased if the next test score is higher than our average test score, thereby raising the average. So as long as we know our MPP (next test score) will be at least higher than our APP (average test score), it is irrational to stop production (not take the next test) until MPP = APP; hence, Stage I is an irrational area of production.

Stage III begins where TPP is at its maximum (MPP = 0), point (B) in Figure 2.9. If we continue to use additional units of input, we actually decrease the total production level. For example, in certain forms of tree production, when growth stops the mature trees become susceptible to disease and insects and if the trees are not harvested, production (TPP) declines. Since output decreases as we increase input use, it is impossible to maximize profits in this production region. Thus, Stage III is obviously an irrational area of production. This decision can be made with no price information since within this stage of production more output can be produced by using fewer units of input.

As shown in Figure 2.9, Stage II begins at point (A), the input use level where APP is maximum (MPP = APP), and ends at point (B), the input use level where TPP is maximum (MPP = 0). Somewhere in this region of production profits will be maximized, making Stage II the rational stage of production. For each additional unit of input that is used within Stage II, TPP is increasing but APP is decreasing, resulting in a trade-off between increased production and decreased productivity for each variable input. This trade-off is the core of economic decision making and requires the skills of the decision maker or manager.

Going back to our sandwich shop example (Table 2.6 and Figure 2.6), we can identify the stages of production and identify the region where profitable production will take place. Stage I begins where $X = 0$ and ends where $X = 3$ (APP is maximum; MPP = APP). Within Stage I both TPP and APP are increasing, and MPP is greater than APP. Stage III begins where TPP is maximum, at $X = 5$. Within Stage III, TPP and APP are decreasing, and MPP is negative. Both Stages I and III are irrational stages of production. Stage II begins where $X = 3$ (APP is maximum; MPP = APP) and ends where $X = 5$ (TPP is maximum). Within Stage II, TPP is increasing, APP is decreasing, and MPP is less than APP. Returning to an earlier example, to maximize profits, the manager of Lou's Sandwich Shop should produce within Stage II using inputs between the range $X = 3$ and $X = 5$. Later we will use prices to determine just where in Stage II Lou should operate.

Examples of firms operating in Stages I and III, while not too common, do exist because our assumption of perfect certainty for the manager doesn't exist. Production in Stage I can occur because of capital shortage or shortage of an input, such as fertilizer or diesel fuel. Stage III production might be seen when a producer, expecting a normal rainfall and not receiving it, adds too much fertilizer for the available moisture.

A Note on Elasticity of Production

The concept of elasticity, or responsiveness of one variable to a change in another, is an important tool in economic analysis and is discussed in greater detail later in this book. But the concept of **elasticity of production** can be associated with the stages of production so it will be examined briefly here.

Elasticity of production (E_p) refers to the amount of change in output we receive for a given change in input. Notationally, it is percent change in output ($\Delta Y \div Y$) divided by percent change in input ($\Delta X \div X$). Since, by definition, MPP indicates the degree of production response to changes in input use, we can express E_p in terms of MPP. Note the following use of simple algebra:

$$E_p = \frac{\%\Delta \text{ in output}}{\%\Delta \text{ in input}} = \frac{\dfrac{\Delta Y}{Y}}{\dfrac{\Delta X}{X}} = \frac{\Delta Y}{Y} \times \frac{X}{\Delta X} = \frac{\Delta Y}{\Delta X} \times \frac{X}{Y} = \frac{\dfrac{\Delta Y}{\Delta X}}{\dfrac{Y}{X}} = \frac{\text{MPP}}{\text{APP}}$$

We can use our knowledge about the relationship between MPP and APP to identify the magnitude of E_p in each stage of production. In Stage I E_p must be greater than one ($E_p > 1$) since MPP > APP in this stage; in Stage III E_p must be less than zero ($E_p < 0$) since MPP is negative in this stage. Since MPP = APP at the beginning of Stage II and MPP = 0 at the end of Stage II, E_p must be less than one ($E_p < 1$) but greater than zero ($E_p > 0$) in this stage. Combining these results indicates that in Stage II, the rational stage of production, $0 < E_p < 1$.

Effect of Technological Change

One of the assumptions we made at the beginning of this section was that there is only one method of producing the good being considered (technology remains constant). Recall that the production function gives the maximum amount of output that can be produced by a firm using a given technology, where technology is defined as "a specific method of producing a product." But what happens if technology changes, and what constitutes **technological change?** If a new technology were to occur, the effect on the production function could be to raise the output level associated with each level of input. This is referred to as "shifting the function upward." The effect of a technological improvement on the firm's production function is illustrated in Figure 2.10.

The production function can shift over time as a result of research and development. In agriculture, farmers are constantly being presented with new seed

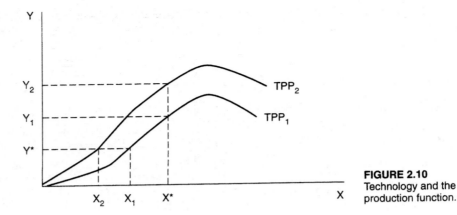

FIGURE 2.10
Technology and the production function.

varieties, improved machinery, and better pesticides to improve TPP on their cropland base. The impact of the technological improvement is the ability to produce more output with the same level of input. In Figure 2.10 if the manager uses input level X^*, under the old technology (TPP_1) the firm could produce Y_1, but the same input use under the new technology (TPP_2) results in a higher output level, Y_2. Conversely, the same output can be produced using less input. To produce output level Y^* using the old technology (TPP_1) required an input use of X_1, but producing Y^* using the new technology (TPP_2) only requires using X_2 units of input. For example, a new hybrid seed corn requires less nitrogen fertilizer to produce the same level of yield; or using the same level of fertilizer the hybrid seed produces a higher yield.

Given an increase in technology, what would the effect be on the APP and MPP curves? The technological change that shifts the TPP curve upward would result in a corresponding upward shift in the MPP and APP curves. Remember, MPP and APP are derived directly from the input–output relationship of the total physical product and vary as the total physical product varies.

From a managerial viewpoint any change in technology or decision making that positively affects the production function has the effect of increasing the net result of the production process. Hence, more goods and services can be produced for the same cost (same units of input); or the same amount of goods and services can be produced for less cost (with fewer units of input). Both situations are appealing to management.

SUMMARY

In this chapter we presented the basics of the physical relationships in the production process. The production function can be expressed in three different ways: tabular, graphically, and mathematically. There are three possible relationships between input use and the resultant output: increasing, decreasing, and constant returns. There are three related concepts used to analyze the production function: TPP, MPP, and APP. Total physical product (TPP) represents the maximum amount of output we can obtain from different levels of one variable input holding other factors fixed. Marginal physical product (MPP) depicts the change in output, or the additional output received from using an additional unit of input, and by definition is the slope of the TPP. Average physical product (APP) depicts the average productivity of the inputs used, or the average number of units of output received per unit of input used. The three curves are related, in that APP and MPP are derived from TPP. Using the relationship between the product curves we were able to identify three stages of production and even without prices, start to answer the question of "How much to produce?" Stage I, which is an irrational stage of production, begins where input use is zero and continues until APP is maximum. Stage II, which begins where APP is maximum and ends where TPP is maximum, is a rational stage of production. Stage III, which begins where TPP is maximum, is an irrational stage of production because each successive unit of input used actually decreases the output level.

QUESTIONS

1. What are the three major questions faced by a firm manager?
2. Distinguish between the following concepts: resource, production, factor, and output.
3. Give a verbal and mathematical representation of a simple production function.
4. Suppose you face a production schedule like that shown in Table 2.1. What would be your production given a resource use of 25 units? What would your factor use be if you needed an output of 48 units? Show your algebra.
5. What are the four possible types of responses possible when factors of production are varied? How do they differ?
6. What is the Law of Diminishing Returns? Illustrate the concept from your own experience.
7. Given a typical production function with three types of returns, where is the inflection point? Is the MPP at that point a maximum or a minimum? Indicate the point in both a tabular example and a graph.
8. Why would a rational decision maker never produce in Stage III? Or in Stage I?
9. What is a technological advance, and why is it an important managerial concept for a farm producer?

CHAPTER 3
Cost Relationships in Production

INTRODUCTION

A manager's goal is to determine how much the firm should produce to reach its objective of maximizing profits. In Chapter 2 we examined the relationships between inputs and output in the physical production process. Using these relationships we identified a rational range of production (Stage II). However, as we found in Chapter 2, information on physical production alone is insufficient for economic decision making. To determine the combination of fixed and variable inputs that will maximize profits, information on costs is necessary.

In This Chapter

The objective of this chapter is to introduce the basic elements of cost relationships in production. To do this, the price of factors utilized in the production process is introduced and used to develop, and analyze in some detail, various cost concepts useful in managerial decision making. In this chapter we combine what we already know about physical production with input price information to examine the relationship between the costs of production and the level of output that is produced.

CONCEPTS AND DEFINITIONS

The most common objective for agricultural firms is to maximize profits. Because profit = revenue − costs, the firm's profit level is affected not only by how much is produced but also by the costs of generating that production. We know from Chapter 2 that the information inherent in a production function is necessary for economic decisions making but by itself is insufficient. We know that as managers seeking to maximize profits we should operate in Stage II of the production function. But to determine the specific level of how much to produce we must have knowledge of how the level of input affects costs of production and revenue from sales of that production. (We discuss revenue concepts in Chapter 4.) In this chapter our focus is on the costs of production. We are asking "How much does it cost to produce a bushel of wheat?" or "How much per-mile does it cost to operate your car?" To derive the costs of production, we combine information on the production function with the relevant price of our inputs, P_x.

Assumptions Used

We want to examine the relationship between the level of physical output and the costs of producing that output. To simplify the analysis, we derive the costs of production utilizing the single-variable input production function introduced in Chapter 2. As we will see later, these cost relationships extend to cases where there are multiple inputs. The assumptions we will be using are as follows:

1. One product, one production method;
2. One variable input (unlimited supply), other inputs are fixed;
3. Firm seeks to maximize profits; and
4. Firm is a **price taker.**

Assumptions 1, 2, and 3 are the assumptions we used in Chapter 2 to examine the production process, and we are still assuming that decisions are made under perfect certainty. We are still talking about a physical production process but now we are interested in the costs of producing the output. Assumption 4 implies something very specific about the behavior of the firm—that the firm can purchase any amount of an input without affecting the price of that input. In other words, the firm can purchase a few units or many units of an input and the price of that input remains unaffected. Thus, in the eyes of the firm the price is fixed, or constant, because within the market the individual firm has no influence over the input's price. The student should be aware that assuming the firm is a price taker does not mean that the market price of the input will never change, just that the firm is not large enough to influence the price for the input. For example, if you went to the feed store to purchase feed grain and the current price for a fifty-pound sack is $9.50, you could purchase 1 sack, 10 sacks, or 100 sacks for $9.50 each. The price per sack remains $9.50 regardless of whether you purchase 1 or 100 sacks of grain. Thus, you are a price taker. The "price taker" assumption is one that is used repeatedly in economic analysis and is a reasonable assumption in many real-world settings. Later on we'll talk about the situation where you might get a "volume discount" on 100 sacks of feed grain.

Definitions

We know that the firm uses factors of production to produce a product. We also know that those factors of production are not "free" but the firm must pay something, often money out of pocket, to use those factors. *The payments that a firm must make to attract inputs (resources) and keep them from being used to produce other outputs* are referred to as *economic costs,* or **costs of production.**

Economic costs can be either implicit or explicit. *Explicit costs* are the normal "out-of-pocket" or "cash" costs of inputs used in production (i.e., seed, fertilizer, wages, fuel, etc.). *Implicit costs* are the costs associated with inputs owned by the firm. Since the firm owns the input there is no need to make a cash payment to be able to use the input, but use is not free. There is an opportunity cost associated with using firm-owned inputs. The **opportunity cost** of using an input is *the value of the contribution that input could make in its highest valued alternative use* (i.e., land used to produce corn versus soybeans; farmer stays on own farm versus working in factory).

Earlier we discussed the distinction between fixed and variable factors of production, a distinction that is dependent on the time horizon of the firm. Associated with fixed and variable inputs, and the time period, are two important cost concepts: fixed costs and variable costs of production.

Fixed costs are *those costs which* **do not** *vary with the level of production, and are the costs associated with the fixed factors of production.* Recall that fixed factors of production must be maintained even if no production occurs; hence, the costs associated with these factors must be paid even if production is zero. **Variable costs** of production **do** *vary as output level changes and are the costs associated with using the variable factors of production.* If no variable input is used, variable costs are zero; as input use increases, resulting in increased output, variable costs increase. The distinction between fixed and variable costs, as with fixed and variable inputs, depends on the time horizon of the firm.

The short-run is a time period such that the firm has both fixed and variable factors of production. Therefore, in the short-run, managerial decisions are affected by the inability to change all variables, such as the size of a plant, location, or amount of land. Because the firm has both fixed and variable factors of production, the firm incurs both fixed and variable costs of production.

The long-run is a time period in which there are no fixed factors of production, thus only variable costs of production are incurred by the firm in this time period. As the time horizon is increased, costs that are fixed in the short-run become variable costs. For example, the typical fixed costs of a business firm include depreciation, interest, rent, taxes, and insurance, often referred to with the acronym *DIRTI.* They become committed costs when the production period, for example, the crop year, is undertaken, and they become variable costs as the decision period increases, for example, "Should I get out of farming and enter another business?" Hence, the long-run is considered to be the situation where any and all changes can be made subject to the availability of capital, information, and manager expertise.

As we indicated in Chapter 2, most economic decisions occur in the short-run because seldom does a manager have perfect control over all the resources necessary for his or her production process. Thus, our discussion focuses on economic decisions made within the short-run.

COST RELATIONSHIPS IN PRODUCTION

When examining the costs of production, seven measures of cost or *cost curves* are of interest. Three of these cost curves measure the costs of production on the basis of the total output produced; the other four measure the costs of production on a per-unit of output basis. We begin our discussion by defining each cost curve, then we use the sandwich shop example from Chapter 2 to illustrate the calculation of and relationship between the cost curves.

Notice that when we graph the cost curves, output (Y) is on the horizontal axis, and cost, measured in dollars ($\$$'s), is on the vertical axis. The cost curves are graphed this way because we are interested in the relationship between the cost level and output, or in how costs of production change as the output level changes. This is different from graphing the product curves where we were interested in how output changes as input use changes; thus, input was on the horizontal axis and output was on the vertical axis.

Costs on Total Output Basis

Three costs are used to measure the costs of production on the basis of total output produced: total fixed costs, total variable costs, and total cost.

Total fixed costs (TFC), sometimes called *overhead costs,* include both *implicit and explicit costs of inputs that are fixed in the short-run and do not change as output level changes.* Because fixed factors of production must be maintained even if production is zero, fixed costs must be paid at all production levels, even zero.

Total variable costs (TVC) include both *implicit and explicit costs of inputs that are variable, or changeable, in the short-run and change as the level of output changes.* Total variable costs represent the total dollar amount paid for each variable input used to produce each output level. Total variable costs are calculated by summing the cost of each variable input used (equal to the units of input used multiplied by the input's price), or:

$$\text{TVC} = (P_{X_1} \times X_1) + (P_{X_2} \times X_2) + \cdots + (P_{x_n} \times X_n)$$

where X_i represent the units of variable input i used, and P_{X_i} is price of variable input X_i. For our single-variable input case,

$$\text{TVC} = P_x \times X$$

Total variable costs are the cost of variable inputs used in production. If no variable inputs are used, TVC is zero and production is zero; as input use increases, output increases and TVC increases.

Total cost (TC) is simply the sum of total fixed costs and total variable costs at each level of output, or

$$TC = TFC + TVC$$

and is the total cost of producing each level of output.

Recall the sandwich shop example from Chapter 2, where the variable input is the number of employees and output is the number of sandwiches prepared per hour. Assume that the store manager can hire an additional worker for $4/hour ($P_x = \$4.00$) and that the TFC of the shop is $10 (TFC = $10.00). The total cost curves associated with this production function are presented in Table 3.1 and graphed in Figure 3.1, where we see that fixed costs do not change as output changes but remain at a constant level of $10. Total variable costs increase as input use, and output levels, increase. At each output level TVCs are calculated by multiplying the units of input used to produce that output level by the price of the input. For example, when $Y = 2$, TVC = $4 ($1 \times \$4 = \$4$); when $Y = 12$, TVC = $20 ($5 \times \$4 = \$20$). This means that the total

TABLE 3.1 Total Cost Curves for Lou's Sandwich Shop:
TFC = $10 and P_x = $4

X	Y	TFC	TVC ($P_x \times X$)	TC (TFC + TVC)
0	0	$10.00	$0.00	$10.00
1	2	10.00	4.00	14.00
2	5	10.00	8.00	18.00
3	9	10.00	12.00	22.00
4	11	10.00	16.00	26.00
5	12	10.00	20.00	30.00
6	11	10.00	24.00	34.00

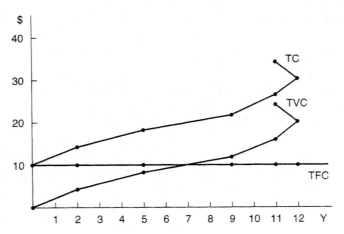

FIGURE 3.1
Total cost curves for Lou's Sandwich Shop: TFC = $10 and P_x = $4.

cost of using the variable input to produce 2 sandwiches is $4; the total cost of using the variable input to produce 12 sandwiches is $20.

Total cost is the sum of total fixed cost and total variable cost at each level of output. For example, when $Y = 2$, the total cost of producing 2 sandwiches is $14 ($10 + $4 = $14); when $Y = 12$, the total cost of producing 12 sandwiches is $30 ($10 + $20 = $30). Graphically, as Figure 3.1 shows, the distance between TVC and TC at each output level is equal to TFC. Figure 3.1 also shows that as we enter Stage III (where $X = 5$ and $Y = 12$) both TVC and TC curve back toward the vertical axis indicating that although output falls as input use increases (Stage III) the costs of using those inputs continue to increase.

Using Smoothed Curves

It is possible, as we just did, to graph the total cost curves using discrete data, although this results in curves that are discontinuous. For teaching purposes it is useful to use continuous and smoothed curves, as we did in Chapter 2, which suggests that small fractions of output would produce corresponding levels of costs. Using smoothed curves allows us to examine more closely the relationships among the three total cost curves. These relationships are presented in Figure 3.2.

Since total fixed costs are constant over all levels of production, even zero, TFC is graphed as a horizontal line as illustrated in Figure 3.2. Where TFC intersects the vertical axis describes the total costs a manager must pay even if production is zero. This represents the costs that must be paid even if we decide to produce nothing in a given year. Since total variable costs are derived from the underlying production process, its shape reflects the increasing/decreasing returns of the TPP curve. Total variable costs start at zero, increase rapidly at first, then continue to increase but at a less rapid rate, then increase rapidly again before beginning to curve back toward the vertical axis as we enter Stage III of the production process. Total cost is the sum of all fixed and variable costs incurred during a decision period; hence, the shape of the TC curve comes directly from the

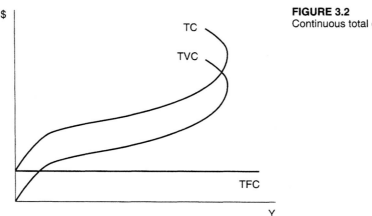

FIGURE 3.2
Continuous total cost curves.

shape of the TVC curve. Graphically, as Figure 3.2 shows, TC is the vertical summation of the TFC and TVC curves; thus, the shapes of the TC and TVC curves are identical because the distance between TC and TVC at each output level is equal to TFC. As output changes, the change in TC is due to changes in TVC because, by definition, TFC does not change with output.

Total costs, or costs measured on a total output basis, are of interest and use; however, a cost measure on a per-unit of output basis is often more beneficial in determining profit maximizing output.

Costs on Per-Unit of Output Basis

Four cost curves are used to measure costs of production on a per-unit of output basis. Just as APP and MPP are derived from TPP, the per-unit costs are derived from the total cost curves. The per-unit costs include three average costs and a related cost concept, which is of great importance in economic analysis, marginal cost.

Average fixed cost (AFC) is the *average cost of the fixed inputs per-unit of output,* and is calculated by dividing the total fixed costs by each output level, or

$$AFC = TFC \div TPP$$

Since Y is synonymous with TPP, we can also write $AFC = TFC \div Y$.

Average variable cost (AVC) is the *average cost of the variable inputs per-unit of output* and is calculated by dividing the total variable costs by each output level, or

$$AVC = TVC \div TPP$$

We can also write $AVC = TVC \div Y$.

Average total cost (ATC) is the *average total cost per-unit of output.* It is calculated by dividing the total cost at each output level by that output level, or

$$ATC = TC \div TPP$$

We can also write $ATC = TC \div Y$, or recognizing that $TC = TFC + TVC$, we can write

$$ATC = AFC + AVC$$

Marginal cost (MC) is the *increase in total cost necessary to produce one more unit of output.* It is calculated as the change in total cost divided by the change in output, or

$$MC = \Delta TC \div \Delta TPP$$

We can also write MC = ΔTC ÷ ΔY. Marginal cost is the addition to total cost incurred to produce an additional unit of output. Since TFC does not change with output levels, TFC does not impact MC. Thus we can express MC as:

$$MC = \frac{\Delta TC}{\Delta Y} = \frac{\Delta(TFC + TVC)}{\Delta Y} = \frac{\Delta TFC + \Delta TVC}{\Delta Y} = \frac{\Delta TVC}{\Delta Y}$$

because as output changes ΔTFC = 0.

We have defined the four per-unit cost measures; we can illustrate their calculation by developing the per-unit cost curves for our sandwich shop example. The per-unit cost curves associated with this production function are presented in Table 3.2 and graphed in Figure 3.3.

As shown in Table 3.2, AFC is calculated by dividing TFC by each output level. For example, when $Y = 2$, AFC = \$5 (\$10 ÷ 2 = \$5); when $Y = 12$, AFC = \$0.83 (\$10 ÷ 12 = \$0.83). So if the sandwich shop produces 2 sandwiches the average cost of the fixed factors per sandwich is \$5; if the shop produces 12 sandwiches the average cost of the fixed factors per sandwich is \$0.83 per sandwich.

Average variable cost, as illustrated in Table 3.2, is calculated by dividing TVC by each output level. For example, when $Y = 2$, AVC = \$2 (\$4 ÷ 2 = \$2); when $Y = 12$, AVC = \$1.67 (\$20 ÷ 12 = \$1.67). Thus, if the sandwich shop produces 2 sandwiches ($Y = 2$) the average cost of the variable inputs ($X = 1$) used in production per sandwich is \$2; if the shop produces 12 sandwiches ($Y = 12$) the average cost of the variable input ($X = 5$) used in production per sandwich is \$1.67.

Average total cost is calculated by dividing TC by each output level, or by adding AFC and AVC at each output level as shown in Table 3.2. For example, when $Y = 2$, ATC = \$7 (\$14 ÷ 2 = \$7; or \$5 + \$2 = \$7); when $Y = 12$, ATC = \$2.50 (\$30 ÷ 12 = \$2.50; or \$0.83 + \$1.67 = \$2.50). So if the sandwich shop produces 2 sandwiches, the average cost (both fixed and variable factors) per sandwich is \$7;

TABLE 3.2 Per-Unit Cost Curves for Lou's Sandwich Shop: TFC = \$10 and P_x = \$4

X	Y	TFC	TVC	TC	AFC (TFC ÷ Y)	AVC (TVC ÷ Y)	ATC (TC ÷ Y)	MC (ΔTC ÷ ΔY)
0	0	\$10	\$0	\$10	\$ —	\$ —	\$ —	
								> \$2.00
1	2	10	4	14	5.00	2.00	7.00	
								> 1.33
2	5	10	8	18	2.00	1.60	3.60	
								> 1.00
3	9	10	12	22	1.11	1.33	2.44	
								> 2.00
4	11	10	16	26	0.91	1.45	2.36	
								> 4.00
5	12	10	20	30	0.83	1.67	2.50	
								> −4.00
6	11	10	24	34	0.91	2.18	3.09	

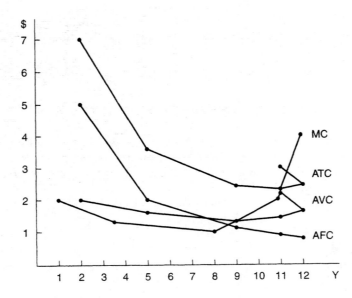

FIGURE 3.3
Per-unit cost curves for Lou's
Sandwich Shop: TFC = $10
and P_x = $4.

when 12 sandwiches are produced the average cost per sandwich is $2.50. Now you can begin to see why volume of sales is so important to a business.

Since marginal cost represents the additional cost incurred to produce another unit of output, it is common to place the values of marginal cost between the two output levels as was done in Table 3.2 and Figure 3.3. For example, when the sandwich shop increases output from 0 to 2 sandwiches ($\Delta Y = 2$), total cost increases from $10 to $14 ($\Delta TC = \$4$). Therefore, the MC, or the additional cost necessary to increase output by one unit ($\Delta TC \div \Delta Y$), associated with increasing output from 0 to 2 is $2 ($4 \div 2 = \$2$), and this value (MC = 2.00) is placed between the two output levels ($Y = 0$ and $Y = 2$). Thus, the shop's production costs will increase an additional $2 per sandwich if output is increased from 0 to 2 units. When the sandwich shop increases production from 11 to 12 ($\Delta Y = 1$) total cost increases from $26 to $30 ($\Delta TC = \$4$). The MC associated with this output change is 4 ($4 \div 1 = \$4$) and its value is placed between the two output levels ($Y = 11$ and $Y = 12$). The sandwich shop will have to bear increased costs of $4 in order to increase output from 11 to 12 sandwiches. Treating marginal cost in this manner reminds us that when we are looking at marginal changes we are looking at the effect on costs of producing an additional unit of output.

Earlier we said that changes in MC are due to changes in TVC and that MC is not affected by TFC. Table 3.2 illustrates this point. Note that as we increase input use from 0 to 1, TVC increases by $4. This increase corresponds exactly to the change in TC as output increases from 0 to 2. As input use increases from 2 to 3, TVC increases by $4, which exactly corresponds to the change in TC as output increases from 2 to 5. As output increases from one level to the next, the associated increase in costs represents the cost of using additional units of the variable input. Thus, changes in MC are the direct result of changes in TVC.

Using Figure 3.3, we can examine the relationships between the per-unit cost curves more closely. Because TFC does not change as total output increases, AFC declines as total output increases. Average variable cost, average total cost, and marginal cost decline initially, reach a minimum, and then begin to increase, generating a U-shaped cost curve. Average variable cost, average fixed cost, and average total cost begin to curve back toward the vertical axis as we enter Stage III where output decreases as input use, and thus costs, increases (since we know we would not rationally produce there, we can ignore this portion of the cost curves). Average total cost reaches its minimum, $2.36, at a higher output level ($Y = 11$) than average variable cost reaches its minimum, $1.33 ($Y = 9$). MC crosses both AVC and ATC from below and at their respective minimums, $1.33 and $2.36.

Using Smoothed Curves

It is possible, as we just did, to graph the per-unit cost curves using discrete data, although this results in curves that are discontinuous. For teaching purposes it is useful to use continuous and smoothed curves, which allows us to examine more closely the relationships between the four per-unit cost curves. These relationships are presented in Figure 3.4 where we can see that average fixed cost is a continuously decreasing function with the shape of a rectangular hyperbola. It reflects the fact that a given amount (TFC) is being divided by an increasing amount (Y). An example is building a 1,000-animal feed lot but having only 100 animals in it. These animals carry the fixed cost of the entire feed lot. When 1,000 animals are put into the feed lot each has 1/1,000 of the fixed cost allocated to it. This is the reason why managers, as we will see, work to make intensive use of capital investments and other fixed costs, sometimes referred to as "spreading the overhead."

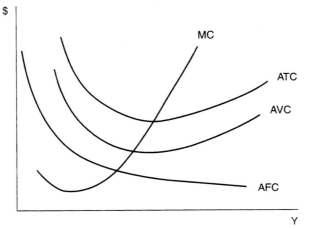

FIGURE 3.4
Continuous per-unit cost curves.

In Figure 3.4 the average total cost, average variable cost, and marginal cost curves are all U-shaped curves, first decreasing at lower levels of output, where increasing returns are typically experienced, and increasing later on when decreasing returns begin to occur. The U-shaped curves reflect the increasing/decreasing returns exhibited by the TPP, from which these curves are derived.

Graphically the average total cost curve is the vertical summation of the average fixed cost and average variable cost curves, thus the vertical distance between ATC and AVC at each output level in Figure 3.4 is equal to AFC. The average total cost curve reaches a minimum (or low point) at a larger output level than AVC because in the output range beyond the AVC minimum AFCs are declining faster than the increase in AVC; hence ATC continues to decrease for a short time. When the increase in AVC is greater than the decrease in AFC, ATC begins to increase. ATC is also called *average cost* or *cost-of-production* by business managers.

Because fixed costs do not change with changes in output, fixed costs do not affect MC. As we will see later, the shape of the MC curve is a direct result of the increasing/decreasing returns exhibited by the TPP. In Figure 3.4 the MC curve crosses the AVC and ATC curves from below and at the minimum point on each of these curves; this is a mathematical relationship to which there are no exceptions. As long as MC is below AVC and ATC, both curves must be falling, because less is being added to the total cost for each successive unit of output than the average of all previous units. As long as MC is above AVC and ATC, both must be rising, because the MC of each added unit of output is larger than the average of the previous unit. This corresponds with our test score example earlier; if our marginal cost (next test score) is lower than our average cost (average test score) the average will decrease. So, until MC = AVC or ATC, the curves are decreasing even though MC might be increasing.

Managers often find the per-unit cost curves particularly useful because most often market prices are quoted or understood by customers on a per-unit basis. Because production costs vary across products produced and even across firms producing the same product (see Box 3.1) per-unit cost curves are also useful tools of comparison. Notice that in Figure 3.4 we did not graph the cost structure for those levels of input where output has decreased (Stage III). This is because, as aspiring managers and decision makers, we now know that it is an irrational area of production.

Costs Related to Production Function

The cost curves are derived directly from the physical production process. Thus, there is a direct relationship between the physical aspect of production (the product curves) and the economic aspects of production (the cost curves). Understanding this difficult concept is very important to the training of a manager. These direct relationships can be shown both graphically, as in Figure 3.6, and mathematically.

BOX 3.1   All Costs of Production Are Not the Same

In the real world of American agriculture, per-unit costs of production vary among producers of the same crop. For example, costs of production vary by region in the United States. Wheat producers in the Northwest and the Central Plains have the lowest average cost of production due to more favorable soil types and rainfall patterns as well as fewer problems with insects and disease than in other regions. The highest cost wheat production areas are the Northeast and Southern Plains.

Costs of production often decrease as the size of a farm increases. USDA survey information has indicated that average variable cash costs of production for all U.S. wheat farms in 1989 was $2.07 per bushel (including costs for seed, fertilizers, chemicals, custom operations, fuel, lubricants, electricity, repairs, and hired labor; see Figure 3.5). Individual wheat farm costs varied from under $1.20 to more than $5 per bushel. But just over half of all wheat farms had variable cash costs at or below the national average of $2.07/bushel.

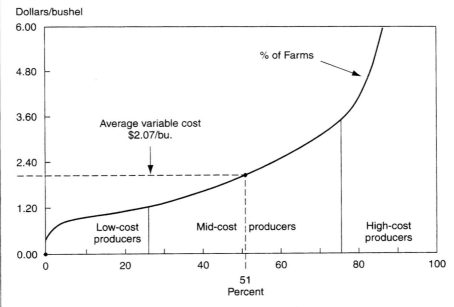

FIGURE 3.5   Cumulative distribution of variable cash expenses for U.S. wheat, 1989. (*Source:* USDA, Agricultural Economic Report No. 712, 1995.)

Can you explain some of the reasons why per unit costs would vary between high-cost and low-cost wheat producers?

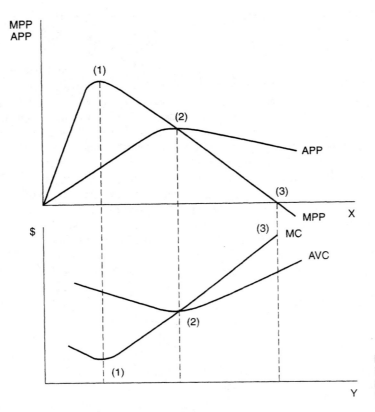

FIGURE 3.6
Relationship between product curves and per-unit cost curves.

Although the axes are different, the physical characteristics of the production function can be transferred directly to the economic values of the cost curves. It is evident from Figure 3.6 that the APP and AVC, and the MPP and MC curves, are mirror images of one another.

The shape of the MC curve is a direct result of the increasing/decreasing returns exhibited by the TPP, and since MPP indicates the slope of the TPP, MPP and MC are related. As indicated in Figure 3.6, when MPP is increasing, that is, we have increasing returns, MC is decreasing. Why? Increasing returns means that each additional unit of input is more productive than the last. Since our firm is a price taker, the cost of using another unit of input is a constant, equal to the input's price. If the cost of using another unit of input is constant, and each additional unit of input produces more output, then the additional cost of producing an additional unit of output (MC) must be decreasing [dividing a constant (ΔTC) by an increasingly larger (ΔY) number]. Thus, when we have increasing returns (MPP increasing), MC will be declining. When MPP reaches its maximum, point (1) in Figure 3.6, MC reaches its minimum. When MPP is decreasing, that is, we have decreasing returns, MC is increasing. Why? Decreasing returns means that each additional unit of input is less productive than the last. Because the cost of using another unit of input is constant, and each additional unit of input produces less output, then the additional cost of producing an additional unit of output (MC) must be increasing [dividing a constant (ΔTC) by an increasingly smaller (ΔY) number]. Thus, when we have decreasing returns (MPP decreasing), MC will be rising. As we enter Stage III, MPP becomes negative, point (3) in Figure 3.6, MC also becomes negative. However, since Stage III is an irrational stage of production, this portion of the curve is typically not depicted.

The APP curve represents the average productivity of each unit of input. As the average productivity of the variable input used increases (APP is increasing), the average cost of using the variable input to produce output decreases (AVC is decreasing) as shown in Figure 3.6. At point (2) where APP reaches its maximum, the point of maximum physical efficiency, AVC reaches its minimum. As APP decreases the average productivity of the variable input is decreasing; thus, the average cost of using the variable input to produce output is increasing (AVC is increasing).

In Figure 3.6 MPP crosses APP from above where APP is at its maximum; MC crosses AVC from below where AVC is at its minimum. This intersection point, point (2), is the beginning of Stage II. Thus, if it is possible to produce at all, the firm will produce up to the minimum point on AVC (maximum point on APP).

The relationships illustrated in Figure 3.6 are summarized in Table 3.3. These same relationships can be shown algebraically. The relationship between AVC and APP is as follows:

$$\text{AVC} = \frac{\text{TVC}}{Y} = \frac{P_x \times X}{Y} = P_x \times \frac{X}{Y} = \frac{P_x}{\dfrac{Y}{X}} = \frac{P_x}{\text{APP}}$$

Since P_x is a constant (our firm is a price taker), as APP increases, AVC will decrease. The shape of AVC depends on the production function, but the amount, or level, of AVC depends on the input price.

The relationship between MC and MPP is

$$\text{MC} = \frac{\Delta \text{TC}}{\Delta Y} = \frac{\Delta (P_x \times X)}{\Delta Y} = \frac{\Delta X \times P_x}{\Delta Y} = \frac{\Delta X}{\Delta Y} \times P_x = \frac{P_x}{\dfrac{\Delta Y}{\Delta X}} = \frac{P_x}{\text{MPP}}$$

TABLE 3.3 Relationships Between Product Curves

Physical or Technical	Economic
1. When MPP > APP, APP ↑.	1. When MC < AVC, AVC ↓.
2. When MPP = APP, APP is maximum.	2. When MC = AVC, AVC is minimum.
3. When MPP < APP, APP ↓.	3. When MC > AVC, AVC ↑.

TABLE 3.4 Relationship Between Per-Unit Cost Curves and Product Curves: Lou's Sandwich Shop (TFC = $10 and P_X = $4)

X	Y	APP	MPP ($\Delta Y \div \Delta X$)		AFC (TFC $\div Y$)	AVC (TVC $\div Y$)	ATC (TC $\div Y$)	MC ($\Delta TC \div \Delta Y$)	
0	0	∞			$—	$—	$—		
			>	2.00				>	$2.00
1	2	2.00			5.00	2.00	7.00		
			>	3.00				>	1.33
2	5	2.50			2.00	1.60	3.60		
			>	4.00				>	1.00
3	9	3.00			1.11	1.33	2.44		
			>	2.00				>	2.00
4	11	2.75			0.91	1.45	2.36		
			>	1.00				>	4.00
5	12	2.40			0.83	1.67	2.50		
			>	-1.00				>	-4.00
6	11	1.83			0.91	2.18	3.09		

Since P_x is a constant (firm is a price taker), as MPP increases, MC will decrease. Remember, the only change in TC that can come about is through change in TVC; thus, marginal cost involves consideration of only variable costs.

To illustrate these relationships, the product curves and the per-unit cost curves for our sandwich shop are presented in Table 3.4. As Table 3.4 illustrates, APP reaches its maximum at $X = 3$ ($Y = 9$), which is the point where AVC reaches its minimum at $Y = 9$ ($X = 3$). The maximum of MPP is between $X = 2$ and $X = 3$ ($Y = 5$ and $Y = 9$), which corresponds to where MC reaches its minimum between $Y = 5$ and $Y = 9$ ($X = 2$ and $X = 3$). The beginning of Stage II where APP is maximum corresponds to where AVC is minimum at $Y = 9$ ($X = 3$). Stage III begins where TPP is maximum, which corresponds to where MC becomes negative at $Y = 12$ ($X = 5$). From the table it is also easy to show that MC $= P_x \div$ MPP and that AVC $= P_x \div$ APP. For example, the average variable cost of producing 2 sandwiches is $2, which is equal to P_x ($4) divided by the APP (2.00) when $X = 1$.

There are specific implications for management in the direct relationship between the production function and costs of production. If little managerial control exists over the price that must be paid for inputs (a situation commonly faced in agriculture or individual firm decisions), then the production function and the method of physically combining inputs should receive managerial emphasis. An increase in the production function (firm productivity) can cause a direct decrease in costs of production and probably an increase in profits. In those cases where different qualities of inputs are available at differing prices, the manager can trade-off, through economic analysis, the lower productivity against a lower price for the input. He or she can identify how much lower the price would have to be to warrant accepting lower productivity.

TABLE 3.5 **Effect of an Input Price Increase**

X	Y	TFC	TVC	TC	AFC	AVC	ATC		MC
				TFC = $10 and P_x = $4					
0	0	$10	$0	$10	$ —	$ —	$ —		
								>	$2.00
1	2	10	4	14	5.00	2.00	7.00		
								>	1.33
2	5	10	8	18	2.00	1.60	3.60		
								>	1.00
3	9	10	12	22	1.11	1.33	2.44		

X	Y	TFC	TVC	TC	AFC	AVC	ATC		MC
				TFC = $10 and P_x = 6					
0	0	$10	$0	$10	$ —	$ —	$ —		
								>	$3.00
1	2	10	6	16	5.00	3.00	8.00		
								>	2.00
2	5	10	12	22	2.00	2.40	4.40		
								>	1.50
3	9	10	18	28	1.11	2.00	3.11		

Changes in Input Price or Production Function

So far in all our examples we have held P_x, the price the manager or farmer must pay for a unit of input, at a single level. In real life, managers must make economic decisions in the face of changing prices. What happens to cost levels and the position of the cost curves if the price of the input changes?

Input Price Increase

If the input price (P_x) increases, the cost of producing each output level increases. We can illustrate this numerically by comparing the cost curves for a situation where TFC = $10 and P_x increases from $4 to $6. This comparison is presented in Table 3.5, where the top portion shows the initial cost structure of the firm when TFC = $10 and P_x = $4. The lower portion of the table shows the impact on production costs when the input price increases to $6 ($P_x$ = $6). Direct comparison of the cost curves shows that the impact of an input price increase is an increase in the cost of producing each output level. At each output level, TVC and TC are higher, as are the per-unit costs AVC, ATC, and MC. Only the TFC and AFC remain unaffected. We illustrate the effect of an input price increase graphically in Figure 3.7.

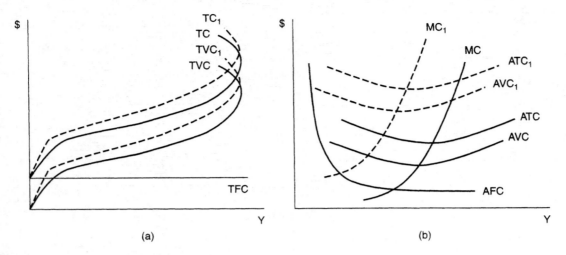

FIGURE 3.7 Effect of an input price increase.

TABLE 3.6   Effect of an Input Price Decrease

| | | | | TFC = $10 and P_x = $4 | | | | | |
X	Y	TFC	TVC	TC	AFC	AVC	ATC		MC
0	0	$10	$0	$10	$ —	$ —	$ —		
								>	$2.00
1	2	10	4	14	5.00	2.00	7.00		
								>	1.33
2	5	10	8	18	2.00	1.60	3.60		
								>	1.00
3	9	10	12	22	1.11	1.33	2.44		
				TFC = $10 and P_x = 2					
X	Y	TFC	TVC	TC	AFC	AVC	ATC		MC
0	0	$10	$0	$10	$ —	$ —	$ —		
								>	$1.00
1	2	10	2	12	5.00	1.00	6.00		
								>	0.67
2	5	10	4	14	2.00	0.80	2.80		
								>	0.50
3	9	10	6	16	1.11	0.67	1.78		

When P_x increases, the cost of producing the same output level is now higher. Thus, as illustrated in Figure 3.7(a), TVC and TC shift upward to the left, to TVC_1 and TC_1, while TFC remains unchanged. When P_x increases, the per-unit cost curves AVC, ATC, and MC all shift upward to the left, to AVC_1, ATC_1, and MC_1, while AFC remains unchanged as illustrated in Figure 3.7(b).

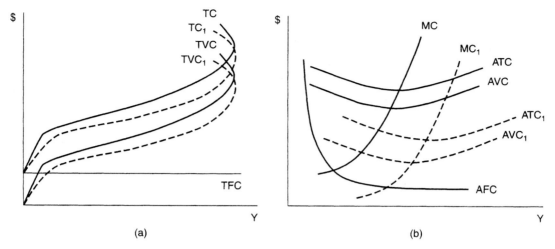

FIGURE 3.8 Input price decrease.

Input Price Decrease

What happens if the input price decreases or if there is an increase in the productivity of inputs due to a technological change? As we discussed earlier, a technological improvement allows the firm to produce the same level of output using fewer units of input, thus producing at a lower cost. If the input price (P_x) decreases, the cost of producing each output level also decreases. We can illustrate this numerically by comparing the cost curves for the situation where TFC = $10 and P_x = $4 with the decreased input price of $2. This comparison is presented in Table 3.6, where the top portion of Table 3.6 shows the initial cost structure of the firm when TFC = $10 and P_x = $4. The lower portion of the table shows the impact on production costs when the input price decreases to $2 ($P_x$ = $2). Direct comparison of the cost curves when P_x = $4 to the cost curves when P_x = $2 illustrates that the impact of an input price decrease is lower costs of producing each output level. Only the TFC and AFC remain unaffected by the price change. Figure 3.8 graphically illustrates the effect of a decrease in P_x on the position of the firm's cost curves.

When P_x decreases or a technological innovation occurs, the cost of producing the same amount of output now costs less. Thus, TVC and TC shift downward to the right, to TVC_1 and TC_1 while TFC remains unchanged as illustrated in Figure 3.8(a). When P_x decreases, the per-unit cost curves AVC, ATC, and MC all shift downward to the right to AVC_1, ATC_1, and MC_1, while AFC remains unchanged as illustrated in Figure 3.8(b).

As illustrated earlier, when input prices increase the cost of producing, each unit of output increases. As a result the total cost, variable cost, and marginal cost curves will shift upward (or inward in appearance) to the left, while fixed cost curves remain unchanged. Conversely, when input prices decrease, or a techno-

After a long controversy, in 1993 the Food and Drug Administration finally approved *bovine somatotropin* or bST for use in dairy production. Sold as "Prosilac" by Monsanto, bST is a hormone produced naturally in dairy cows but synthetically manufactured with new biotechnology. University researchers believe bST is "safe" and will not affect the quality or safety of milk and, after four years of use, this appears to be true.

Although there was concern about consumer acceptance of milk produced with bST, producers are beginning to make use of this technology. Currently, about 25% of all dairy cows receive bST injections about every two weeks once they are 60 to 120 days into lactation. These injections cost about $3 per week and milk output increase is dependent on dairy nutrition (i.e., the "balanced ration"). However, with a well-balanced ration a conservative estimate of the expected increase in milk production is about 6 to 9 pounds per day, or an additional 48 to 63 pounds per cow every week.

For $3 per week American dairy farmers can use a new technology that will increase milk production **from the same feed input** for each cow. At 1996 milk prices of $12.50 per hundredweight (net of hauling costs), the additional farmer income with bST injections is $6 to $7.87 per cow every week. Thus, bST represents a technology that increases the output of a given level of inputs and lowers the overall costs of production.

logical innovation occurs (see Box 3.2), the cost of producing each unit of output also decreases. As a result the total, variable, and marginal cost curves will shift downward (or outward in appearance) to the right, while the fixed cost curves remain unchanged.

What will happen to the curves if fixed costs increase or decrease? Will MC change if fixed costs increase or decrease? No, since fixed costs do not affect MC. If fixed costs increase or decrease, then the TFC and TC will change. The curves will shift upward or downward depending on the direction of change in TFC (since TC = TFC + TVC). If fixed costs change then AFC and ATC will change. The curves will shift with the direction of the shift depending on whether fixed costs increase or decrease (since ATC = AFC + AVC).

SUMMARY

In this chapter we learned the basics of the economic relationships in the production process, or the costs of producing output. Seven cost measures are of interest to the manager; three measure costs of production on a total output basis, and four measure the per-unit output cost of production. Because the cost curves are based on physical production, they are related to the product curves of the firm and a direct relationship between them can be shown. Using the cost curves, it is possible to identify a range of output in which profits will be maximized, that is, Stage II of production. The question of how much to produce cannot be fully answered until we consider the price of the output, but we are now one step closer to being able to make an economic decision.

QUESTIONS

1. Why don't total fixed costs vary with the level of output?
2. Show that marginal cost is equal to the price of the input divided by the marginal physical product.
3. How do technological improvements in factors of production affect costs of production, and why is it an important managerial concept for a farm producer?
4. Is it reasonable to say that marginal costs include only variable costs? Explain.
5. Prove that average variable cost is equal to the price of the input divided by the average physical product.
6. Using the production function presented in Table 3.1, calculate the total cost curves (TFC, TVC, TC) if TFC = $15 and P_x = $6. Graphically represent these curves.
7. Using the production function presented in Table 3.1, calculate the per-unit cost curves (AFC, AVC, ATC, MC) if TFC = $15 and P_x = $6. Graphically represent these curves.
8. Using the cost curves derived in Question 6, what will happen if P_x increases to $10? If P_x decreases to $3? Show these changes both numerically and graphically.
9. Using the cost curves derived in Question 7, what will happen if P_x increases to $10? If P_x decreases to $3? Show these changes both numerically and graphically.

CHAPTER 4

Management Decisions: How Much to Produce?

INTRODUCTION

Recall that the question we are trying to answer is "How much to produce?" Thus far we have examined the physical aspects of production (the product curves) and the costs associated with that production (the cost curves). Using this information we have been able to identify a range of input use, which translates into a range of output, in which profit maximization will occur. However, to determine the specific output level that will maximize profits, we need one last piece of information: the price of our output, P_y.

The output price, P_y, is the price received by the producer or manager for each unit of product that is sold. By using the output price, the manager is able to determine the value of output produced by each alternative input level. This value is revenue and when costs are subtracted from revenue we can identify profits. Thus, combining revenue information with costs of production we can finally determine the output level that will maximize profits and answer our question: How much to produce?

In This Chapter

The objective of this chapter is to introduce the student to the decision rules appropriate to answer the question "How much to produce?" Since the producer's objective is to maximize profits, deciding how much to produce is based on profit maximization. The profit maximization decision utilizes our knowledge of the physical production process and the costs of production, combined with the prices of products, to evaluate the costs and revenue at alternative levels of production, allowing an economic decision to be made. This decision, often referred to as the **factor–product decision,** is the cornerstone of economic decision making. In this chapter you will learn formal decision rules that will be useful for the rest of your life.

Assumptions Used

Let's restate the question we've been seeking to answer and apply it in two situations. First, given that a farmer has available all the variable input he or she might desire, how much variable input should be used to maximize profits? Secondly, if only limited amounts of variable resource are available, how should this "fixed amount" of resource be used in order to maximize profits? To analyze the profit maximization decision, we continue to use the single-variable input production function introduced in Chapter 2. As we will see later, the decision rules developed in this chapter extend quite easily to cases where there are multiple inputs and outputs.

Before examining the profit maximization decision more closely, let's review the assumptions we are using:

1. One product is produced, one production method.
2. One variable input, others are fixed.
3. The firm wants to maximize profits.
4. The firm is a price taker.
5. All firms produce a homogeneous product.

As in previous chapters we are assuming that all decisions are made under perfect certainty. Assumptions 1, 2, and 3 were introduced in Chapter 2. Assumption 4, which was introduced in Chapter 3, has an expanded meaning in this chapter. In addition to the inability to influence input prices, the assumption that the firm is a price taker also implies that the firm can sell as much of its product as it wants without affecting the market price. That is, within the market the individual firm has no influence over price, thus in the firm's eyes the price is fixed or constant. As in Chapter 3, assuming the firm is a price taker does not mean that the price of the output will never change, just that the individual firm has no influence over the price received for its product. Assumption 5 implies that the market does not distinguish between the products of individual firms, thus the firm must accept whatever the market price is for its product.

PROFIT MAXIMIZATION: THE DECISION RULES

We will examine the profit maximizing decision from two perspectives: the input basis and the output basis. Under the input basis approach the objective is to determine the input use level that will maximize the firm's profits; the output basis approach seeks to determine the output level, on either a total output or per-unit output basis, that will maximize profits. As we will see later, both of these approaches are exactly equivalent to each other and equivalent to the input basis approach. That is, all three approaches yield the same answer (the same input use level and production level). To illustrate the decision rules we will be developing, we will continue to use the sandwich shop example from Chapters 2 and 3.

Profit Maximization Decision: Input Basis

The objective in the input basis situation is to determine the level of input use that will maximize profits. We can achieve this objective by comparing the value of the product(s) produced to the value of the input(s) used to produce them.

Up to this time we have used the production function to describe the physical relationships between inputs and output in a given production process. However, to determine what input level will maximize profits, we need a measure of revenue on the basis of input use level. The product curves, presented in Chapter 2, show the relationship between input use and physical production. By multiplying each product curve by the output price we can translate production measured on a physical basis to output measured on a value, or monetary, basis and thus establish the relationship between input use and the value of production. The resulting information can be graphically presented as curves, or the revenue concepts: total revenue product, average revenue product, and marginal revenue product.

Total revenue product (TRP) shows the *dollar value of the output produced from alternative levels of variable input.* It is calculated by multiplying TPP at each input level by the output price, P_y, or

$$\text{TRP} = \text{TPP} \times P_y$$

In some textbooks TRP is called total value product (TVP).

Average revenue product (ARP) shows the *average value of output per-unit of input at each input use level.* It is calculated by multiplying APP at each input level by the output price, P_y, or

$$\text{ARP} = \text{APP} \times P_y$$

Alternatively, ARP = TRP ÷ X, or ARP = (P_y × TPP) ÷ X. Average revenue product is also called average value product (AVP).

Marginal revenue product (MRP) shows the *additional (marginal) value of output obtained from each additional unit of the variable input.* It is calculated by multiplying the MPP for each input use level by the output price, P_y, or

$$\text{MRP} = \text{MPP} \times P_y$$

Alternatively,

$$\text{MRP} = \frac{\Delta \text{TRP}}{\Delta X} = \frac{\Delta(\text{TPP} \times P_y)}{\Delta X} = \frac{P_y \times \Delta \text{TPP}}{\Delta X} = P_y \times \frac{\Delta Y}{\Delta X} = P_y \times \text{MPP}$$

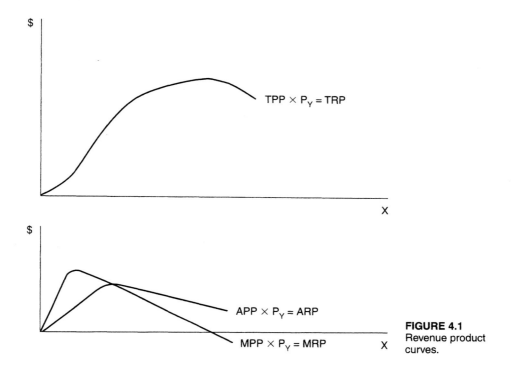

FIGURE 4.1
Revenue product
curves.

since P_y is a constant and does not change with output levels. This is sometimes called marginal value product (MVP) or value of the marginal product (VMP).

The revenue product curves represent the value of production obtained from alternative levels of the variable input combined with the fixed factors and are illustrated in Figure 4.1.

As illustrated in Figure 4.1, when we graph these revenue concepts, we must now change the vertical axis from physical units of output to dollars ($'s) to reflect the information now contained in the figure. Since we have multiplied each product curve by a constant, the relationships between the curves have not changed, only what the curves represent. Thus, Stage II, the rational production stage, begins where MRP = ARP (ARP is maximum) and ends where TRP is maximum (MRP = 0).

These revenue concepts provide more information than the product curves. We can now associate the revenue received, or the value of output, with each alternative level of input. We are gradually moving toward the ability to answer our question of how much input should be used in order to maximize profits.

Our objective is to determine the level of input use that will maximize profits. Because profits are equal to revenue minus costs of production, we need to compare the value of production in terms of the units of input used to the cost of using that input level. We have a measure of output in value terms, now we need a measure of cost that measures the cost of production in the same manner. To do so, we introduce a new cost concept: marginal factor cost.

Marginal factor cost (MFC) is the *cost of an additional unit of input* and is the amount added to total cost from using one more unit of variable input. Because our firm is a price taker marginal factor cost is simply P_x, the cost of purchasing one more unit of input. For example, in hiring labor for our firm it is the wage we must pay per person; in purchasing fertilizer it is the cost of an additional (marginal) unit of fertilizer.

We can use marginal factor cost to measure cost on the same basis that marginal revenue product measures the value of output, at the margin. But how does the producer use this information to determine how much input to use?

Since profits are the returns after the costs of production have been paid, the producer determines how much input to use by comparing the value of output produced by each input level to the cost of using that input level. The producer should continue to use additional units of input until the additional value of output received from using an additional unit of input is equal to (or slightly more than) the cost of using that unit of input. This is the economic decision rule that maximizes profit for a single-variable resource. In other words, the producer determines how much input to use by directly comparing the MRP (the additional value of output) of the input to the MFC (the cost of the input). Thus, profit maximizing managers will *use additional input until MRP $\geq$ MFC*. This decision rule states that unless the additional value of output received from using the next unit of input is greater than or equal to the additional cost of using that unit of input, the firm will not use that unit of input. This decision is shown graphically in Figure 4.2.

In a competitive market a firm is a price taker, so the additional cost of using an additional unit of input is equal to that input's price, P_x. Thus when graphed, MFC is a horizontal line. To find the input use level that maximizes profits, we find the point at which MFC = MRP. In Figure 4.2 the input use level that maximizes profit is X_1. If the firm uses an input level of less than X_1, for example, X_0, the firm would not be maximizing profits. Why? Because each additional unit of input that is used up to the level X_1 adds more to the value of output than to the cost of using the input (MRP > MFC); thus, revenue is increasing faster than costs. If the firm uses inputs beyond X_1 the firm will not maximize profits. At input use beyond X_1, the costs of using additional units of the input exceed the additional value of output from using those units (MRP < MFC), hence costs are increasing faster than revenue. Therefore, the firm will continue to increase production (use additional units of input) until MRP = MFC. If X_1 is the input level that maximizes profit, what is the output level? To find the output level produced, we simply find the TRP level produced by the input level X_1 (TRP$_1$ in Figure 4.2) and divide by the output price P_y (since TRP = $P_y \times$ TPP; then TPP = TRP $\div P_y$).

We can illustrate the rationale behind this decision rule even more clearly using our sandwich shop example. If the manager can sell each sandwich for $3.00 ($P_y$ = \$3), what is the profit maximizing level of input use, or how many workers should we hire to maximize profits? The appropriate curves are shown in Table 4.1, where TRP, ARP, and MRP are simply the TPP, APP, and MPP at each input level multiplied by P_y. For example, when $X = 1$ ($Y = 2$), TRP = $6 ($3 \times 2 = \$6$); when APP = 2, ARP = $6 ($3 \times 2 = \$6$); and when MPP = 2, MRP = 6 ($3 \times 2 = \$6$). We

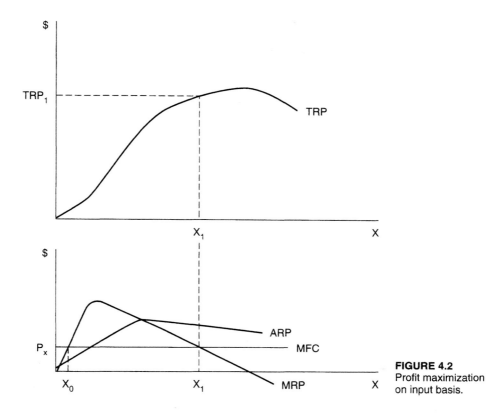

FIGURE 4.2
Profit maximization
on input basis.

know that it is profitable for a firm to use additional units of an input as long as the value of output received from the additional unit of input (MRP) is greater than the cost of using that unit of input (MFC). If we use the third unit of input, increasing the number of employees from two to three, we receive $12 as MRP (MPP $\times P_y$, or $4 \times \$3 = \$12$) and pay $4 for the input, so we have added $8 ($12 − $4 = $8) to the net returns of our firm.

Since MRP > MFC for the third unit of input, we examine the use of an additional unit of input. If we use the fourth unit of input, increasing employees from three to four, we receive $6 as MRP (MPP $\times P_y$, or $2 \times \$3 = \$6$) and pay $4 for the input, so we have added $2 ($6 − $4 = $2) to the net returns of our firm. Since MRP > MFC for the fourth unit of input, we look to use an additional unit of input. Looking at the possibility of hiring the fifth unit of input (increasing from four to five employees) we decide not to do so because we must pay $4 for it and it only increases our revenue by $3, hence net return would be decreased.

Thus, using our decision rule, MRP ≥ MFC, the profit maximizing input use level is four workers. The fourth worker adds $6 in revenue but only costs $4 to use. With four workers the output produced is eleven sandwiches (profit maximizing output). To see what our total net return is while operating at the profit maximizing level of four units of input, we simply add the net return received for

TABLE 4.1   Decision Information on an Input Basis: Lou's Sandwich Shop
(TFC = $10, Px = $4, and Py = $3)

X	Y	APP	MPP	ARP	TRP	MRP	MFC	Net Return per-unit[a]	Net Total Return[b]
0	0	∞		$0	$0				
			> 2			> $6.00	$4.00	$2.00	$2.00
1	2	2.00		6.00	6.00				
			> 3			> 9.00	4.00	5.00	7.00
2	5	2.50		7.50	15.00				
			> 4			> 12.00	4.00	8.00	15.00
3	9	3.00		9.00	27.00				
			> 2			> 6.00	4.00	2.00	17.00
4	11	2.75		8.25	33.00				
			> 1			> 3.00	4.00	−1.00	16.00
5	12	2.40		7.20	36.00				
			> −1			> −3.00	4.00	−7.00	9.00
6	11	1.83		5.49	33.00				

[a]Net return per-unit is MRP − MFC for each individual unit.
[b]The sum of the net per-unit returns is net total return at each input level.

each of the units, yielding a net return over variable costs of $17. This net return is not considered profit until fixed costs are considered. Assuming that the fixed costs for our sandwich shop remain at $10, as in Chapter 3, the profit for our firm is $7 ($17 − $10 = $7).

The student should be aware that MFC ≠ MC. Why? Marginal cost, which is the change in total cost divided by the change in output, is calculated on the basis of a change in output. Marginal factor cost, which is the change in total cost divided by the change in input use, is calculated on the basis of a change in input use and is always P_x.

At this point students often ask, "Is this really used in the real world?" The answer is *yes!* Although the terminology and method of presentation may differ, the basic underlying principle of the decision rule MRP ≥ MFC is taught in many of the management workshops and seminars presented by university faculty and extension economists across the country.

For example, the University of Kentucky holds workshops for new and inexperienced farmers on a regular basis. The faculty who teach these workshops try to illustrate basic profit maximization principles with straightforward examples in order to teach these farmers to apply the same principles in the input decisions on their farms or other businesses. For profit maximization decisions, the faculty gives participants this **basic rule**: *Continue to add input as long as the value of the output is greater than the cost of the input.* This basic rule is illustrated with examples such as the orchardgrass hay response to nitrogen fertilizer illustrated in Figure 4.3.

As Figure 4.3 illustrates, the first 40 lbs of nitrogen applied gives a 1.10-ton response, the second 40 lbs generates an additional 0.2-ton response, and the third 40 lbs results in a 0.05-ton loss in hay yields. The faculty often ask participants the

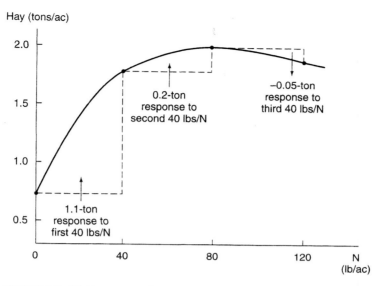

Hay (tons/ac)

0.2-ton
response to
second 40 lbs/N

−0.05-ton
response to
third 40 lbs/N

1.1-ton
response to
first 40 lbs/N

FIGURE 4.3 Yield response of orchardgrass hay to nitrogen fertilizer. (*Source:* University of Kentucky Farm Management Workshop, 1996.)

following question: "If the price of orchardgrass hay is $80/ton and the price of ammonium nitrate (nitrogen) is $0.35/lb of actual nitrogen, then how much fertilizer (40 lbs? 80 lbs? 120 lbs?) should the farmer topdress on his or her hay?" Using the basic rule, the answer is 80 lbs of nitrogen. Why? The first 40 lbs of nitrogen costs $14 to use and generates an $88 increase in returns, the second 40 lbs of nitrogen costs $14 to use and results in a $16 increase in returns, whereas the third 40 lbs of nitrogen costs $14 to use but results in a $4 decrease in returns. In this example, 80 lbs of nitrogen (or where MRP ≥ MFC) will maximize profits given these prices. Although the faculty can teach the principle of the decision rule MRP ≥ MFC without using terminology like "marginal factor cost" and "marginal revenue product," the underlying principle remains the same. Why would a farmer probably not want to increase the topdressing from 80 to 120 lbs of nitrogen, even if the extra 40 lbs of nitrogen were *free?*

Decision Under Changing Input Prices

Up until now we have been assuming that the P_x remains constant. In real life managers often make economic decisions in the face of changing prices. Figure 4.4 illustrates the impact changing input prices have on a firm's input use decision.

Because a rational producer only produces in Stage II, and because profits can be maximized in Stage II, only the Stage II portion of the marginal revenue product curve is considered in Figure 4.4. We know that to maximize profits the firm will continue to use additional units of input until MRP ≥ MFC. When input prices

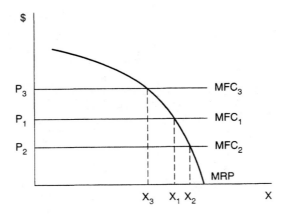

FIGURE 4.4
Input decision under changing input prices.

change, the firm must make a new decision about how much input to use. As input price (P_x) increases, from P_1 to P_3 in Figure 4.4, the rational manager decreases input use from X_1 to X_3, since the revenue at the initial input use level, X_1, is less than the cost of using X_1, at the new price level P_3 (MFC$_3$). Therefore, the firm reduces input use to X_3, where MRP = MFC$_3$. Conversely, if input prices were to decrease, the firm would find it economical to increase input use. If input price (P_x) decreases, from P_1 to P_2 in Figure 4.4, the firm would increase its input use from X_1 to X_2. Why? At the initial input use level, X_1, the marginal revenue product now exceeds the cost of using an additional unit of input, P_2. Therefore, the firm can afford to use more units of input and will increase input use from X_1 to X_2 where MRP = MFC$_2$. In the face of changing prices, the firm applies the same decision rule, MRP $\geq$ MFC, to determine the new profit maximizing level of input use.

Firm's Demand Curve for Variable Input

The rational manager will buy and use each input just up to the point where the marginal revenue product of using each input is equal to its marginal factor cost. Thus, the price of the input (MFC) is a major factor in determining how much of an input the firm will purchase or how much of an input the firm demands. The marginal revenue product curve provides us with the information of how much of an input, for example, fertilizer, will be utilized or demanded at various prices for that input. This is referred to by economists as **factor demand** and is useful to such individuals as fertilizer dealers for identifying producer needs (and sales opportunities) in an area.

The demand for an input reflects the value that input adds to revenue, or its MRP. As the price of an input changes, the quantity of the input used, or demanded, by the firm also changes. Thus, using changes in the price level for an input and the decision rule MRP $\geq$ MFC, we can trace the firm's demand schedule for that input, holding all other inputs and their prices constant. This is illustrated in Figure 4.5.

Since the rational manager operates in Stage II, the firm's demand curve for factors, or factor demand, is represented by the MRP of that input within Stage II.

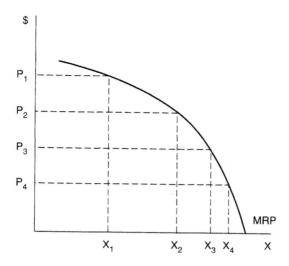

FIGURE 4.5
Firm's demand curve for variable input.

The Stage II portion of the marginal revenue product curve is downward sloping, depicting the real-world behavior of firms: As the price of the input increases, use of the input will decrease; as the price of the input decreases, use of the input will increase. As shown in Figure 4.5, as input price decreases from P_1 to P_2, input use by the firm increases from X_1 to X_2; as input price decreases from P_2 to P_3 input use increases from X_2 to X_3; as price decreases from P_3 to P_4 input use increases from X_3 to X_4. Conversely, if input prices increase from P_4 to P_1, input use by the firm decreases correspondingly from X_4 to X_1 as the firm reestablishes input use where MRP = MFC to maximize profits.

Using this new knowledge, we can trace the sandwich shop's demand curve for workers, its variable input. Using the basic decision rule, MRP ≥ MFC, and the production information presented in Table 4.1, we can determine how many workers will be hired at different wage rates. When P_x = $4, four employees are hired. If the wage rate increases to $7 (MFC = $7), the sandwich shop will hire three employees (MRP of third worker is $12, MFC is $7). If the wage rate decreases to $2 (MFC = $2), the sandwich shop will hire five employees (MRP of fifth worker is $3, MFC is $2). Thus, as the input price increases (from $4 to $7) the sandwich shop hires fewer employees (from X = 4 to X = 3); as the input price decreases (from $4 to $2) the sandwich shop hires additional employees (from X = 4 to X = 5). Using the decision rule, MRP ≥ MFC, and different input prices we are able to determine the firm's demand for workers, which is the MRP curve within Stage II.

Position of Firm's Demand Curve for Variable Input

Since marginal revenue product is derived by multiplying the marginal physical product by the output price, the position of the firm's demand curve for a variable input depends on both the output price, P_y, and the underlying production function.

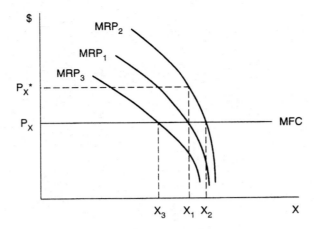

FIGURE 4.6 Input decision under changing output prices.

Thus, a change in either P_y or TPP will change the position of the marginal revenue product curve, as illustrated in Figure 4.6.

When a technological innovation occurs, the TPP and MPP are shifted upward to the right as the input becomes more productive. This shift results in MRP shifting upward (or outward) to the right, as illustrated in Figure 4.6 where marginal revenue product shifts from MRP_1 to MRP_2. As MRP shifts upward, input use increases from X_1 to X_2 as it becomes more profitable for the firm to use additional units of input at the same input price, P_x. The upward shift also implies that at the same input use level (X_1) the firm can now afford to pay more per unit of input, P_x^* instead of P_x.

When output prices change, the MRP will shift in the direction of the change in output price. This is the case because MRP shifts when multiplying the same MPP curve times a larger constant, in the event of an increase in P_y, or a smaller constant, in the case of a decrease in P_y. If the output price increases, the MRP will shift upward to the right, from MRP_1 to MRP_2 in Figure 4.6. At the same input price level, P_x, if P_y increases it is now more profitable to use additional input, so input use increases from X_1 to X_2, resulting in increased output levels. If P_y decreases, the MRP will shift downward (or inward) to the left, from MRP_1 to MRP_3 in Figure 4.6, indicating that at the same input price (P_x) it is less profitable to use the variable input. Hence, input use decreases from X_1 to X_3 and production levels fall.

By comparing the value of production obtained from using an additional unit of input (MRP) to the cost of using that unit of input (MFC), we are able to determine the input use level that maximizes profit, and to develop a decision rule for this economic decision. We now turn our attention to determining the output level that will maximize profits.

Profit Maximization Decision: Total Output Basis

Under this approach, the manager or producer compares the total revenue and total cost at each production level to determine which output level will maximize profits. **Total revenue (TR),** also referred to as total sales or gross income, is sim-

TABLE 4.2   Total Revenue

Y	TR (P_y = $2)	TR (P_y = $3)	TR (P_y = $1)
1	$2	$3	$1
2	$4	$6	$2
3	$6	$9	$3

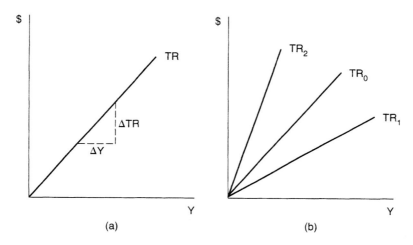

FIGURE 4.7   Revenue on a total output basis: total revenue curve

ply the *amount of money received when the producer sells the product.* It is calculated by multiplying each potential output level by the output price, P_y, or

$$TR = TPP \times P_y$$

Note that TR is synonymous with TRP, and so total revenue is equal to the dollar value of the output. Table 4.2 illustrates this concept.

From Table 4.2 it is obvious that total revenue is dependent on the level of the output and the price of the output. When output price increases from $2 to $3, the revenue at each output level also increases; conversely when output price decreases from $2 to $1, the revenue at each output level also decreases. How do we express the firm's total revenue graphically? We use a total revenue curve, which is presented in Figure 4.7.

Since the firm is a price taker, the firm sells each additional unit of output for the same price. Therefore, the added revenue from each additional unit sold is the same as the output price, P_y. Thus, TR is linear (straight line), as shown in Figure 4.7(a), with a slope ($\Delta TR \div \Delta Y$) equal to the output price, P_y. As illustrated in Table 4.2, as the output price changes, the firm's total revenue also changes since the firm is a price taker. The impact of changing output prices on the firm's total revenue

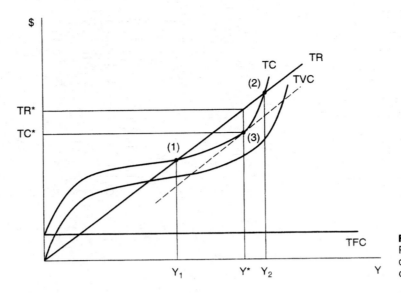

FIGURE 4.8
Profit maximization
decision: total
output basis.

curve is illustrated graphically in Figure 4.7(b). At the initial output price (for example, $2) the firm's total revenue curve is represented by TR_0. If P_y decreases (for example, to $1), the total revenue curve rotates downward, from TR_0 to TR_1, indicating that less revenue is received for each output level. If P_y increases (for example, to $3), the total revenue curve rotates upward, from TR_0 to TR_2, indicating that more revenue is received for each output level.

Combining this new knowledge of total revenue with the total cost concepts introduced in Chapter 3, we can now determine the output level that will maximize profits. The firm will continue to expand production as long as the increase in revenue is greater than the increase in costs. This is illustrated in Figure 4.8.

Profits are equal to total revenue minus total costs. Therefore, maximum profits will occur where the distance between the total revenue curve and the total cost curve is greatest. Using Figure 4.8, we can find the output level that will maximize profits. Points (1) and (2), corresponding to output levels Y_1 and Y_2, are **break-even points** because *at these output levels total revenue exactly equals total costs* (TR = TC). Thus producing output at levels less than Y_1 or greater than Y_2 will result in negative profits (TR < TC), or losses, for the firm. Profits are made in the range of output between Y_1 and Y_2 because total revenue exceeds total costs (TR > TC) in this output range.

Maximum profit is found graphically by taking a line parallel to the total revenue curve (line has same slope as TR) and finding where it is just tangent to the total cost curve (dashed line in Figure 4.8). Two curves are tangent at a particular point if the curves touch (without crossing) at that point. At the point of tangency, point (3) in Figure 4.8, the added cost of producing a unit of output is exactly equal to the added revenue from selling that unit of output and is the point where the vertical distance between TR and TC is the greatest. The optimal output level, the

TABLE 4.3 Decision Information on a Total Output Basis: Lou's Sandwich Shop (TFC = $10, P_x = $4, and P_y = $3)

X	Y	TR ($Y \times P_y$)	TFC	TVC	TC	Profit (TR − TC)
0	0	$0.00	$10.00	$0.00	$10.00	$−10.00
1	2	6.00	10.00	4.00	14.00	−8.00
2	5	15.00	10.00	8.00	18.00	−3.00
3	9	27.00	10.00	12.00	22.00	5.00
4	11	33.00	10.00	16.00	26.00	7.00
5	12	36.00	10.00	20.00	30.00	6.00
6	11	33.00	10.00	24.00	34.00	−1.00

[handwritten margin note: Profit max with arrows pointing to rows 3 and 4]

output level that maximizes profit, is Y^*. The total revenue from selling Y^* is TR^* and the total cost of producing Y^* is TC^*, with the profit earned from producing Y^* equal to $TR^* - TC^*$.

We can illustrate this decision more clearly by using our sandwich shop example. The TR and TC curves for the sandwich shop are presented in Table 4.3.

Profit at each output level is calculated by subtracting the total cost of that output level from the total revenue. Table 4.3 shows that maximum profit, $7, is made when eleven units are produced even though some profit would be made at the ninth, eleventh, and twelfth levels of output. The point where TR first equals TC, referred to as the break-even point, is between the output level of five and nine units. Another break-even point occurs between the twelfth and last level of output when TR slips below TC. For our sandwich shop, the output level that maximizes profit is eleven sandwiches. Since this output level is produced using four workers ($X = 4$), the profit maximizing output level found using the TR/TC approach is identical to the profit maximizing output level found using MRP analysis (Table 4.1).

Decision Under Changing Prices

Thus far we have only looked at this decision under a single price level. However in real life the manager must adjust his or her output levels in the face of changing input and output prices. What happens to the profit maximizing decision if the input or output price changes?

Decision under changing output prices. If the output price, P_y, changes, the level of revenue earned at each output level also changes. This is illustrated in Figure 4.9.

If the output price (P_y) increases, each unit of output produced by the firm earns more revenue. For the same level of production costs, an increase in P_y makes it less profitable to produce at the initial level, so production levels will increase. This is illustrated in Figure 4.9, where the firm maximizes profits initially by producing output level Y_1. As output price increases the TR rotates upward, from TR_1

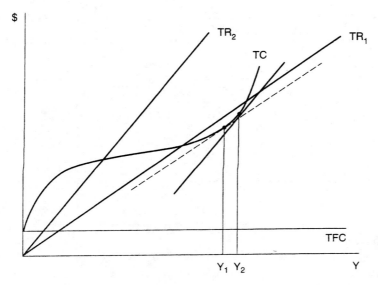

FIGURE 4.9
Total output decision under changing output prices.

to TR_2 in Figure 4.9, indicating that more revenue is earned at each output level. Given the new output price, the initial production level, Y_1, no longer maximizes profits. The firm determines its new production level, Y_2, by taking a line parallel to the new total revenue curve, TR_2, and finding where it is tangent to the total cost curve (dashed line in Figure 4.9). Since the firm's costs of production have not changed, the increased revenue means more profit is earned at each output level, enabling the firm to expand production from Y_1 to Y_2. As output price increases, the firm reallocates its resources in order to maximize profits, given the new output price, resulting in increased production. Conversely, as P_y decreases (for example, from TR_2 to TR_1), less revenue is earned at each output level and so production levels will decrease (for example, from Y_2 to Y_1).

Decision under changing input prices. If the input price changes, the cost structure of the firm also changes. As the costs of production change, so will the output level that maximizes profits. The profit maximizing decision under changing input prices is illustrated in Figure 4.10.

If the input price increases, the cost of producing each output level increases. At the same output price, an increase in P_x makes it less profitable to produce at the initial level, so profit maximizing production levels will decrease. This is illustrated in Figure 4.10, where for simplicity only TFC and TC curves are drawn. Initially, the firm produces Y_1 to maximize profits. As TC increases from TC_1 to TC_2 due to an increase in P_x, the profit maximizing output level decreases from Y_1 to Y_2. The decrease in output occurs because at the same level of revenue the increased costs of production mean less profit is earned at each output level. Thus, the firm must reallocate its resources in order to maximize profits under the new cost structure. Conversely, as total costs decrease (for example, from TC_2 to TC_1) due to a decrease in input prices, the firm would find it profitable to increase pro-

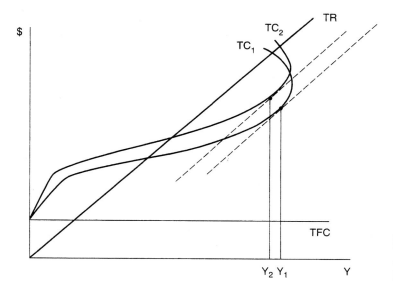

FIGURE 4.10
Total output decision under changing input prices.

duction (for example, from Y_2 to Y_1) since each output level earns the same total revenue but costs less to produce.

Most producers and managers do not make decisions looking at total revenue versus total costs; most decisions are made "at the margin," for example, "What small changes or additions should be made to our entire enterprise?" Accordingly, examining the marginal or incremental unit to be added or deleted from a production schedule is valuable for learning appropriate managerial decision making.

Profit Maximization Decision: Per-Unit Output Basis

Under the per-unit approach the output level that will maximize profits is determined by comparing the amount that each additional unit of output adds to total revenue and to total cost. Recall marginal cost (MC), by definition, represents the additional cost of producing an additional unit of output (MC = ΔTC ÷ ΔY). To compare added cost to added revenue, we need a measure of revenue on a per-unit of output basis. We can use the concept of total revenue, introduced in the last section, to derive average revenue and marginal revenue.

Average revenue (AR) is the *average dollar amount received per-unit of output sold (produced).* It is calculated by dividing the total revenue at each output level by the output level, or

$$AR = \frac{TR}{TPP} = \frac{TPP \times P_y}{TPP} = P_y$$

Since Y is synonymous with TPP we can also write AR = TR ÷ Y.

Marginal revenue (MR) is the *addition to total revenue from selling (producing) one more unit of output.* It is calculated by dividing the change in total revenue by the change in output level, or

$$MR = \frac{\Delta TR}{\Delta TPP} = \frac{\Delta(TPP \times P_y)}{\Delta TPP} = \frac{\Delta TPP \times P_y}{\Delta TPP} = P_y$$

Since our firm is a price taker, P_y does not change with output levels. We can also write $MR = \Delta TR \div \Delta Y$. Marginal revenue is always equal to P_y in this stage of our discussion and is extremely useful in economic analysis. Table 4.4 illustrates this revenue concept.

Given our assumption that the firm is a price taker, the firm sells each additional unit of output for the same price. Thus, as illustrated in Table 4.4, $MR = AR = P_y$, and if the firm can sell all it wants at the same price, it will sell all it produces. Just as with TR, the level of MR and AR depends on the output price level. For example, if output price increases from \$2 to \$3 (as in Table 4.2) the MR and AR both increase to \$3. If output price decreases from \$2 to \$1 (as in Table 4.2) the MR and AR decrease correspondingly to \$1. The AR and MR curves are illustrated in Figure 4.11.

Since the firm is a price taker, the additional revenue from selling an additional unit of output is equal to that output's price, P_y. Thus when graphed, MR (and AR) is a horizontal line, as illustrated in Figure 4.11(a). If the output price changes, the position of the MR curve changes. The impact of changing output prices on the firm's marginal revenue curve is illustrated graphically in Figure 4.11(b). At the initial output price (for example, \$2) the firm's marginal revenue curve is represented by MR_0. If P_y increases from P_0 to P_2 (for example, to \$3), the marginal revenue curve shifts upward from MR_0 to MR_2 indicating that more revenue is received for each unit of output sold. If P_y decreases from P_0 to P_1 (for example, to \$1) the marginal revenue curve shifts downward, from MR_0 to MR_1, indicating that less revenue is received for each unit of output sold.

Combining the per-unit cost curves with the per-unit measure of revenue will allow us to determine the output level that will maximize the firm's profits. The firm will continue to expand production until the added revenue received from an additional unit of output sold is equal to the additional cost of producing that unit

TABLE 4.4 Per-Unit Revenue: Marginal Revenue and Average Revenue

Y	TR (P_y = \$2)	AR (TR $\div$ Y)	MR ($\Delta TR \div \Delta Y$)
1	\$2	\$2	
			> \$2
2	\$4	\$2	
			> \$2
3	\$6	\$2	

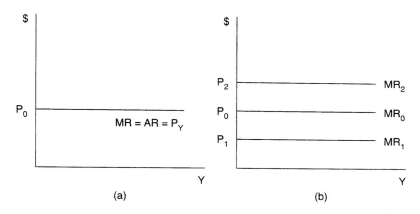

FIGURE 4.11 Revenue on a per-unit output basis: marginal revenue curve.

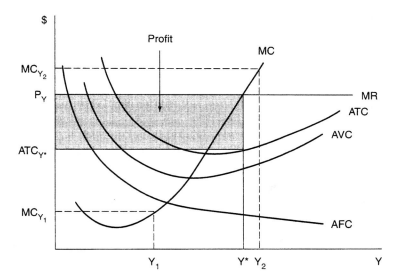

FIGURE 4.12 Profit maximization decision: per-unit output basis.

of output. Thus, the firm compares MR and MC to determine what output level to produce. This decision is illustrated graphically in Figure 4.12.

Profits (represented by a Greek pi, π) will be maximized where the added cost of producing an additional unit of output (MC) is equal to the additional revenue from selling that unit of output (MR). Thus, the firm will continue to produce additional units of output up to the point where MR = MC. As illustrated in Figure 4.12, the firm will produce Y^* units of output (intersection of MR and MC) to maximize profits. How do we calculate the profits at this production level? The per-unit profit of producing Y^* is equal to the per-unit revenue from selling Y^* less the cost of producing Y^*. The per-unit revenue is the output price (P_y), and production costs comes directly from the ATC (ATC$_{Y^*}$), or

$$\text{Per-unit } \pi = P_y - \text{ATC}_{Y^*}$$

The total profit of producing Y^* is equal to the per-unit profit times the number of units produced, or

$$\pi = (P_y - \text{ATC}_{Y^*}) \times Y^*$$

The intersection of MR and MC is a unique point and is the basis for the general profit maximizing decision rule: MR = MC. In Figure 4.12, Y^* is the output level that maximizes profit. Why? At the output level Y^*, the additional cost of producing the Y^*'th unit of output is exactly equal to the added revenue from selling the Y^*'th unit of output (MR = MC). If the firm produces an output level less than Y^*, for example Y_1 in Figure 4.12, the marginal revenue from selling another unit of output (P_y) is greater than the additional cost of producing another unit of output (MC$_{Y_1}$), and the firm can increase profits by expanding production. If the firm produces an output level greater than Y^*, for example Y_2 in Figure 4.12, the marginal revenue from selling another unit of output (P_y) is less than the added cost of producing it (MC$_{Y_2}$) and the firm's profit will decrease if it expands production. As illustrated in Figure 4.12, the firm maximizes profits only if it produces where the added cost equals added revenue, the point where MR = MC, and if the firm produces at this point there is no incentive for the firm to change its production level.

We can illustrate the MR = MC rule numerically using our sandwich shop example. The appropriate per-unit cost curves and marginal revenue are presented in Table 4.5.

TABLE 4.5 **Decision Information on a Per-Unit Output Basis: Lou's Sandwich Shop (TFC = $10, P_x = $4, and P_y = $3)**

X	Y	AFC	AVC	ATC		MC	MR
0	0	$—	$—	$—			
					>	$2.00	$3.00
1	2	5.00	2.00	7.00			
					>	1.33	3.00
2	5	2.00	1.60	3.60			
					>	1.00	3.00
3	9	1.11	1.33	2.44			
					>	2.00	3.00
4	11	0.91	1.45	2.36			
					>	4.00	3.00
5	12	0.83	1.67	2.50			
					>	−4.00	3.00
6	11	0.91	2.18	3.09			

The firm will continue to produce additional sandwiches until MR ≥ MC or, stated another way, if an additional unit of output costs more to produce than the additional return received for that unit it is not economical to produce it. In Table 4.3, the MR (P_y = $3) is greater than the MC of producing another unit of output until the twelfth unit of output is reached. At this point, increasing production from eleven to twelve sandwiches (the twelfth unit) would add $4 in expenses (MC = $4) to our firm, but we would only receive $3 in additional revenue (MR = $3). Obviously it makes no economic sense to increase production to twelve units. So our firm would use four workers to make eleven sandwiches per hour as the profit maximizing level of production. The per-unit profit is equal to $0.64 ($P_y$ − ATC, or $3 − 2.36) and the total profit is equal to $7.04 ($0.64 × 11), which is equal, barring a small rounding error, to the total profit found using the previous two approaches.

Decision Under Changing Prices

Thus far we have been examining this economic decision under a single price level. However, in real life, managers must make production decisions in the face of changing output prices and input prices. We can use the decision rule developed in the preceding section to analyze the profit maximizing decision, given changes in either product prices or input prices.

Decision under changing output prices. If the output price changes, the profit maximizing output level will change depending on the direction of the price change. This is illustrated graphically in Figure 4.13, where the firm maximizes profit given the initial output price P_1 by producing Y_1, where MR_1 = MC. If the output price, P_y, changes, the point where MR = MC changes and production changes accordingly. In this situation, an increase in product price is reflected in an increase in MR; a decrease in product price is reflected in a decrease in MR. If output price

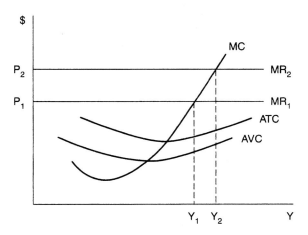

FIGURE 4.13
Per-unit output decision under changing output prices.

increases from P_1 to P_2 in Figure 4.13, the additional revenue per-unit of output also increases (from MR_1 to MR_2). As output price increases, the manager is economically justified to produce additional units of output because revenue per added unit is greater than the added costs of producing it (MR > MC). Thus, as P_y increases from P_1 to P_2 the firm expands production from Y_1 to Y_2 to maximize profits. Conversely, if output price decreases (say, from P_2 to P_1), production levels would decrease (say, from Y_2 to Y_1).

Decision under changing input prices. If input price changes, the optimal production level will also change due to changes in the firm's cost structure. This is illustrated in Figure 4.14. For simplicity, only the MC and ATC curves are used in this figure to show the impact input price changes have on managerial decisions. At the initial output price (P_y) and input price (P_x) levels in Figure 4.14, the firm maximizes profits by producing Y units (where MR = MC) of output. If the input price increases, both the MC and ATC curves will shift upward to the left, from MC to MC_1 and ATC to ATC_1 in Figure 4.14, indicating it now costs more to produce each unit of output. At the same output price, P_y, the firm will decrease production from Y to Y_1 because higher costs make it less profitable to maintain the initial production level. If the input price decreases, or if a technological improvement occurs that lowers input costs, the MC and ATC curves will shift downward to the right, from MC to MC_2 and ATC to ATC_2 in Figure 4.14, indicating that it is now less expensive to produce each unit of output. At the same output price, the firm will expand production from Y to Y_2 as the lower costs make it economical to increase production levels. Thus, as price of input increases, production decreases; as input price decreases, production increases.

The decision rules we have developed (MRP ≥ MFC; TR − TC; MR = MC) allow managers to understand how profit is maximized given changes in prices and costs. You, as the student, now have tools that will enable you to evaluate and make economic decisions under alternative management situations.

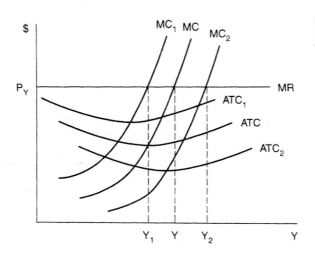

FIGURE 4.14
Per-unit output decision under changing input prices.

Equivalence of Profit Maximization Decision Rules

When we began talking about profit maximization we stated that the three methods of determining how much to produce were equivalent. Not only do these approaches yield the same decision and profit picture for the producer, the decision rules can be shown to be mathematically equivalent. In essence, the same decision is being made regardless of whether the manager is looking for the input level to use to maximize profits or the output level to produce in order to maximize profits. This equivalence is shown algebraically in Table 4.6.

Recall that when using the TR/TC approach we determined the optimal output level by equating the slope of the total revenue curve to the slope of the total cost curve by using a line parallel to TR (same slope) and tangent to TC. Since the slope of the total revenue curve ($\Delta TR \div \Delta Y$) is equal to MR, and the slope of the total cost curve ($\Delta TC \div \Delta Y$) is equal to MC, we were in fact equating MR and MC to determine the profit maximizing output level.

The equivalence of these three methods can also be illustrated using our sandwich shop example. Using the decision rule MRP $\geq$ MFC (Table 4.1), we determined that the profit maximizing input use was $X = 4$, which produced eleven sandwiches with a net return of $17, which, after subtracting fixed costs of $10, yields a profit of $7. Comparing total revenue to total cost (Table 4.3), we determined that the output level of eleven sandwiches (produced using four employees) maximized profits, which equal $7. Using the decision rule MR = MC (Table 4.5), we determined the optimal output level to be eleven sandwiches. The net profit at this level is derived by multiplying the per-unit profit times the output level: ($3.00 $-$ $2.36) $\times$ 11 = $7.04, which, except for some rounding error, is the same profit estimated using the other two approaches.

TABLE 4.6 Equivalence of Profit Maximization

Input Decision Rule	Output Decision Rule
MRP = MFC	MC = MR
MRP = P_x	MC = P_y
MPP $\cdot P_y = P_x$	$\dfrac{P_x}{MPP} = P_y$
	$\dfrac{P_x}{\left(\dfrac{\Delta Y}{\Delta X}\right)} = P_y$
$\left(\dfrac{\Delta Y}{\Delta X}\right) \times P_y = P_x$	$\left(\dfrac{\Delta X}{\Delta Y}\right) \times P_x = P_y$
$\Delta Y \times P_y = \Delta X \times P_x$	$\Delta X \times P_x = \Delta Y \times P_y$
(Added output value must equal added input cost)	(Added input cost must equal added output value)

SELECTED APPLICATIONS

The material we have covered thus far in our attempt to be able to perform economic analysis and become managers is so important that it will be used again and again in this textbook. Many circumstances warrant use of this basic microeconomic theory and decision rules. To show you how these rules can be utilized in different situations, consider the following applications.

Break-Even Analysis

It is very common for a manager to be concerned about when, or if, he or she reaches the break-even point at which revenue covers all costs as production or sales increase. Over the long run, if break-even of total revenue over total costs isn't achieved, the firm will or should go out of business. Figure 4.15 indicates the difference between the short-run versus the long-run and gives us some idea about how to use **break-even analysis.**

 Figure 4.15 shows that until point (1), where TR = TVC, we aren't even covering the variable costs of production, and it isn't until point (2), where TR = TC, that we've covered all costs of production and can then start potentially earning a profit. Obviously we would never increase production past point (3) because no profit is made at that output level. In the short-run, we would stay in production as long as the production (sales level) we can achieve is at least greater than point (1). Any return we receive over variable costs, for example, amount A at point (4),

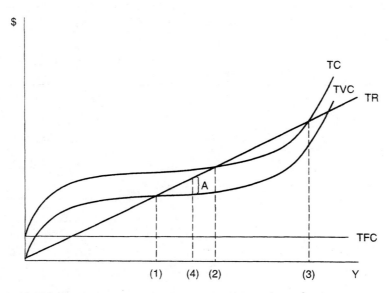

FIGURE 4.15 Break-even analysis: total output basis.

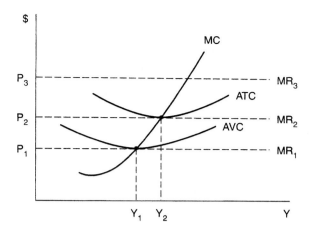

FIGURE 4.16
Break-even analysis: per-unit output basis.

can be used to pay partially for some of the fixed costs. In the long-run, where all factors are variable and we can decide to go into or out of business, the manager must expect to achieve sales of at least output level (2) before a decision in favor of continuing production would take place.

The same analysis can be performed using the per-unit or average cost curves, as illustrated in Figure 4.16, where the break-even price is P_2, where $P_y = MC = min$ ATC (since $P_y = MR$, at this point $MR = MC = min$ ATC). At this price level, all costs of production are just being covered by revenue. The firm is not earning a profit, nor is it suffering a loss (firm produces output Y_2 when $P_y = P_2$); it is breaking even. As long as product price, P_y, is expected to be above P_1 in the short-run, the manager would continue producing the product because some contribution to paying fixed costs is available. Prices between P_1 and P_2 in Figure 4.16 (the range between where $MR = min$ AVC and $MR = min$ ATC) are often referred to as **loss minimization prices.** In this price range revenues are covering the variable costs of production but not all of the fixed costs so the firm is operating at a loss. However, if the firm shuts down it must still pay its fixed costs. So the choice is between paying fixed costs and producing nothing, or producing an output level between Y_1 and Y_2 covering its variable costs and a portion of the fixed costs.

If the firm cannot make a profit, the next best choice is to minimize the loss. In the short-run if product prices fall below where $MR = min$ AVC (P_1 in Figure 4.16), the firm is not covering its variable costs of production. In this case it pays for the firm to shut down in the short-run even though it must still pay for its fixed factors. Prices below P_1 (where $MR = min$ AVC) are often referred to as the **shut-down case** or **close-down case.** In the long-run, where all factors of production are variable (ATC and AVC are the same), an average price of P_2 is necessary to keep that manager and his or her firm in production. Obviously a price of P_2 or greater in Figure 4.16 (where $MR \geq min$ ATC) would generate smiles and profits for the manager. Since P_2 is the break-even price level, where revenues are just covering costs, then at prices greater than P_2, say, P_3, profits can be made.

We personally saw this principle illustrated in 1962 at Thorne, North Dakota. A drought hit the farms in this area and, at harvest time, the following situation held. Eighteen dollars per acre had been invested in the crop at that time; harvesting the crop (the variable costs) was going to cost another $6 per acre; so total costs were $24 per acre. The wheat yield was only six bushels per acre and market price was about $2, so total revenue was only $12 per acre. The question to be answered was "Should we harvest the wheat even though we will lose $12 per acre?" The decision was that if we didn't harvest the crop, we would lose $18 per acre (all the fixed costs) instead of $12 so we went ahead and harvested. In this case, as we minimized our loss in the short-run, we were also maximizing our profits because our maximum profit was a loss. Thus, in the short-run the manager need not worry about the fixed costs. In the longer run, crop yields improved in North Dakota.

Maximizing Yield

Agricultural researchers, chemical companies, and many other agribusiness firms are constantly talking about "improving yield" or "maximizing yield." Farmers speak with pride about the Washington State wheat field that yielded 141 bushels per acre or the Wisconsin dairy cow that produced 26,000 pounds of 3.5% milk last year. Until you became at least a beginning manager you probably agreed with these quite reasonable statements. Until now, right?

Returning to Table 4.1 we see the effect of the law of diminishing returns: Each unit of input has differing effectiveness in producing output. Thus, profits are maximized at eleven units of output even though yield is maximized at twelve units. It costs the manager $1 in lost profits to maximize yield rather than profits! The manager must decide if the ego trip of "having a higher yield than my neighbors" is worth the lost profit!

Limited Resources

Up to now, we have been determining output levels under the assumption that the supply of the variable input is unlimited. However, in some cases the supply of the variable input is limited (i.e., capital − dollars, machinery, etc.). When the supply of input is limited, the opportunity cost of using the input becomes very important. Recall that *opportunity cost* is defined as *the return that the input could have earned in another use, its next best alternative.* Thus, opportunity costs represent the returns foregone, or given up, by using the input for one use instead of another. A good example is a college education. A student is painfully aware of the out-of-pocket costs of tuition, room and board, books, clothes, and entertainment. But the same student might not be aware of the non-out-of-pocket expense caused by the potential income lost because, while attending school, he or she is not earning an income. These costs are just as real as the cash costs associated with attending school. So, when a student pays $5,000 to $7,000 per year to attend school he or she is also paying $8,000 to $10,000 per year in foregone income or opportunity cost. Total cost per year might be $13,000 to $17,000 in actuality.

The effect of our new knowledge of the concept of opportunity cost is to remind us as managers to minimize opportunity cost when resources are limited. When the supply of inputs is limited, we want to employ those inputs in uses where marginal returns are greatest. In other words, if input supply is limited we want to minimize the opportunity cost of using the input, or we want to employ resources in uses where marginal returns are highest. In the words of former Secretary of Defense Robert McNamara, "Let's get the biggest bang for the buck!"

A specific concept embodying the preceding discussion is the **equi-marginal return principle,** which states that *returns will be maximized when scarce resources are employed so as to have equal (or as nearly equal as possible) marginal returns in each enterprise, product, or activity.* The hypothetical example in Table 4.7 is of a farm supply store that has four opportunities to expand its offerings to its customers. The expected returns to its investment (and sales efforts) are presented as the marginal revenue products associated with each additional $1,000 investment. In this case the $1,000 incremental investment might be considered to be the marginal factor cost.

According to Table 4.7, if a manager had unlimited capital resources, he or she would only invest in each enterprise up to the point where the marginal revenue product of an enterprise was equal to the $1,000 investment costs. Thus, $8,000 would be invested in chemicals because any further investment would yield a return less than the $1,000. The investment in fertilizer, garden tools, and diesel would be $5,000, $2,000, and $5,000, respectively.

If only $9,000 is available, the manager should invest where the return is highest. The first $1,000 should be invested in chemicals because $2,100 is greater than the returns possible in the other enterprises. The second and third $1,000 increment

TABLE 4.7   Alternative Enterprises for a Farm Supply Store

Capital Investment		Marginal Revenue Product			
Total investment	Marginal investment (or MFC)	Chemicals	Fertilizer	Garden tools	Diesel
$1,000	$1,000	$2,100	$1,450	$1,080	$1,800
2,000	1,000	2,000	1,340	1,050	1,680
3,000	1,000	1,850	1,210	980	1,450
4,000	1,000	1,640	1,090	890	1,150
5,000	1,000	1,450	1,040	780	1,010
6,000	1,000	1,220	960	690	960
7,000	1,000	1,140	880	660	820
8,000	1,000	1,010	720	620	700
9,000	1,000	940	650	450	590
10,000	1,000	850	570	400	500

should also be invested in chemicals because the return would be $2,000 and $1,850 as contrasted with a possible $1,450, $1,080, and $1,800 in the other three alternatives. The fourth $1,000 increment would be invested in diesel because the return would be $1,800 as contrasted with the possible $1,640, $1,450, and $1,080 in the other three alternatives. This procedure, investing each $1,000 increment of our $9,000 available capital where marginal returns are highest, results in $5,000 in chemicals, $1,000 in fertilizer, no investment in garden tools, and $3,000 in diesel. Note that the marginal revenue products happen to be equal at $1,450 at this distribution. Obviously this is not expected in day-to-day decisions faced by managers.

Before leaving this discussion of opportunity costs, equi-marginal return principle, and limited inputs, several points can be made. First, when inputs are limited, the cost we use in economic analysis should not automatically be what we paid for them, P_x, but it should be whatever return we give up by using the input in this production process. If that return foregone is less than P_x, then—and only then—is P_x the appropriate price to use in our managerial analysis.

Second, the concept of limited resources has a physical and economic aspect to it, just as did the concepts of production function and costs of production. Resources might be limited, physically, in number, for example, 1,000 men and women in the labor force available for hire. But, if our decision-making rules of comparing revenue and costs dictate that only 750 people be employed in our firm, these resources are not economically limited. The question is not whether you can hire more than 1,000 people, but rather can you hire enough people to attain all profitable levels of production?

Discrete Units of Input and Output

For simplicity in teaching, the figures we have been using are smooth, continuous curves. As will be seen in many of the applied examples in this book, decisions are not made in the extremely small units suggested by continuous curves. Rather indivisible, discrete, or even lumpy decisions are commonly made. We cannot decide to purchase one-half of a tractor or one-half of a person to use on the farm. But these difficulties do not negate the importance of economic analysis; rather, they increase it. These "lumpy" inputs are extremely expensive, and the question of whether to use one more in our firm is even more critical. We do know that we can vary the size of tractors and the number of hours of overtime paid to labor as a means of smoothing the lumpiness of our investment purchases. Here again, appropriate decisions will be made based on rules developed in this chapter.

SUMMARY

This is the most important chapter in this book. It has introduced you to, and hopefully eliminated fear of, terminology useful in economic analysis. Several concepts useful throughout this book were introduced; these concepts, such as opportunity cost and marginal returns, are based on common sense and were merely stated more formally in this discussion. We emphasized the decision rules necessary to answer the question "How much to produce?" both in cases of limited or unlimited resources. In addition, we now have the tools to analyze the profit maximizing decision for factors outside our control, such as product or input prices, change over time. Yes, this chapter did introduce you to a wealth of new material—let's hope you will develop a good relationship with it.

Remember, though, that only three pieces of information were necessary to make the proper decisions: production function, price of inputs, and price of outputs. These data allow us to decide the appropriate level of input. In our definition of economics and management in Chapter 1, we emphasized the problem of allocation and the role of the manager as the decision maker. In this chapter the first producer question was examined as an application of managerial economics. In the next chapter the second question of how to produce will be emphasized. Much of the terminology developed in this discussion will be reused, and some new concepts will be introduced. The theme and goal of this and succeeding chapters continue to be development of management in agricultural economics, both production agriculture and agribusiness management.

QUESTIONS

1. What do we mean by a *factor–product decision?* Give a specific example of such a decision.
2. Under a situation of unlimited resources the manager will maximize profits by equating MFC and MRP. How does this situation change when resources are limited?
3. Graphically show the effects of an increase in factor price on factor input usage, using MRP analysis.
4. Using the information provided in Table 4.5, numerically show what happens to the profit maximizing output level if P_x increases to $6; if P_x decreases to $2. In each case identify the new output level and the per-unit and total profit earned.
5. Graphically show the effects of an increase in product price on factor input usage using an MRP analysis.
6. Graphically show the effects of an increase in factor price on profit maximizing output levels using MR/MC analysis.
7. Using the information provided in Table 4.7, if you had $7,000 to invest how would you allocate it across the four enterprises to maximize returns? If you had $14,000 to invest? Be sure to explain why you would invest in each alternative enterprise.
8. Using the information provided in Table 4.3, numerically show what happens to the profit maximizing output level if P_y increases to $5; if P_y decreases to $1. In each case identify the new output level and the total profit earned.

CHAPTER 5

Management Decisions: How to Produce?

INTRODUCTION

Thus far we have examined the production of one commodity where only one input was variable and all others were fixed at some amount. The question we were addressing was "How much to produce?" Instances of only one variable input are indeed rare. In reality firms use more than a single-variable input to produce their products, which leads to different—and more complex—decision-making criteria.

This chapter addresses the second of our producer questions: "How to produce?" This approach reflects the existence of more than one variable input in the production activities of a firm or farm and emphasizes the question of substitution of inputs in the production process. Because most products can be produced with varying quantities of inputs (think of how many combinations are possible in producing beef), different input combinations have different costs associated with them. Hence, the manager must decide which combination of inputs to use in order to achieve the firm's economic objective. This decision-making process is called the **factor–factor decision** and is slightly more complex than the factor–product decision. For example, when cleaning the barn on a farm, the farmer can use different combinations of labor and capital. A tractor and loader with only one driver (high capital/low labor) or a large number of people with pitchforks (low capital, high labor) are two possible alternatives. Once the price of each is known, the lowest cost alternative can be chosen. A term often associated with factor–factor decisions is *cost minimization,* or, given that a certain level of output is desired (cleaning one barn every Saturday), what combination of inputs will produce that output at the lowest cost?

In This Chapter

The objective of this chapter is to introduce the basic decision rules appropriate for proper managerial decision making and economic analysis in situations where the firm combines multiple variable inputs in the physical production process of its output. Since the inputs are combined, it is possible, to a certain extent, to substitute one input for another in the production process. For example, hay versus concentrate in feeding livestock, fertilizer versus amount of seed per acre in crop production, or capital versus labor in manufacturing. When multiple inputs are available, a firm or producer must decide which combination of variable inputs to use with the fixed factors to achieve the firm's production objectives. We assume the firm will act as a rational producer, therefore the production objective is to combine the variable inputs in the proportion that results in the highest profit. Most products can be produced using varying combinations of inputs, and each input combination will have a cost associated with it. Therefore, the decision regarding which combination of inputs to use to maximize profits also involves deciding which combination of inputs will minimize cost.

Assumptions Used

The factor–factor decision involves finding the least-cost combination of inputs, from the different input combinations possible, that will produce a given level of output. To examine this decision, we use the following assumptions:

1. The firm produces a single product.
2. The firm has two variable inputs, others are fixed.
3. The firm is a price taker.

As in previous chapters we are assuming all decisions are made under perfect certainty and we are examining a single product production function. However, as assumption 2 states, the firm now has two variable inputs, which can be combined in different ways, to use with the fixed factors in producing this product. As in Chapter 3, the firm has no influence over the price it pays for inputs; that is, the firm is a price taker.

To examine the factor–factor decision, we begin by developing a measurement or tool to identify how one input substitutes for another within the firm's production process. Then we determine what combination of two variable inputs to use to maximize profit at each level of output, recognizing that for a given level of output, maximizing profit and cost minimization are the same goal. Finally, we determine some decision rules to identify the most profitable level of output using two variable inputs. As we will see, the decision rules developed in this chapter can be used in the same conceptual manner for more complex situations (i.e., those with more than two variable inputs).

PHYSICAL RELATIONSHIPS

We begin our discussion of the factor–factor decision by examining the physical production aspects of the decision. The firm has two variable inputs that it can combine in different combinations along with its fixed factors of production to produce its product. Notationally, we are now using a production function that has two variable inputs, X_1 and X_2, with the other factors fixed at some level. Thus the firm's production function is expressed as:

$$Y = f(X_1, X_2 \mid X_3, \ldots, X_n)$$

where Y is the product, or output, produced; X_1 and X_2 are the variable inputs; and $X_3, \ldots, X_n$ are the fixed factors of production.

In determining a measurement for describing how the two inputs physically substitute for each other, we need to introduce two new concepts. The first is a production surface; the second is a theoretical tool called an isoquant.

Production Surface

The concept of a **production surface** describes *the "hill" of increased production obtained as we increase use of the variable inputs.* This is nothing more than a production function in three dimensions (i.e., Y, X_1, and X_2). This concept is graphically depicted in Figure 5.1, where the height of the surface is the production level. As we move up the production surface the level of production is increasing. If we were to stop at any point on the production surface (or hill) and walk around it at the same elevation, the production level would be the same at every point and we could identify all the possible input combinations that would produce that production level. This is illustrated in Figure 5.1 by lines A, B, and C (later we will learn these are isoquants). The production level at line B is greater than at line A, and the production level at C is greater than the production at both lines B and A. Why? Because C is higher up on the production surface, or at a higher elevation, than lines B or A, and as we move up the production surface production levels are increasing. At each point along line B the output level remains constant, or the same. Thus, as we move along line B, we can identify all the combinations of the two vari-

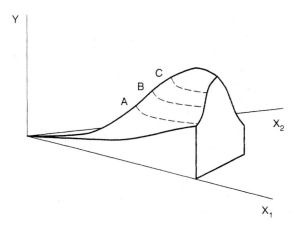

FIGURE 5.1
Production surface with two variable inputs.

able inputs, X_1 and X_2, that will produce this output level. We can do the same for production levels A and C, or any other production level on the production surface. Using the production surface allows us to identify all the combinations of the two variable inputs that will produce a given level of output. This knowledge will enable us to identify the costs associated with each input combination and allow us to identify the least-cost combination to produce a given level of output.

Isoquant

If we identify all the combinations of X_1 and X_2 that produce exactly the same total output, we would have an **isoquant** (sometimes called an **isoproduct**). An isoquant is *a line indicating all combinations of two variable inputs that will produce a given, or constant, level of output. Iso* means "the same"; hence, isoquant means that the same quantity of output is reflected everywhere along the line, which exactly describes lines A, B, and C in Figure 5.1. Thus, lines A, B, and C are isoquants.

Deriving an Isoquant

An isoquant identifies all the combinations of two variable inputs that will produce the same level of output. To illustrate how to derive an isoquant, consider the following example. A dairy farmer can combine grain (X_1) and hay (X_2) for his herd's feed ration. From past production records the dairy farmer obtains the combinations of the variable inputs (hay, grain) that will produce 11,000 lbs of milk, as listed in Table 5.1. Each combination of grain and hay shown produces the same level of output, thus they are all on the same isoquant. For example, the first combination (720 lbs grain, 600 lbs hay) produces 11,000 lbs of milk as does the third combination (600 lbs grain, 700 lbs hay) or the fifth combination (530 lbs grain, 800 lbs hay). As indicated in Table 5.1, if we increase the use of one variable input (hay), the other variable input can be decreased (grain). That is, hay is substituting for grain in the production of milk.

To derive the isoquant graphically, we simply plot the input combinations, as illustrated in Figure 5.2. The downward slope of the isoquant indicates that grain

TABLE 5.1 Input Combinations Producing 11,000 Pounds of Milk

X_1 (Pounds of Grain)	X_2 (Pounds of Hay)
720	600
650	650
600	700
560	750
530	800
510	850
500	900

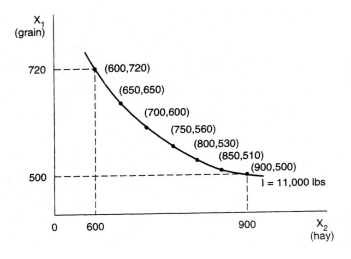

FIGURE 5.2
An isoquant for 11,000
pounds of milk.

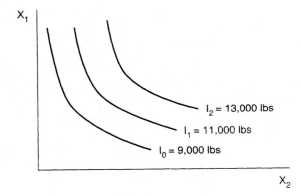

FIGURE 5.3
Isoquants for alternative output levels of
milk.

(X_1) can be decreased, or substituted for, by increasing the use of hay (X_2). Economic convention (the way we usually do things) places the input being replaced, grain in this example, on the vertical axis (labeled X_1), and the input being added, hay in this example, on the horizontal axis (labeled X_2). The designation of which input is to be replaced and added is arbitrary, but is important in terms of specifying the marginal rate of substitution (MRS) and the inverse price ratio (IPR), terms we discuss later.

The isoquant concept is commonly used in economic analysis. The isoquant is used not only to indicate the different combinations of two inputs capable of producing a given level of output, but it can represent many levels of output as well. Since each isoquant represents a unique output level, theoretically there is an isoquant for each level of output a firm is capable of producing, as illustrated in Figure 5.3. This figure shows that the distance away the origin indicates the level of output, or the height of the production surface. The closer the isoquant is to the origin, I_o, the lower the level of output. The farther away from the origin, I_2, the higher

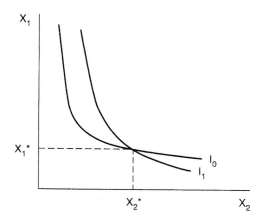

FIGURE 5.4
Intersecting isoquants (this is not possible; see
discussion in text).

the output level. The use of distance from the origin as an indication of higher production arises because of the difficulty of portraying production functions in three-dimensional space.

Because each isoquant represents a unique output level, isoquants can never intersect. Why? If isoquants intersected, it would mean that a single combination of inputs would be capable of producing more than one level of output. This is depicted graphically in Figure 5.4. If the isoquants I_0 and I_1 intersected, as in Figure 5.4, it would mean that the input combination X_2^* and X_1^* could produce two different levels of output. Given the unique relationship between an input use combination and the resultant output in the physical production process, this is not possible.

Marginal Rate of Substitution

As you can see from Figure 5.2, as we move along an isoquant the output level remains constant but the quantities of the variable inputs are changing as one input replaces, or substitutes, for the other. As managers we are interested in how well the inputs substitute for each other. *The rate at which one variable input can physically substitute for another variable input in the physical production process of the firm* is measured by the **marginal rate of substitution (MRS),** also referred to as marginal rate of technical substitution (MRTS) in some textbooks.

The marginal rate of substitution describes the rate at which one input can be decreased as use of the other input increases and, by definition, is the slope of the isoquant. Since economic convention places the input being replaced on the vertical axis and the input being added on the horizontal axis, MRS is calculated by dividing the change in the replaced input (Δ replaced) by the change in the added input (Δ added). Thus the computational formula and the interpretation of MRS depend on which input is being replaced and which is being added.

For example, if X_1 is the variable input being replaced and X_2 is the variable input being added, then the **marginal rate of substitution of X_2 for X_1,** written

$\text{MRS}_{X_2 X_1}$, is calculated by dividing the ΔX_1 (input replaced) by the ΔX_2 (input added), or

$$\text{MRS}_{X_2 X_1} = \frac{\Delta X_1}{\Delta X_2} = \frac{\text{replaced}}{\text{added}}$$

Graphically, as shown in Figure 5.5, the variable input being replaced, X_1 in this case, is on the vertical axis and the variable input being added, X_2 in this case, is on the horizontal axis. The slope of the isoquant in Figure 5.5 is the change in X_1 divided by the change in X_2 ($\Delta X_1 \div \Delta X_2$), which corresponds exactly to the definition of $\text{MRS}_{X_2 X_1}$.

Conversely, if X_2 is the variable input being replaced and X_1 is the variable input being added, then the **marginal rate of substitution of X_1 for X_2**, written $\text{MRS}_{X_1 X_2}$, is calculated by dividing the ΔX_2 (input replaced) by the ΔX_1 (input added), or

$$\text{MRS}_{X_1 X_2} = \frac{\Delta X_2}{\Delta X_1} = \frac{\text{replaced}}{\text{added}}$$

Graphically, as shown in Figure 5.6, X_2, the variable input being replaced, is on the vertical axis and X_1, the variable input being added, is on the horizontal axis. The

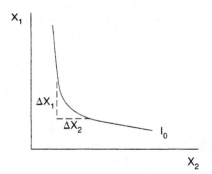

FIGURE 5.5
Marginal rate of substitution of X_2 for X_1.

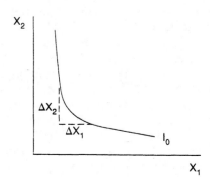

FIGURE 5.6
Marginal rate of substitution of X_1 for X_2.

slope of the isoquant in Figure 5.6 is equal to the change in X_2 divided by the change in X_1 ($\Delta X_2 \div \Delta X_1$), which corresponds exactly to $\text{MRS}_{X_1 X_2}$.

Note that the subscripts on the definition of MRS match the axes labels on the graphs in Figures 5.5 and 5.6. Since the order of the subscripts on MRS tells us which input is being added and which input is being replaced, it is very important that the graphical presentation be consistent with stated MRS, and vice versa. Again, conventionally, we usually put X_1 on the vertical axis but we do not have to.

We indicated and illustrated earlier that the designation of which input is to be replaced and which input is to be added is an arbitrary designation. This is because

$$\text{MRS}_{X_2 X_1} = \frac{1}{\text{MRS}_{X_1 X_2}}$$

Why?

$$\text{MRS}_{X_2 X_1} = \frac{\Delta X_1}{\Delta X_2} = \frac{1}{\dfrac{\Delta X_2}{\Delta X_1}} = \frac{1}{\text{MRS}_{X_1 X_2}}$$

As illustrated, $\text{MRS}_{X_2 X_1}$ is the inverse of $\text{MRS}_{X_1 X_2}$; thus, deciding which input is replaced and which input is added for analysis is a management decision. Although they represent the same information, it is important to be aware of which input is being replaced and added for computation and for correct graphical presentation.

Using our dairy example, we can now illustrate how to calculate the marginal rate of substitution numerically. As shown in Table 5.1, the variable input hay (X_2) is being added and the variable input grain (X_1) is being replaced. Therefore, the correct formula to calculate the marginal rate of substitution between these inputs is $\text{MRS}_{X_2 X_1} = \Delta X_1 / \Delta X_2$, as shown in Table 5.2.

The formula, which states the rate at which X_2 (is added) substitutes for X_1 (is replaced), is equal to the change in X_1 divided by the change in X_2, and, as illustrated in Figure 5.7, is the slope of the isoquant. Because one input can be decreased by adding the other input, MRS as a numerical coefficient is always negative, as indicated in Table 5.2 and Figure 5.7. The MRS of -1.40 ($-70/50$) as we go from the first input combination (720, 600) to the second combination (650, 650) indicates that 1.4 units of grain (X_1) are replaced as we add one unit of hay (X_2). This -1.40 also describes the slope of the isoquant at this point, since, by definition, MRS measures the slope of the isoquant. As shown in Table 5.2, initially hay (X_2) substitutes very well for grain (X_1) in the feed ration because a one-unit increase in X_2 allows a greater than one-unit decrease in use of X_1. Later, as decreasing returns (remember our discussion from Chapter 2) affect the trade-offs, the two inputs substitute less freely. As we substitute hay for grain in our dairy example, the MRS decreases in magnitude (in terms of absolute value). Thus, an isoquant exhibits diminishing MRS. Graphically, as in Figure 5.7, it can be stated that when the isoquant has a steep slope (MRS > |1|, where |1| is the absolute value of 1) the

TABLE 5.2   Calculating the Marginal Rate of Substitution of Hay (X_2) for Grain (X_1)

X_1 (Grain)	X_2 (Hay)	ΔX_1 (Replaced)	ΔX_2 (Added)	$MRS_{x_2 x_1}$	=	$\Delta X_1 \div \Delta X_2$
720	600					
		−70	50	−1.40	=	−70/50
650	650					
		−50	50	−1.00	=	−50/50
600	700					
		−40	50	−0.80	=	−40/50
560	750					
		−30	50	−0.60	=	−30/50
530	800					
		−20	50	−0.40	=	−20/50
510	850					
		−10	50	−0.20	=	−10/50
500	900					

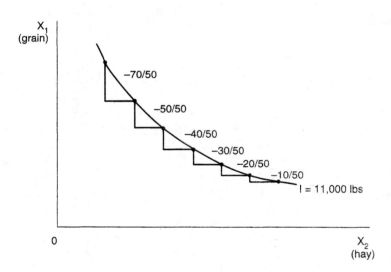

FIGURE 5.7
Isoquant and marginal rate of substitution of X_2 for X_1.

input on the horizontal axis, usually X_2, is a good substitute for X_1; as the slope of the isoquant decreases (MRS < |1|), X_2 becomes a less efficient substitute for X_1.

The slope of the isoquant is the MRS between the two inputs and the **substitutability** between two inputs depends on how responsive production is to changes in each input, or in the productivity of the individual inputs.

Possible Rates of Input Substitution

The example developed for the dairy producer had inputs that were imperfect substitutes for each other, or the isoquant exhibited diminishing marginal rate of substitution. Diminishing MRS occurs where one unit of an input can be exchanged for another, but at a decreasing rate due to diminishing returns in the production function. As illustrated in Figure 5.7, initially X_2 substitutes rather freely for X_1 [a large decrease in grain (X_1) for a small increase in hay (X_2)]; but as decreasing returns affect the trade-off between X_1 and X_2, the two inputs substitute less freely [a small decrease in grain (X_1) for a large increase in hay (X_2)]. Two other possibilities for substitutability between inputs exist, as discussed later.

Constant, or perfect, substitutability occurs when one unit of an input can be exchanged for another input on a consistent basis of one-to-one or some other unchanging ratio. An example of nearly perfect input substitutes is nitrogen (N) fertilizer use on corn. A producer can get a specified amount of actual N on corn using ammonia nitrate (33% N) or anhydrous ammonia (84% N). The producer's choice is determined by personal preferences, soil types, availability, and other factors but in any case, the rate of substitution is almost linear. In this constant case, diminishing marginal rate of substitution does not occur, and the isoquant is linear, as illustrated in Figure 5.8(a).

Fixed proportions (or perfect **complementarity**) occurs when inputs must be used in a fixed ratio. There is no choice regarding what proportion of each input to use. For example, every time a truck is needed, a driver is also required. In such a case, the substitutability is zero, thus the isoquant is a single point as illustrated in Figure 5.8(b).

In either of these cases, constant substitutability or fixed proportions, the role of the manager is not very active since little choice is available. The most commonly seen relationship between inputs is one of diminishing MRS. Why? Because MRS is the ratio of the marginal products of the two inputs, and decreasing returns are evident in most production functions (i.e., the law of diminishing returns; see Chapter 2).

One application for isoquants is to illustrate the options available to producers in the Corn Belt should nitrogen fertilizer use be restricted in the face of groundwater

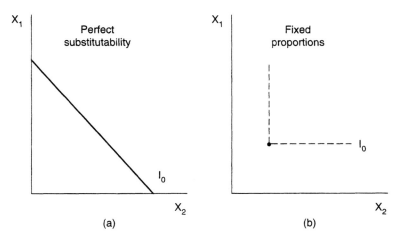

FIGURE 5.8 Constant (perfect) substitutes and fixed proportions isoquants.

TABLE 5.3 Cropland and Nitrogen Combinations to Produce 14,000 Bushels of Corn

Nitrogen Applied (lbs/acre)	Acres Planted	Total Corn Production (bu)	Yield (bu/acre)
40	127.3	14,000	110
80	100.0	14,000	140
120	90.3	14,000	155
160	87.5	14,000	160

quality problems. For example, research conducted at the University of Minnesota suggests that corn producers in southern Minnesota could use the alternative combinations of cropland and nitrogen shown in Table 5.3 to produce 14,000 bushels of corn. So, if environmental regulators restricted nitrogen use on corn acreage, farmers could maintain output by expanding acreage. Expanded corn acreage may not be the outcome environmentalists are seeking, but it would be one adjustment possible for corn producers.

ECONOMIC RELATIONSHIPS

Up to this point, we have only been comparing physical units of two different inputs and how they substitute for each other in a technical, noneconomic relationship. To calculate minimum cost combinations of inputs we need factor prices to build on the technical relationships. Input prices are appropriate since we are examining the factor–factor decision.

Isocost Line

A useful tool for describing the value of the two inputs is an **isocost line,** which is defined as *a line indicating all combinations of two variable inputs that can be purchased for a given, or same, level of expenditure.* It is calculated by adding up the total money spent on each input, or

$$TE = (P_{x_1} \times X_1) + (P_{x_2} \times X_2)$$

where TE is the total dollar expenditure on inputs (price of each input multiplied by the units of each input), P_{x_1} and P_{x_2} are the prices of the variable inputs, and X_1 and X_2 are the units of the variable inputs purchased. This is sometimes referred to as an equal expenditure line in economic literature.

The amount of each input that can be purchased depends on the amount of money to be spent (total outlay) and the respective prices of the inputs. Going back to our dairy example, if the dairy farmer only has $100 to purchase inputs, and he can purchase grain for $0.20/lb and hay for $0.10/lb, how many units of each input could he purchase? Several possible input combinations that could be pur-

TABLE 5.4 Input Combinations with Same Total Expenditure of $100 with Price of Grain (P_{x_1}) = $0.20 and Price of Hay ($P_{x_2}$) = $0.10

Grain Purchased (X_1)	Expenditure on Grain ($P_{x_1} \times X_1$)	Hay Purchased (X_2)	Expenditure on Hay ($P_{x_2} \times X_2$)	Total Expenditure ($P_{x_1} \times X_1 + P_{x_2} \times X_2$)
500	$100	0	$ 0	$100
400	$ 80	200	$ 20	$100
300	$ 60	400	$ 40	$100
200	$ 40	600	$ 60	$100
0	$ 0	1,000	$100	$100

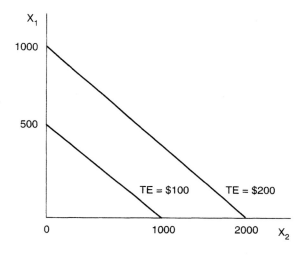

FIGURE 5.9
Isocost lines for $100 and $200 total expenditure.

chased for $100 are listed in Table 5.4. Each of the combinations listed requires the same level of expenditure. For example, $100 would purchase the first combination (500 lbs grain, 0 lbs hay) or the third combination (300 lbs grain, 400 lbs hay) or even the fifth combination (0 lbs grain, 1,000 lbs hay). Thus, with $100 to spend on inputs, the dairy farmer could use any of the grain/hay (X_1, X_2) combinations given in Table 5.4.

The isocost line for $100 is depicted graphically in Figure 5.9. If the dairy farmer spent the entire $100 on X_1 he could purchase 500 units of X_1 ($100 ÷ $0.20 = 500). If he spent the entire $100 on X_2 he could purchase 1,000 units of X_2 ($100 ÷ $0.10 = 1,000). These points are the end points, or intercept terms (points), of the isocost line. Connecting the two endpoints yields the isocost line (TE = $100 in Figure 5.9) for $100. Any combination of inputs on this line requires the same level of total expenditure ($100) to purchase. Thus, all of the combinations listed in Table 5.4 lie on the isocost line for $100 (TE = $100) in Figure 5.9.

If the dairy farmer has $200 and only purchases X_1, he can purchase 1,000 units ($200 ÷ $0.20 = 1,000); if he purchases only X_2, he can purchase 2,000 units ($200 ÷ $0.10 = 2,000). The isocost line for $200 (TE = $200) is also drawn in

Figure 5.9. As illustrated, the distance from the origin indicates the level of expenditure. The closer to the origin, the lower the level of total expenditure and the fewer inputs can be purchased; the farther away from the origin the higher the level of total expenditure and the more inputs can be purchased.

As you are probably aware, an infinite number of isocost lines could have been drawn in Figure 5.9, each reflecting a different expenditure amount. For illustrative purposes we have drawn only two, one for $100 and one for $200.

The slope of the isocost line drawn in Figure 5.9 is $-P_{x_2}/P_{x_1}$, and is referred to as the inverse price ratio. It is derived from the isocost line equation:

$$TE = (P_{x_1} \times X_1) + (P_{x_2} \times X_2)$$

Since X_1 is on the vertical axis, we solve the equation for X_1. First we subtract $(P_{x_2} \times X_2)$ from both sides:

$$(P_{x_1} \times X_1) = TE - (P_{x_2} \times X_2)$$

and then we divide both sides by P_{x_1} to obtain

$$X_1 = \frac{TE}{P_{x_1}} - \frac{P_{x_2}}{P_{x_1}} \times X_2$$

which is the familiar formula for a straight line $Y = a + bX$, where the intercept term (a) in this case is TE/P_{x_1} (which identifies the X_1 intercept on the X_1 axis) and the slope (b) is $-P_{x_2}/P_{x_1}$. The slope of the isocost line is called an **inverse price ratio (IPR)** because it is *the price of the input on the horizontal axis divided by the price of the input on the vertical axis.* (Later on we'll have another good way to use this term.) Thus, the way the graph axes are labeled will dictate which input price is in the numerator and denominator of the price ratio. The slope of the isocost line tells us the relative value of the inputs, or how they substitute for one another economically; this information will allow us to identify the least-cost combination we've been looking for.

Effects of a Price Change

Because the slope of the isocost line depends on the relative price ratio of the two inputs, if the price of one of the inputs changes, the slope of the isocost line also changes.

Figure 5.10 illustrates the impacts of a change in P_{x_2}. Recall our previous example where TE = $100, P_{x_1} = $0.20, and P_{x_2} = $0.10. At these prices, if the dairy farmer purchased only X_2, he could purchase 1,000 units; or if he purchased only X_1, he could purchase 500 units (TE_0 in Figure 5.10). If the price of X_2 changes, the number of units of X_2 that he can purchase with $100 also changes. If P_{x_2} increases,

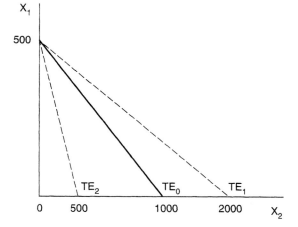

FIGURE 5.10
Effect of price of X_2 changes on isocost line.

X_2 is now more expensive to use relative to X_1, and with the price increase, $100 will now purchase fewer units of X_2. For example, if P_{x_2} increases to $0.20 the dairy farmer can only purchase 500 units of X_2 with $100 instead of the initial 1,000 units. The isocost line for $100 given the new prices (P_{x_1} = $0.10; P_{x_2} = $0.20) is TE_2 in Figure 5.10. On the other hand, if P_{x_2} decreases X_2 is now less expensive to use relative to X_1, and with the price decrease $100 purchases more units of X_2. For example, if P_{x_2} decreases to $0.05 the farmer can purchase 2,000 units of X_2 with $100 instead of the initial 1,000 units. The isocost line for $100 given the new prices (P_{x_1} = $0.10; P_{x_2} = $0.05) is TE_1 in Figure 5.10.

As illustrated in Figure 5.10, if the price of the variable input X_2 changes, the number of units of X_2 that can be purchased for the same expenditure level also changes. If P_{x_2} increases, the isocost becomes steeper, or rotates inward, because the same level of expenditure now buys fewer units of X_2. If P_{x_2} decreases, the isocost becomes flatter, or rotates outward, because the same level of expenditure now buys more units of X_2. However, note that the X_1 intercept does not change and the isocost line rotates around the axis intercept of the input whose price has not changed (X_1 in this case). This is because with the same level of expenditure ($100) the dairy farmer can still purchase 500 units of X_1.

Figure 5.11 illustrates the impacts of a change in P_{x_1}. Recall our previous example where TE = $100, P_{x_1} = $0.20, and P_{x_2} = $0.10. At these prices, if the dairy farmer purchased only X_2, he could purchase 1,000 units; or if he purchased only X_1, he could purchase 500 units (TE_0 in Figure 5.11). If the price of X_1 changes, the number of units of X_1 that he can purchase with $100 also changes. If P_{x_1} increases, X_1 is now more expensive to use relative to X_2, and with the price increase $100 purchases fewer units of X_1. For example, if P_{x_1} increases to $0.40 he can only purchase 250 units of X_1 with $100 instead of the initial 500 units. The isocost line for $100 given the new prices (P_{x_1} = $0.40; P_{x_2} = $0.10) is TE_2 in Figure 5.11. On the other hand, if P_{x_1} decreases, X_1 is now less expensive to use relative to X_2, and with

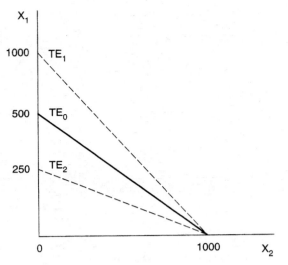

FIGURE 5.11
Effect of price of X_1 changes on isocost line.

the price decrease $100 purchases more units of X_1. For example, if P_{x_1} decreases to $0.10 we can purchase 1,000 units of X_1 with $100 instead of the initial 500 units. The isocost line for $100 given the new prices (P_{x_1} = $0.10; P_{x_2} = $0.10) is TE_1 in Figure 5.11.

As illustrated in Figure 5.11, if the price of the variable input X_1 changes, the number of units of X_1 that can be purchased for the same expenditure level also changes. If P_{x_1} increases the isocost line becomes flatter, or rotates downward, because the same level of expenditure now buys fewer units of X_1. If P_{x_1} decreases, the isocost line becomes steeper, or rotates upward, because the same level of expenditure now buys more units of X_1. However, note that the isocost line rotates around the axis intercept of the input whose price has not changed (X_2 in this case). This is because with the given level of expenditure ($100) the dairy farmer can still purchase 1,000 units of X_2.

We can summarize the important points to remember about the isocost line as follows:

1. The end points of the isocost line show how many units of the variable input could be purchased if all the expenditure were spent on that one input.
2. A change in total outlay causes a parallel shift in the isocost line.
3. A change in one input price causes the isocost line to rotate around the intersection of the isocost line and the axis of the other input.
4. A reduction in the use of one input is necessary to increase use of the other input.
5. The slope of the isocost, termed the inverse price ratio (IPR), is the price of the input on the horizontal axis divided by the price of the input on the vertical axis.

In briefer summary, the two main points to remember about an isocost line are distance from the origin (dollars being spent) and the slope (relative value of the inputs).

FACTOR–FACTOR DECISION

When the manager's goal is to maximize profits, then he or she is interested in minimizing costs. Thus, the objective of the producer is to find the **least-cost combination of inputs** to produce a given level of output. Three equivalent approaches can be used to determine the least-cost combination of inputs, all of which are based on the same economic trade-offs. These are the total outlay, numerical, and graphical approaches.

Total Outlay Approach

Under the total outlay approach, the manager or producer simply calculates the total expenditure associated with each input combination and chooses the combination with the smallest total expenditure or dollar outlay in order to minimize costs.

We can illustrate this approach using our dairy example. Assuming that the dairy farmer can purchase grain for $0.20/lb and hay for $0.10/lb, the total expenditure required for each combination of hay and grain that will produce 11,000 lbs of milk is presented in Table 5.5. Examining Table 5.5 we find that the least-cost combination of inputs to produce 11,000 lbs of milk is 530 units of grain (X_1) and 800 units of hay (X_2) with a total expenditure of $186.

The total outlay approach is the simplest of the three. As indicated in Table 5.5, identifying the least-cost input combination is simply a matter of adding the money spent on each input to determine the total expenditure required to purchase each input combination and then choosing the smallest cost.

TABLE 5.5 Determining Least-Cost Combination of Inputs to Produce 11,000 Pounds of Milk: Total Outlay Approach with Price of Grain (P_{x_1}) = $0.20 and Price of Hay ($P_{x_2}$) = $0.10

X_1 (Grain)	X_2 (Hay)	Total Expenditure ($P_{x_1} \times X_1 + P_{x_2} \times X_2$)				
720	600	$144	+	$60	=	$204
650	650	$130	+	$65	=	$195
600	700	$120	+	$70	=	$190
560	750	$112	+	$75	=	$187
530	800	$106	+	$80	=	$186
510	850	$102	+	$85	=	$187
500	900	$100	+	$90	=	$190

Numerical (or Mathematical) Approach

Under the numerical approach, the manager equates the marginal rate of substitution (the slope of the isoquant) to the inverse price ratio (the slope of the isocost) to determine the least-cost combination of inputs. The marginal rate of substitution describes the physical (technical) substitution between the two inputs, while the inverse price ratio (IPR) describes the value (cost) substitution between the two inputs. In other words, we are determining the point at which the rate of physical substitution (MRS) is equal to the rate of economic substitution (IPR) for the two variable inputs, or for our dairy example, where

$$MRS_{X_2X_1} = -\frac{P_{x_2}}{P_{x_1}}$$

Why? We know that $MRS_{X_2X_1} = \Delta X_1/\Delta X_2$, and since the inverse price ratio is negative and MRS is always negative, the negative signs will cancel so we can write

$$\frac{\Delta X_1}{\Delta X_2} = \frac{P_{x_2}}{P_{x_1}}$$

Multiplying both sides by P_{x_1} and ΔX_2 gives

$$P_{x_1} \times \Delta X_1 = P_{x_2} \times \Delta X_2$$

which can be described as **cost of replaced input (X_1) = cost of added input (X_2)**.

So the producer continues to substitute one input (adds X_2) for the other input (replaces X_1) until the cost of adding additional units of input is equal to (or less than) the cost of the input being replaced. In other words, the producer examines the input combinations and:

If $(P_{x_1} \times \Delta X_1) > (P_{x_2} \times X_2)$, then X_2 should be added because the cost savings from replacing X_1 is greater than the cost of adding X_2.

If $(P_{x_1} \times \Delta X_1) < (P_{x_2} \times \Delta X_2)$, then X_1 should be added because the cost of adding X_2 is greater than the cost savings from replacing X_1.

This comparison is the basis for the factor–factor decision rule, which tells us that the least-cost combination of inputs to produce a given level of output will be found where the physical rate of substitution (MRS) between the inputs is equal to the economic rate of substitution (IPR) between the inputs, or where MRS = IPR.

We can illustrate this approach using our dairy example. The appropriate information is presented in Table 5.6. Using our decision rule, $MRS_{X_2X_1} = -P_{x_2}/P_{x_1}$ (replaced/added = inverse price ratio), and the information in Table 5.6, the least-cost combination of inputs to produce 11,000 lbs of milk is 530 units of grain (X_1) and 800 units of hay (X_2). Note that this is the same input combination identified using the total outlay approach.

TABLE 5.6 Determining Least-Cost Combination of Inputs to Produce 11,000 Pounds of Milk: Numerical Approach with Price of Grain (P_{x_1}) = $0.20 and Price of Hay (P_{x_2}) = $0.10

X_1 (Grain)	X_2 (Hay)	ΔX_1	ΔX_2	MRS$_{X_2 X_1}$ $(\Delta X_1/\Delta X_2)$	$-\dfrac{P_{x_2}}{P_{x_1}}$ $(-0.10/0.20)$	Cost Savings $(\Delta X_1 \times P_{x_1})$	Added Costs $(\Delta X_2 \times P_{x_2})$
720	600						
		−70	50	−1.40	−0.50	−$14	$5
650	650						
		−50	50	−1.00	−0.50	−$10	$5
600	700						
		−40	50	−0.80	−0.50	−$8	$5
560	750						
		−30	50	−0.60	−0.50	−$6	$5
530	800						
		−20	50	−0.40	−0.50	−$4	$5
510	850						
		−10	50	−0.20	−0.50	−$2	$5
500	900						

Why is this input combination (530 units of X_1 and 800 units of X_2) the least-cost combination when MRS$_{X_2 X_1}$ > $-P_{x_2}/P_{x_1}$ at this point? Since each input combination in Table 5.6 produces the same output level we can compare the cost savings from replacing X_1 to the costs of adding X_2 to illustrate why this combination minimizes costs for the dairy producer. The last two columns in Table 5.6 show the cost savings from replacing X_1 (equal to $\Delta X_1 \times P_{x_1}$) and the added costs from adding units of X_2 (equal to $\Delta X_2 \times P_{x_2}$). A simple comparison shows that the cost savings from replacing X_1 are greater than the costs of adding X_2 up to the input combination of 530 units X_1 and 800 units of X_2. Since the producer saves more by replacing X_1 than he spends on adding X_2 the producer can reduce costs by replacing grain (X_1) and adding hay (X_2) until this combination is reached. If the producer were to use the next input combination (510 units of X_1 and 850 units of X_2) the cost of adding X_2 would be $5 and the cost savings from replacing X_1 would be $4. Because the producer has to spend more money ($5) to add X_2 (from 800 to 850 units) than he would save ($4) by replacing X_1 (from 530 to 510 units) he would continue to use the input combination of 530 units of X_1 and 800 units of X_2 to produce 11,000 lbs of milk.

Graphical Approach

Using the graphical approach, the least-cost combination of inputs is located at the point of tangency between the isocost line and the isoquant. Recall that the point of tangency between two curves is the point where the slopes of the two curves are equal to each other. Since the slope of the isocost line is the inverse price ratio and the slope of the isoquant is the MRS, graphically making the two curves tangent to each other is equivalent to equating the two numerically. Thus, the graphical approach is simply a picture of our decision rule $MRS_{X_2X_1} = -P_{x_2}/P_{x_1}$ and is illustrated in Figure 5.12.

Since we assume that the producer is rational and wants to maximize profits, the producer wants to reach the highest possible level of production for the lowest possible total expenditure. Therefore, the producer wants to find the tangency between the highest obtainable isoquant and the lowest possible isocost line. The isocost lines TE_0, TE_1, and TE_2 in Figure 5.12 represent three of the many possible expenditure levels that could have been drawn. The output level for which we are trying to minimize costs is represented by the isoquant I_0 in Figure 5.12. The lowest expenditure depicted, isocost line TE_0, is so low that no input combination capable of producing the desired level of output can be purchased; that is, TE_0 does not intersect the isoquant at any point. The highest expenditure isocost line, TE_2, intersects the isoquant at two points, B and C, indicating that this level of expenditure would allow us to purchase either of these input combinations. However, our goal is to produce at the lowest cost so we would like to use the lowest valued isocost line possible. Since they are on the same isoquant, I_0, the input combinations at point B and C produce the same level of output as the input combination at point A. However, since the input combinations at points B and C are on a higher isocost, TE_2, they cost more to purchase. Thus, the input combination at point A produces the desired output level but at a lower expenditure than either of the in-

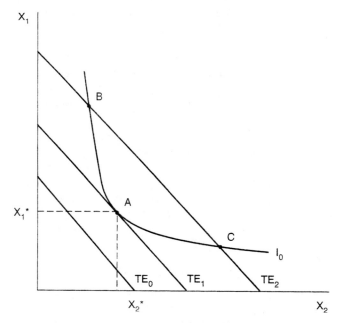

FIGURE 5.12 Factor–factor decision: determining the least-cost combination of inputs.

put combinations at points B and C. The least-cost combination of inputs, X_1^* units of X_1 and X_2^* units of X_2, is at point A in Figure 5.12 where the slope of the highest obtainable isoquant (I_0) is tangent to the lowest possible isocost line (TE$_1$), or where $\text{MRS}_{X_2X_1} = -P_{x_2}/P_{x_1}$.

We can illustrate this approach using our dairy example. This factor–factor decision is illustrated in Figure 5.13. We know from Table 5.5 that the least-cost combination of inputs requires an outlay of $186. If the producer spent all $186 on X_1, he could purchase 930 units of X_1 (intercept point on vertical axis in Figure 5.13). If the producer spent all $186 on X_2, he could purchase 1,860 units of X_2 (intercept point on horizontal axis in Figure 5.13). At point A, the point of tangency between the isoquant for 11,000 lbs and the isocost line for $186, the slopes of the two curves are equal. The least-cost combination of inputs to produce 11,000 lbs of milk is 530 units of X_1 (grain) and 800 units of X_2 (hay).

The three methods of determining the least-cost combination of inputs are equivalent. The total outlay approach is useful when the choices are simple and straightforward. The graphical approach displays relationships between prices, factors, and production levels. But in the complex problems of the real world, computers solve these problems using spreadsheet or database software. In our dairy example, the least-cost combination of inputs, as determined under all three approaches, is 530 units of grain (X_1) and 800 units of hay (X_2) with a total expenditure of $186. But what happens to the decision, or determination of the least-cost-combination, if the price of one of the variable inputs changes? We examine this in the next section.

Impact of Changes in Input Prices

Managers are often faced with changing input prices when making factor–factor decisions. When the relative prices of inputs change, the least-cost combination of

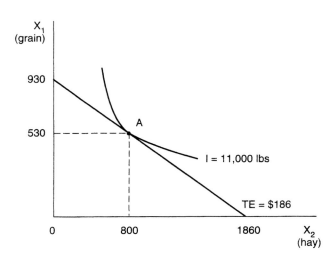

FIGURE 5.13
Least-cost combination of inputs to produce 11,000 lbs of milk.

inputs will also change. The decision the producer makes will depend on whether or not the funds available to purchase inputs are limited.

Input Price Increase

When the price of one variable input increases, using that input becomes more expensive relative to the other input. The manager or producer must then reevaluate the factor–factor decision to find the new least-cost combination of inputs. This decision is depicted in Figure 5.14.

Initially the producer is at point A in Figure 5.14 where the slope of the isoquant ($MRS_{X_2 X_1}$) is equal to the slope of the isocost line ($-P_{x_2}/P_{x_1}$), and the least-cost combination of inputs to produce the desired output level (I_0) is X_1^* units of X_1 and X_2^* units of X_2. If P_{x_2} increases to P'_{x_2}, the isocost line becomes steeper, rotating inward from TE to TE' in Figure 5.14, because the slope is now equal to $-P'_{x_2}/P_{x_1}$. The producer has two options available given the price change. The first option is to maintain the same level of production at a higher level of expenditure. Thus, the producer must have funds available to meet increased expenditure levels. The second option, if the producer has limited funds to purchase inputs, is to maintain the same expenditure level and reduce production.

Unlimited funds. If the producer has unlimited funds with which to purchase inputs, she can choose to maintain her current level of production (stay on isoquant I_0) by recombining the variable inputs such that the slope of the new isocost line (TE') is equal to the slope of the initial isoquant (I_0). In this case the least-cost combination of inputs, X'_1 units of X_1 and X'_2 units of X_2, is found at point B in

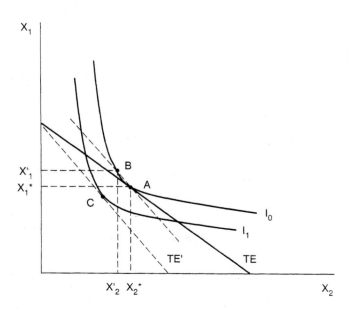

FIGURE 5.14
Factor–factor decision under changing prices: price of X_2 increase.

Figure 5.14. Point B is found by locating the point of tangency between the initial isoquant (I_0) and a line with the same slope as (parallel to) the new isocost line (TE′). When the price of X_2 increases, X_2 becomes more expensive to use relative to X_1. Thus, the new least-cost combination of inputs incorporates more X_1 ($X_1' > X_1^*$) and less X_2 ($X_2' < X_2^*$). Since the inputs substitute for one another, the producer can produce the same level of output using more X_1 (the relatively cheaper input) and less X_2 (the relatively more expensive input). However, because the new combination of inputs (X_1', X_2') at point B lies on a higher isocost line, it is more expensive to purchase than the initial combination (X_1^*, X_2^*) at point A in Figure 5.14.

Limited funds. If the producer is constrained by limited funds with which to purchase inputs, the producer can maintain the same level of total expenditure on inputs by moving to a lower isoquant (I_1), thus reducing production. If the producer makes this decision she would operate at point C in Figure 5.14, where the slope of the lower isoquant (I_1), is equal to the slope of the new isocost line (TE′). At point C the producer is using less of both variable inputs; hence, the production level is lower than the initial production level.

We can illustrate the effect of an input price increase on the factor–factor decision using our dairy example. What happens to the least-cost combination of hay and grain if the price of hay increases to $0.25/lb? The total outlay required to purchase the input combinations that will produce 11,000 lbs of milk is presented in Table 5.7. At the initial prices of $P_{x_1} = \$0.20$ and $P_{x_2} = \$0.10$, the least-cost combination of inputs is 530 units of grain (X_1) and 800 units of hay (X_2) with

TABLE 5.7   Determining Least-Cost Combination of Inputs Under Changing Prices: Dairy Example, Price of Hay Increases

X_1 (Grain)	X_2 (Hay)	Total Expenditure $P_{x_1} = \$0.20$ $P_{x_2} = \$0.10$	Total Expenditure $P_{x_1} = \$0.20$ $P_{x_2} = \$0.25$	$MRS_{X_2X_1}$ $\left(\dfrac{\Delta X_1}{\Delta X_2}\right)$	$-\dfrac{P_{x_2}}{P_{x_1}}$ $\left(-\dfrac{0.10}{0.20}\right)$	$-\dfrac{P'_{x_2}}{P_{x_1}}$ $\left(-\dfrac{0.25}{0.20}\right)$
720	600	$204.00	$294.00			
				−1.40	−0.50	−1.25
650	650	$195.00	$292.50			
				−1.00	−0.50	−1.25
600	700	$190.00	$295.00			
				−0.80	−0.50	−1.25
560	750	$187.00	$299.50			
				−0.60	−0.50	−1.25
530	800	$186.00	$306.00			
				−0.40	−0.50	−1.25
510	850	$187.00	$314.50			
				−0.20	−0.50	−1.25
500	900	$190.00	$325.00			

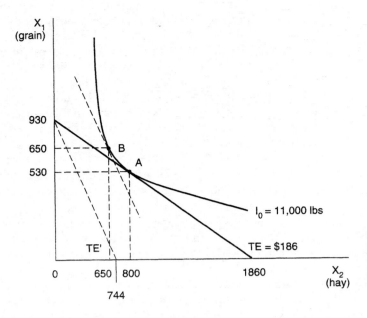

X_1
(grain)

930

650 B

530 A

I_0 = 11,000 lbs

TE' TE = $186

0 650 | 800 1860 X_2
(hay)

744

FIGURE 5.15
Least-cost combination of
inputs to produce 11,000
lbs of milk: price of hay
increase with unlimited
expenditure.

a total expenditure of $186. When the price of hay increases, the dairy producer
has two options, as examined next.

With unlimited funds to purchase inputs, the producer can continue to produce the same output level (11,000 lbs) but at a higher level of expenditure. Using
our decision rule, we locate the point of tangency between the slope of the new isocost line (-1.25) and the slope of the isoquant ($\text{MRS}_{X_2X_1}$). With input prices $P_{x_1} =$
$0.20 and P'_{x_2} = $0.25 the new least-cost combination of inputs to produce 11,000
lbs of milk is 650 units of grain (X_1) and 650 units of hay (X_2) with a total expenditure of $292.50. This decision is graphically depicted in Figure 5.15.

When the price of hay increases to $0.25/lb the isocost line steepens, rotating
inward from TE to TE' in Figure 5.15, showing that hay (X_2) is now more expensive
to use relative to grain (X_1). Due to the higher price, if the producer spent the entire $186 on hay ($X_2$) he can only purchase 744 units of X_2 instead of the initial 1,860
units. To maintain production at 11,000 lbs the producer recombines the variable
inputs until the slope of the new isocost line (TE') is equal to the slope of the isoquant (I_0). So the producer moves from point A in Figure 5.15, using 530 units of X_1
and 800 units of X_2, to point B, using 650 units of X_1 and 650 units of X_2, to maintain production at 11,000 lbs of milk. After the price of hay increases the producer
is substituting grain (X_1), the relatively cheaper input, for hay (X_2), the relatively
more expensive input. This substitution of grain (adding X_1 from 530 to 650 units)
for hay (replacing X_2 from 800 to 650 units) represents movement along the isoquant so production remains at the same level (11,000 lbs of milk). As you can see
in Table 5.7, at the new price levels our initial least-cost combination of inputs (530
units of X_1 and 800 units of X_2) costs $306, which is more than the new least-cost
combination of inputs (650 units of X_1 and 650 units of X_2), which costs $292.50.

TABLE 5.8 Input Combinations Producing 9,000 Pounds of Milk with Price of Grain (P_{x_1}) = \$0.20 and Price of Hay (P'_{x_2}) = \$0.25

X_1 (Grain)	X_2 (Hay)	ΔX_1	ΔX_2	$MRS_{X_2 X_1}$	$-\dfrac{P'_{x_2}}{P_{x_1}}$	Total Expenditure
405	423					\$186.75
		−35	25	−1.40	−1.25	
370	448					\$186.00
		−25	25	−1.00	−1.25	
345	473					\$187.25
		−20	25	−0.80	−1.25	
325	498					\$189.50
		−15	25	−0.60	−1.25	
310	523					\$192.75
		−10	25	−0.40	−1.25	
300	548					\$197.00
		−5	25	−0.20	−1.25	
295	573					\$202.25

Thus, by increasing the use of grain, from 530 to 650 lbs, the producer is able to maintain the same level of production but at a higher total expenditure on inputs.

With limited, or constrained, funds to purchase inputs, the producer cannot exceed the initial expenditure level of \$186. With input prices P_{x_1} = \$0.20 and P'_{x_2} = \$0.25, Table 5.7 indicates that with only \$186 to spend the producer cannot afford any of the input combinations required to produce 11,000 lbs of milk. Hence the only option available to the producer is to reduce production, move to a lower isoquant, and maintain the same level of expenditure. The producer obtains the input combinations that will produce 9,000 lbs of milk from past production records. These input combinations, and their total expenditure, are listed in Table 5.8.

In Table 5.8 the least-cost combination of inputs is found at the point of tangency between the isocost line and the highest obtainable isoquant (in this case 9,000 lbs of milk). Using our decision rule, we locate the point of tangency between the slope of the new isocost line (−1.25) and the slope of the new isoquant ($MRS_{X_2 X_1}$). The least-cost combination of inputs to produce 9,000 lbs of milk is 370 units of grain (X_1) and 448 units of hay (X_2) with a total expenditure of \$186. At this lower production level, use of both variable inputs declines. This decision is depicted graphically in Figure 5.16.

If the price of hay increases to \$0.25/lb the isocost line steepens, rotating inward from TE to TE′ in Figure 5.16. Since the producer is constrained by the available funds to purchase inputs, he finds the point of tangency between the highest obtainable isoquant (I_1) and the isocost line (TE′). Due to the price increase, combined with limited funds, the producer is forced to move to a lower isoquant, thus reducing production. The producer moves from point A in Figure 5.16, using 530 units of X_1 and 800 units of X_2 producing 11,000 lbs of milk, to point B, using 370 units of X_1 and 448 units of X_2 producing 9,000 lbs of milk. The net impact of an input price increase when the

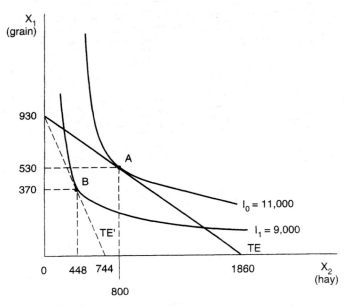

FIGURE 5.16 Least-cost combination of inputs to produce 11,000 lbs of milk: price of hay increase with limited expenditure.

producer is constrained by available funds to purchase inputs is a reduction in the use of both inputs, resulting in lower production levels.

In the face of increasing input prices, producers try to locate, and use, lower cost substitutes to maintain production levels, as illustrated in Box 5.1, and they try to minimize production costs. As demonstrated in our earlier example, if the price of a variable input increases the producer or manager has two options: The producer can produce the same output level (remain on initial isoquant) requiring a higher level of expenditure on inputs, or he can maintain the same level of expenditure and reduce output level (move to a lower isoquant). But in either case the decision rule, which equates the slope of the isoquant (MRS) to the slope of the isocost (IPR), is used to determine the new least-cost combination of inputs.

As an exercise to help you understand these concepts better, recalculate the preceding example of Table 5.8 and Figure 5.16 assuming the price of hay is $0.10/lb and the price of grain increases to $0.50/lb.

Input Price Decrease

When the price of a variable input decreases it becomes less expensive to use relative to the other input. The manager must then reevaluate the factor–factor decision to find the new least-cost combination of inputs. This decision is illustrated in Figure 5.17.

BOX 5.1   *Dairy Farmers Adjust to Higher Grain Prices*

In 1996 grain prices exploded as world stocks reached all-time low levels. Dairy farmers were faced with serious decision making about least-cost rations to maintain production. As corn prices reached $5.50 per bushel and soybean prices reached $7 per bushel, feed mills and dairy farmers scrambled to find balanced rations with lower cost feedstuffs. Most producers attempted to maintain a "nutritionally balanced ration" in order to maintain milking goals by finding substitutes that supplied equivalent protein, carbohydrates, and roughage. They found some substitutes in soy hulls, corn gluten feed (a by-product of corn syrup production), and corn distiller's dried grain (a by-product of distilled spirits production). For example, a typical Southeastern dairy farmer's balanced ration formulated to maintain a 19,000-lb herd average yearly production (i.e., isoquant = 19,000 lbs) in early 1996 and in late 1996 is illustrated in Table 5.9.

TABLE 5.9   Illustration of Dairy Farmers Adjustment to Higher Grain Prices in 1996

Feed	Early 1996 Ration			Late 1996 Ration		
	Cost/ton	lbs/day	Daily cost	Cost/ton	lbs/day	Daily cost
Alfalfa silage	$ 45	35.0	$0.79	$ 45	40.0	$0.90
Corn silage	25	35.0	0.43	25	40.0	0.50
Whole cottonseed	120	6.4	0.38	200	6.4	0.38
Ground corn	80	8.9	0.36	200	4.0	0.39
Mineral	470	0.3	0.07	470	0.3	0.07
Protein supplement	225	4.8	0.54			
Soy hulls				145	3.0	0.22
Corn gluten feed				250	2.0	0.25
Distiller's dried grain				170	1.5	0.13
Soybean meal				250	1.5	0.19
Total daily feed cost			**$2.57**			**$3.29**

In adjusting the feed ration, dairy producers modestly increased forage use, slashed corn use, and changed the composition of protein supplements (but maintained a "balanced" ration) to maintain production goals. Thus, dairy farmers tried to adjust to dramatically higher grain prices (corn prices increased from $80/ton to $200/ton) by substituting lower cost, but less desirable feeds. However, even with the substitutions, daily feed cost (per cow) still increased more than 25% to $3.29 in order to meet the goal of 19,000 lbs of milk per cow. This is just one of many examples of the factor–factor decision process used by producers in the real world.

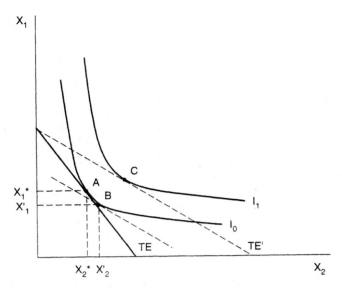

FIGURE 5.17 Factor–factor decision under changing prices: price of X_2 decrease.

Initially the producer is at point A in Figure 5.17, where the slope of the isoquant ($MRS_{X_2X_1}$) is equal to the slope of the isocost line ($-P_{x_2}/P_{x_1}$), and the least-cost combination of inputs to produce the desired output level (I_0) is X_1^* units of X_1 and X_2^* units of X_2. If P_{x_2} decreases to P'_{x_2}, the isocost line becomes flatter, rotating outward from TE to TE′ in Figure 5.17, because the slope is now equal to $-P'_{x_2}/P_{x_1}$. The producer has two options available given the price change. The first option is to maintain the same level of production at a lower level of expenditure. The second option is to maintain the same level of expenditure and expand production.

Reduce expenditure. The producer can choose to maintain his level of production (stay on isoquant I_0) by recombining the variable inputs until the slope of the initial isoquant (I_0) is equal to the slope of the new isocost line (TE′). In this case the least-cost combination of inputs, X_1' units of X_1 and X_2' units of X_2, is found at point B in Figure 5.17. Point B is found by locating the point of tangency between the initial isoquant (I_0) and a line with the same slope as (parallel to) the new isocost line (TE′). When the price of X_2 decreases, X_1 becomes more expensive to use relative to X_2; as a result the new least-cost combination of inputs (X_1', X_2') incorporates more X_2 ($X_2' > X_2^*$) and less X_1 ($X_1' < X_1^*$). This is because the producer can produce the same output level using more X_2 (the relatively cheaper input) and less X_1 (the relatively more expensive input). However, because the new combination of inputs (X_1', X_2') at point B lies on a lower isocost line, it is less expensive to purchase than the initial combination (X_1^*, X_2^*) at point A in Figure 5.17.

TABLE 5.10 **Determining Least-Cost Combination of Inputs Under Changing Prices: Dairy Example, Price of Hay Decreases**

X_1 (Grain)	X_2 (Hay)	Total Expenditure $P_{x_1} = \$0.20$ $P_{x_2} = \$0.10$	$P_{x_1} = \$0.20$ $P_{x_2} = \$0.25$	$MRS_{X_2X_1}$ $\left(\dfrac{\Delta X_1}{\Delta X_2}\right)$	$-\dfrac{P_{x_2}}{P_{x_1}}$ $\left(-\dfrac{0.10}{0.20}\right)$	$-\dfrac{P'_{x_2}}{P_{x_1}}$ $\left(-\dfrac{0.05}{0.20}\right)$
720	600	$204.00	$174.00			
				-1.40	-0.50	-0.25
650	650	$195.00	$162.50			
				-1.00	-0.50	-0.25
600	700	$190.00	$155.00			
				-0.80	-0.50	-0.25
560	750	$187.00	$149.50			
				-0.60	-0.50	-0.25
530	800	$186.00	$146.00			
				-0.40	-0.50	-0.25
510	850	$187.00	$144.50			
				-0.20	-0.50	-0.25
500	900	$190.00	$145.00			

Expand production. After the price decrease, the same level of expenditure purchases more. Therefore, the producer can maintain the same level of total expenditure on inputs by moving to a higher isoquant (I_1), thus increasing production. If the producer makes this decision he would operate at point C in Figure 5.17, where the slope of the higher isoquant, I_1, is equal to the slope of the new isocost line, TE′. At point C the producer is using more of both variable inputs, hence the production level is higher than the initial production level.

We can illustrate the effect of an input price decrease on the factor–factor decision using our dairy example. What happens to the least-cost combination of hay and grain if the price of hay decreases to $0.05/lb? The total expenditure required to purchase the input combinations that will produce 11,000 lbs of milk is presented in Table 5.10.

At the initial prices of $P_{x_1} = \$0.20$ and $P_{x_2} = \$0.10$, the least-cost combination of inputs is 530 units of grain (X_1) and 800 units of hay (X_2) with a total expenditure of $186. When the price of hay decreases the dairy producer has two options as examined next.

The producer can continue to produce the same level of output (11,000 lbs) but at a lower level of expenditure. Using our decision rule we locate the point of tangency between the slope of the new isocost line (-0.25) and the slope of the isoquant ($MRS_{X_2X_1}$). With input prices $P_{x_1} = \$0.20$ and $P'_{x_2} = \$0.05$ the new least-cost combination of inputs is 510 units of grain (X_1) and 850 units of hay (X_2) with a total expenditure of $144.50. This decision is depicted graphically in Figure 5.18.

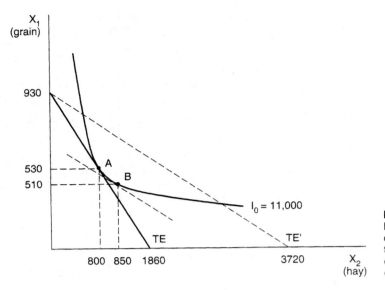

FIGURE 5.18
Least-cost combination of inputs to produce 11,000 lbs of milk: price of hay decrease.

When the price of hay decreases to $0.05/lb the isocost line flattens, rotating outward from TE to TE' in Figure 5.18, showing that hay (X_2) is now less expensive to use relative to grain (X_1). Due to the lower price, if the producer spends the entire $186 on X_2 he can purchase 3,720 units of X_2 instead of the initial 1,860 units. To maintain production at 11,000 lbs, the producer recombines the variable inputs until the slope of the new isocost line (TE') is equal to the slope of the isoquant (I_0). So the producer moves from point A in Figure 5.18, using 530 units of X_1 and 800 units of X_2, to point B, using 510 units of X_1 and 850 units of X_2, to maintain production at 11,000 lbs. After the price of hay decreases, the producer is substituting the relatively cheaper input hay (X_2) for the relatively more expensive input grain (X_1). This substitution of hay (adding X_2 from 800 to 850 units) for grain (replacing X_1 from 530 to 510 units) represents movement along the isoquant so production remains at the same level. As you can see in Table 5.10, at the new price level our initial least-cost combination (530 units of X_1 and 800 units of X_2) costs $146, which is more than the new least-cost combination of inputs (510 units of X_1 and 850 units of X_2), which costs $144.50. Thus by increasing the use of hay, from 800 to 850, the producer is able to maintain the same level of production with less total expenditure on inputs after the price decrease.

Table 5.10 shows that with input prices P_{x_1} = $0.10 and P'_{x_2} = $0.05 and $186 to spend on inputs, the producer can afford any of the input combinations that will produce 11,000 lbs of milk, with money left over. Thus, due to the decrease in the price of hay the producer can now afford to purchase additional inputs and expand his production by moving to a higher isoquant. The producer obtains the input combinations that will produce 13,000 lbs of milk from past production records. These input combinations, and their total expenditure, are listed in Table 5.11.

TABLE 5.11 Input Combinations Producing 13,000 Pounds of Milk with Price of Grain (P_{x_1}) = \$0.20 and Price of Hay (P'_{x_2}) = \$0.05

X_1 (Grain)	X_2 (Hay)	ΔX_1	ΔX_2	$MRS_{X_2 X_1}$	$-\dfrac{P'_{x_2}}{P_{x_1}}$	Total Expenditure
1,030	780					\$245
		−140	100	−1.40	−0.25	
890	880					\$222
		−100	100	−1.00	−0.25	
790	980					\$207
		−80	100	−0.80	−0.25	
710	1,080					\$196
		−60	100	−0.60	−0.25	
650	1,180					\$189
		−40	100	−0.40	−0.25	
610	1,280					\$186
		−20	100	−0.20	−0.25	
590	1,380					\$187

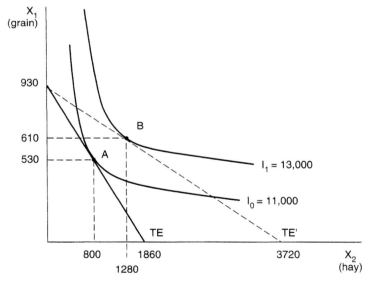

FIGURE 5.19 Least-cost combination of inputs to produce milk under changing prices: price of hay decrease with increased production.

The least-cost combination of inputs is found at the point of tangency between the isocost line and the highest obtainable isoquant (in this case 13,000 lbs of milk). Using our decision rule, we locate the point of tangency between the slope of the new isocost line (−0.25) and the slope of the new isoquant ($MRS_{X_2 X_1}$). From Table 5.11, the least-cost combination of inputs to produce 13,000 lbs of milk is 610 units of grain (X_1) and 1,280 units of hay (X_2) with a total expenditure of \$186. At this increased production level, use of both variable inputs increases. This decision is depicted graphically in Figure 5.19.

If the price of hay decreases to \$0.05/lb the isocost line flattens, rotating outward from TE to TE' in Figure 5.19. Since the producer wants to maintain the same level of expenditure, he finds the point of tangency between the highest obtainable isoquant (I_1) and the isocost line (TE'). Due to the price decrease, the producer can afford to purchase additional units of the inputs, enabling him to expand production and move to a higher isoquant. The producer moves from point A in Figure 5.19, using 530 units of X_1 and 800 units of X_2 producing 11,000 lbs of milk, to point B, using 610 units of X_1 and 1,280 units of X_2 producing 13,000 lbs of milk. When the producer desires to spend all of his available funds to purchase inputs, an input price decrease enables the producer to use more of both inputs and increase production.

As demonstrated earlier, if the price of a variable input decreases, the producer or manager has two options. The producer can produce the same output level (remain on the initial isoquant) requiring a lower level of expenditure; or maintain the same level of expenditure and increase output level (move to a higher isoquant). But in either case the decision rule, which equates the slope of the isoquant (MRS) to the slope of the isocost (IPR), is used to determine the new least-cost combination of inputs.

As an exercise to help you understand these concepts more thoroughly, rework the preceding example of Table 5.11 and Figure 5.19 assuming the price of hay is \$0.10/lb and the price of grain decreases to \$0.10 lb.

Decision Rule Etiquette

As we discussed earlier, the order of the subscripts on MRS indicates which input is being replaced and which input is being added. Since economic convention places the input being replaced on the vertical axis, the order of subscripts on MRS also indicates the labeling on the graphs. Thus, it is very important to have the graphical presentation of the decision rule match the written form, and vice versa.

When the input being replaced is X_1 and the input being added is X_2, the decision rule is

$$MRS_{X_2X_1} = -\frac{P_{x_2}}{P_{x_1}}$$

since X_1 is on the vertical axis and X_2 is on the horizontal axis.

However, if X_2 is the input being replaced, and X_1 is being added, the decision rule would be

$$MRS_{X_1X_2} = -\frac{P_{x_1}}{P_{x_2}}$$

because in this case X_2 will be on the vertical axis and X_1 will be on the horizontal axis.

Maximizing Profits Using Two Variable Inputs

When the least-cost combination of inputs to use in producing a given level of output is found, the manager has greatly improved his or her possibility of success. However, the manager is also concerned with what level of production will maximize profits. In our dairy example, if demand were to increase beyond 11,000 lbs of milk, which output level should our manager produce? Even though his capacity is greater than 11,000 pounds we know that output (yield) maximization is not the same as profit maximization; hence, we need price information and a decision rule to make the appropriate choice.

In a one-variable production function we found that profit was maximized when MRP was equated to MFC or the added revenue from employing another unit of input was equal to or greater than the cost of purchasing that input unit. The same decision rule holds for the two- or many-variable input situation, namely, $MRP_{X_1} = P_{x_1}$ and $MRP_{X_2} = P_{x_2}$. For profit maximization the general rule

$$\frac{MRP_{x_1}}{P_{x_1}} = \frac{MRP_{x_2}}{P_{x_2}} = \cdots = \frac{MRP_{x_n}}{P_{x_n}} = 1$$

holds, which, remembering our equi-marginal return principle from Chapter 4, states that resources should be employed so that the marginal returns are equal in each case. Recognizing that $MRP/P_x = 1$ means that $MRP = P_x$, the criterion can be rewritten as

$$\frac{MRP_{x_1}}{P_{x_1}} = \frac{MRP_{x_2}}{P_{x_2}} = \cdots = \frac{MRP_{x_n}}{P_{x_n}}$$

Thus, profits will be maximized when each input is employed so that the marginal returns are equal.

Another way of finding the profit maximizing level of product is to compare total costs (total outlay) at each output level with the total revenue received for that output. At each output level, the cost of production is $(P_{x_1} \times X_1 + P_{x_2} \times X_2)$ and revenue is $(P_y \times Y)$. As the decision to increase production is evaluated, it is easy to compare the marginal, or added, cost of going from one level of output to the next with the marginal, or added, revenue received from use of more inputs. Actually, if we want to look at the inputs, we can equate the MFC of the combined least-cost combination of inputs to the MRP of achieving that least-cost level of output for each alternative level.

Expansion of Business Production Level

As we indicated earlier, the production surface for a firm contains an isoquant for each output level the firm is capable of producing. In deciding how to expand production or the size of a business, it is useful to introduce the concept of an isocline.

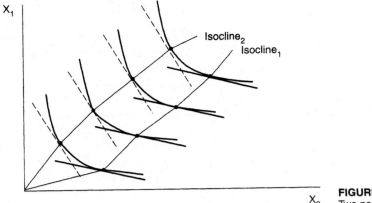

FIGURE 5.20
Two possible isoclines.

An **isocline** is defined as *a line connecting the least-cost combinations of inputs for all output levels at a specific price ratio.* Two isoclines, each representing a different price ratio, are graphically depicted in Figure 5.20.

The isocline shows how, at a given price ratio, inputs should be added as production increases. The specific direction for expanding production is given by an "expansion path." An **expansion path** is defined as *a line connecting least-cost combinations of inputs for each level of output at "the specific price ratio the manager believes relevant."* The important point is that, of all the isoclines available due to the many possible price ratios, when a manager chooses that price ratio considered to be most probable, this isocline then becomes the expansion path for the firm. Thus, while every expansion path is an isocline, not every isocline is an expansion path; only the one representing the chosen, relevant price ratio is the expansion path.

In Figure 5.20, isocline$_2$ represents the situation where X_2 is relatively more expensive than X_1; that is, the isocost lines associated with this price ratio have a steep slope. Thus, as the firm expands production along this isocline, the firm will use relatively more of X_1 in the production process than X_2. On the other hand, isocline$_1$ represents the situation where X_1 is relatively more expensive than X_2; that is, the isocost lines associated with this price ratio have a flat slope. Thus, as the firm expands along this isocline, the firm will use relatively more of X_2 in the production process than X_1.

SUMMARY

In this chapter we have been concerned with the basic production question "How to produce?" This decision is concerned with a production function where two inputs are variable, and production results from different combinations of these two inputs. Commonly referred to as factor–factor decision making, this analysis determines the least-cost combination of inputs to produce a given output level. In developing the cost-minimization criterion we introduced the concepts of an iso-quant and isocost lines. The isoquant describes the physical, or technical, manner in which inputs substitute for each other in the production process; the isocost line describes the economic relationships between relative factor prices. When the firm has more than one variable input to use in the production process, the decision criteria are used to find the least-cost combination of inputs that will produce the output level desired by the firm. The decision rule developed equates the slope of the isoquant (MRS), the rate of physical substitution, to the slope of the isocost line (IPR), the rate of economic substitution.

The firm faces additional decisions if the price of one of the inputs changes. Being a price taker the firm must pay the market price for its variable inputs. If an input price decreases, the producer can either produce more output (move to higher isoquant) with the same total outlay, or produce the same output (stay on same isoquant) with a lower expenditure level. If an input price increases, the producer can either produce less output (move to lower isoquant) with the same total outlay, or produce the same output (stay on same isoquant) and increase expenditure on inputs.

We continued to study the question of allocation and producer decisions developed in the first four chapters of the book. Price relationships combined with physical input–output relationships have allowed us to answer two producer questions, "How much to produce?" and "How to produce?" Many products, enterprises, or activities are available to every manager. In the next chapter we will develop rules for managerial economic decision making in choosing between products to produce.

QUESTIONS

1. Utilizing Figure 5.1, what is the difference in output of milk achieved by using 650 units of grain and 650 units of hay versus using 600 units of grain and 700 units of hay?
2. True or false? Isoquants are bowed in toward (convex) the origin, represent equal output along any given isoquant, and increase in output value as you move closer to the origin. Explain.
3. Suppose $MRS_{X_2 X_1}$ is -5. Does this mean that input X_2 substitutes fairly well for X_1 or vice versa? In general, what can you say about an MRS equal to -1.0?
4. Explain the law of diminishing marginal rate of substitution. How is this law related to the steepness of isoquants? How is it related to the law of diminishing returns?
5. Suppose a wheat producer has $1,000 to purchase inputs, and can purchase wheat seed for $100 a ton and nitrogen fertilizer for $25 per unit. What is the slope of this isocost line? Illustrate with a graph.
6. Using Figure 5.9 show the effect on the slope of each isocost line from a $0.10 increase in the price of X_1 and then a $0.05 decrease in the price of X_2.

7. From Table 5.6 suppose that the price of X_1 dropped to $0.10. What combination of grain and hay would now be optimal? What if the price of X_1 doubled from the original price?
8. Distinguish between isoclines and expansion paths. What determines the manager's selection of an expansion path as being most optimal?
9. Define isoquant and isocost line. What concepts do we use to define the slopes of each?

CHAPTER 6

Management Decisions: What to Produce?

INTRODUCTION

The preceding chapters developed concepts and decision rules aimed at helping the manager answer these two questions: "How much to produce?" and "How to produce?" Both of these decisions, the factor–product and the factor–factor decision, involve using input(s) to produce a single product. But few operators or decision makers are concerned with producing just one product. In fact, most firms and farms are concerned with producing many alternative products. The need for trained managers with good knowledge of managerial economics becomes particularly acute in this situation.

This chapter addresses the third of the basic producer questions: "What to produce?" This approach, called a **product–product decision** reflects the reality and complexities of life regarding the many choices that we all must make. On a farm, the manager is faced with choices of which enterprises (crops, livestock, etc.) and what amount of each should be produced to achieve the goal of profit maximization. A student has to decide what discipline to major in and which employment opportunity to accept. A farm-supply store or book publisher must decide how extensive and complete a product line to offer its customers. All people have to make the choice continually between work or leisure. Each of these choices reflects the process of allocating resources between competing ends to achieve the stated managerial goal.

In This Chapter

The objective of this chapter is to introduce the basic decision rules appropriate for proper managerial decision making and economic analysis for situations in which the firm produces multiple products from a given set of resources. Because more than one product can be produced, it is possible, to a certain extent, to substitute one product for another. That is, the firm can move resources from the production of one output to the production of another output in various combinations. For example, a farmer may produce oats instead of barley on his or her acreage, a clothing factory may produce jackets instead of pants from its cloth, or a bakery may produce cakes instead of cookies from its baking supplies. When multiple products can be produced, a firm or producer must decide whether to specialize by producing a single product, or which combination of products will meet the firm's economic objectives. We assume that the firm will act as a rational producer; therefore, the objective is to combine the outputs produced in the proportion that results in the highest revenue, given a resource constraint.

Assumptions Used

The product–product decision involves finding the revenue maximizing combination of products, from the different product combinations possible, that can be produced using a given set of resources. To examine this decision we use the following assumptions:

1. The firm produces two products.
2. The firm has a fixed set of resources.
3. The firm is a price taker (both inputs and outputs).

The firm is capable of producing different combinations of two products from the same set of fixed or given level of resources. Since the firm is a price taker, neither its input use or production levels have an impact on the price it pays for inputs or on the price it receives for production. As in previous chapters we assume that all decisions are made under perfect certainty.

The approach in this chapter is very similar to the factor–factor discussion of the previous chapter. First, a measurement or tool is developed to identify how one product substitutes for another in the physical production activities of the firm. Second, a decision rule is developed to identify the combination of products that will maximize the firm's revenue for a given level of input use. Finally, the most profitable level of output and associated input usage will be identified. As we will see, the decision rules developed in this chapter can be used in the same conceptual manner for more complex situations where more than two products are involved.

PHYSICAL RELATIONSHIPS

We begin our discussion of the product–product decision by examining the physical production aspects of the decision. The firm can produce two products from its set of available resources. Notationally, we are now using a production function that has two alternative products, Y_1 and Y_2, that can be produced from a given level of resources available, or fixed, in the short-run. For example, in a given crop year a farmer can produce either corn or soybeans, or some combination, given the available land, machinery, and fertilizer; on a given day a bakery can produce either bread or cakes, or some combination, with the available baking supplies, employees, and oven space. Thus the firm's production function is expressed as

$$Y_1, Y_2 = f(\mid X_1, \ldots, X_n)$$

where Y_1 and Y_2 are the two products produced by the firm, and $X_1, \ldots, X_n$ are the factors of production fixed in the short-run. In the long-run, all resources are variable in the production of these two products. Because limited resources, especially in time and space, exist in most decisions, the objective of the product–product decision is revenue maximization. With resources fixed at some level, revenue maximization is identical to profit maximization. The product–product decision involves determining how the two products substitute for one another in the firm's physical production process. In determining a measurement for describing how the two products physically substitute for one another, we introduce the concept of a production possibilities frontier.

Production Possibilities Frontier

The firm is capable of producing two products from a given set of resources. How do we represent the firm's production possibilities in a form that is easy to understand? A convenient device for depicting two production functions in one figure is the **production possibilities frontier (PPF),** which is defined as *a curve depicting all the combinations of two products that can be produced using a given level of inputs (or expenditure).* The production possibilities frontier, also referred to as the production possibilities curve (PPC), identifies all the possible product combinations the firm is capable of producing from a given, or fixed, level of inputs. Since both products are produced using the same resources, we are talking about two separate production functions that are no longer independent because expanding production of one output requires decreasing production of the other output.

Deriving a Production Possibilities Frontier

To illustrate how to derive a production possibilities frontier, consider the following example. Assume a farmer can use nitrogen fertilizer (the input, X) to produce either corn or grain sorghum, or a combination of both crops, on her farm. Since we are talking about physical production, we begin by defining, or specifying, the production function (TPP) of each individual product the firm can produce. The agronomic relationship between fertilizer and each of the two crops is presented in Table 6.1. This agronomic information reveals two things: (1) Both production functions exhibit decreasing returns over the relevant range. (We'd never go past eight units of fertilizer, would we? Remember Stage III of production?) (2) Corn (Y_1) is more physically responsive in yield to the addition of fertilizer (X) than sorghum (Y_2). Note that the production function for each crop exhibits the characteristics of a typical production function (three Stages of Production).

TABLE 6.1　Production Functions for Corn and Sorghum

Fertilizer (X)	Corn in Bushels (Y_1)	Fertilizer (X)	Sorghum in Bushels (Y_2)
0	0	0	0
1	10	1	8
2	25	2	25
3	50	3	44
4	70	4	56
5	85	5	65
6	95	6	72
7	100	7	76
8	101	8	78
9	95	9	77
10	85	10	73

TABLE 6.2　Production Combinations Possible for X = 4

X in Y_1 (Corn)	X in Y_2 (Sorghum)	Total X	Production Y_1 (corn)	Production Y_2 (sorghum)
4	0	4	70	0
3	1	4	50	8
2	2	4	25	25
1	3	4	10	44
0	4	4	0	56

From its definition we know that the production possibilities frontier illustrates the different combinations of two products that can be produced using a given level of input. So the second step in deriving the production possibilities frontier is to specify the level of input use. Consider the production possibilities frontier for an input use level of four units ($X = 4$), presented in Table 6.2, and for an input use level of 8 units ($X = 8$), presented in Table 6.3.

As indicated in Tables 6.2 and 6.3, the total input use between the two products always sums to the specified input use level. The relationship that exists between the two crops, corn and sorghum, is considered competitive. That is, with limited resources available, to be able to increase the production of one product (sorghum) resources must be removed from the other product (corn). Thus, in a competitive relationship, the production of one product is increased by decreasing production of the other product.

The production possibilities frontier for an input use of $X = 4$ and $X = 8$ are presented in Figure 6.1. Notice how Figure 6.1 is labeled; economic convention plots the output that is being replaced (Y_1 in this example) on the vertical axis, and

TABLE 6.3 Production Combinations Possible for X = 8

X in Y_1 (Corn)	X in Y_2 (Sorghum)	Total X	Y_1 (corn)	Y_2 (sorghum)
			Production	
8	0	8	101	0
7	1	8	100	8
6	2	8	95	25
5	3	8	85	44
4	4	8	70	56
3	5	8	50	65
2	6	8	25	72
1	7	8	10	76
0	8	8	0	78

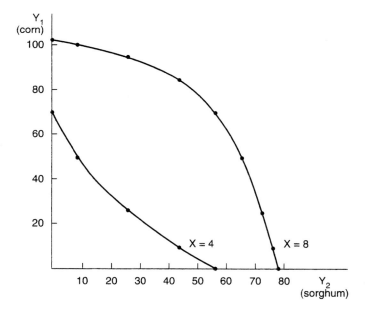

FIGURE 6.1
Production possibilities frontiers for X = 4 and X = 8.

the output being added (Y_2 in this example) on the horizontal axis. This designation of which output is to be replaced and added is arbitrary, but is important in terms of specifying the marginal rate of product substitution (MRPS) and the inverse price ratio (IPR), terms that are discussed later.

The production function for each crop exhibits decreasing returns. This is also indicated by the shape of the production possibilities frontier. The production possibilities frontier for eight units of fertilizer (X = 8) in Figure 6.1 is bowed out (concave) indicating that to increase production of one product, we must decrease production of the other product. Note that the shape of the production possibilities frontier for four units of fertilizer ($X = 4$) is more convex (bowed inward) than the production possibilities frontier for eight units. When we only use four units of fertilizer, we are operating within the increasing returns portion of the production functions for each crop, thus the production possibilities frontier is bowed inward. It is evident now why production possibilities frontiers are considered convenient devices to depict two production functions. They do, as in this example, identify the different combinations of corn and sorghum that can be produced using a specified input use level.

Since each production possibilities frontier represents one level of input use, theoretically there is a production possibilities frontier for each level of input the firm is capable of using. The distance from the origin is an indication of the level of input (or level of expenditure on input) being used. Generally, the farther away from the origin, the higher the input use level; the closer to the origin, the lower the input use level. This is illustrated in Figure 6.1, where the PPF for $X = 8$ is farther from the origin than the PPF for $X = 4$.

Because each production possibilities frontier represents a unique input use level, it also represents a unique expenditure level on inputs. That is, along each production possibilities frontier input use is held at a constant level; therefore, expenditure on (or the cost of using) the inputs is also held at a constant level.

Marginal Rate of Product Substitution

When resources are limited, increased production of one output can only come about if production of the other product is decreased, since inputs must be transferred from producing one product to another. As you can see from Figure 6.1 and Table 6.3, as we move along the production possibilities frontier the input use level remains constant but the quantities of the two products produced change as one output replaces, or substitutes for, the other. As managers we are interested in how the outputs substitute for one another. The rate at which one product must be reduced to increase the other product by one unit, or how the two products physically substitute for one another, is measured by the marginal rate of product substitution (MRPS). Note that the P in MRPS is useful to remind us that we are working in the product–product decision.

The **marginal rate of product substitution (MRPS)** describes *the rate at which one output must be decreased as production of the other product is increased, and, by definition, is the slope of the production possibilities frontier.* Since economic convention places the product being replaced on the vertical axis and the product being added on the horizontal axis, MRPS is calculated by dividing the change in the product being replaced (Δ replaced) by the change in the product being added (Δ added). Thus, the computational formula and interpretation of MRPS depend on which product is being replaced and which is being added.

For example, if Y_1 is the product being replaced and Y_2 is being added, then the **marginal rate of product substitution of Y_2 for Y_1**, written $MRPS_{Y_2Y_1}$, is calculated by dividing the ΔY_1 (product replaced) by the ΔY_2 (product added), or

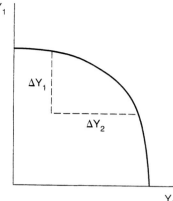

FIGURE 6.2
Marginal rate of product substitution of Y_2 for Y_1.

$$\text{MRPS}_{Y_2 Y_1} = \frac{\Delta Y_1}{\Delta Y_2} = \frac{\text{replaced}}{\text{added}}$$

Graphically, as shown in Figure 6.2, Y_1 is on the vertical axis and Y_2 is on the horizontal axis. The slope of the production possibilities frontier in Figure 6.2 is equal to the change in Y_1 divided by the change in Y_2 ($\Delta Y_1 \div \Delta Y_2$), which corresponds exactly to the definition of $\text{MRPS}_{Y_2 Y_1}$.

Conversely, if Y_2 is the product being replaced and Y_1 is the product being added, then the **marginal rate of product substitution of Y_1 for Y_2**, written $\text{MRPS}_{Y_1 Y_2}$, is calculated by dividing the ΔY_2 (product replaced) by the ΔY_1 (product added), or

$$\text{MRPS}_{Y_1 Y_2} = \frac{\Delta Y_2}{\Delta Y_1} = \frac{\text{replaced}}{\text{added}}$$

Graphically, as shown in Figure 6.3, Y_2, the product being replaced, is on the vertical axis, and Y_1, the product being added, is on the horizontal axis. The slope of the production possibilities frontier in Figure 6.3 is equal to the change in Y_2 divided by the change in Y_1 ($\Delta Y_2 \div \Delta Y_1$), which corresponds exactly to $\text{MRPS}_{Y_1 Y_2}$.

Note that the subscripts on the definition of MRPS match the axes labels on the graphs in Figures 6.2 and 6.3. Since the order of the subscripts on MRPS tells us which product is being added and which product is being replaced, it is very important that the graphical presentation be consistent with the stated MRPS, and vice versa.

We indicated and illustrated earlier that the designation of which output is to be replaced and which output is to be added is an arbitrary designation. This is because

$$\text{MRPS}_{Y_2 Y_1} = \frac{1}{\text{MRPS}_{Y_1 Y_2}}$$

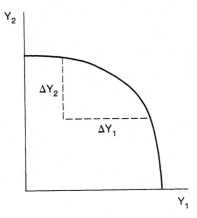

FIGURE 6.3
Marginal rate of product substitution of Y_1 for Y_2.

Why?

$$\text{MRPS}_{Y_2 Y_1} = \frac{\Delta Y_1}{\Delta Y_2} = \frac{1}{\frac{\Delta Y_2}{\Delta Y_1}} = \frac{1}{\text{MRPS}_{Y_1 Y_2}}$$

As illustrated earlier, $\text{MRPS}_{Y_2 Y_1}$ is the inverse of $\text{MRPS}_{Y_1 Y_2}$, and although they represent the same information, each represents a specific method of presenting the information. Therefore, it is important to be aware of which product is being replaced and added for computation and for correct graphical presentation.

Using our crop example we can illustrate how to calculate the marginal rate of product substitution numerically. As shown in Table 6.3, sorghum (Y_2) is being added and corn (Y_1) is being replaced. Therefore, the correct formula to calculate the marginal rate of product substitution between these products is $\text{MRPS}_{Y_2 Y_1} = \Delta Y_1 / \Delta Y_2$, as presented in Table 6.4.

The formula, which states the rate at which Y_2 (is added) substitutes for Y_1 (is replaced), is equal to the change in Y_1 divided by the change in Y_2, and, as graphically depicted in Figure 6.4, is the slope of the production possibilities frontier. When two products are competing for limited resources the firm must decrease production of one product to be able to increase production of the other product; thus, MRPS is negative as indicated in Table 6.4 and Figure 6.4. The MRPS of -0.125 ($-1/8$) as we go from the first combination (101,0) to the second combination (100, 8) indicates that only one-eighth of a unit of corn has to be given up (replaced) to gain (add) a unit of sorghum. This -0.125 also describes the slope of the production possibilities frontier at this point, since, by definition, MRPS measures the slope of the production possibilities frontier. As we substitute sorghum for corn in our crop example, the MRPS increases in magnitude (in terms of absolute value). Thus, this production possibilities frontier exhibits increasing MRPS.

TABLE 6.4 Calculating the Marginal Rate of Product Substitution of Sorghum (Y_2) for Corn (Y_1)

Y_1 (Corn)	Y_2 (Sorghum)	ΔY_1 (Replaced)	ΔY_2 (Added)	$MRPS_{Y_2 Y_1}$	=	$\Delta Y_1 \div \Delta Y_2$
101	0					
		−1	8	−0.125	=	−1/8
100	8					
		−5	17	−0.294	=	−5/17
95	25					
		−10	19	−0.526	=	−10/19
85	44					
		−15	12	−1.250	=	−15/12
70	56					
		−20	9	−2.222	=	−20/9
50	65					
		−25	7	−3.571	=	−25/7
25	72					
		−15	4	−3.750	=	−15/4
10	76					
		−10	2	−5.000	=	−10/2
0	78					

Types of Competitive Substitution

The example developed for the crop producer has products that are imperfect substitutes for each other, or the production possibilities frontier exhibits increasing marginal rate of product substitution. This phenomenon, which is one of three possible competitive situations (increasing, constant, and decreasing MRPS), arises because the production functions for corn and sorghum both exhibit decreasing returns. Thus, when we decrease the amount of fertilizer employed in corn (Y_1) production from eight to seven, we decrease corn production by only one unit since the marginal physical product (MPP_{Y_1}) at this level of input usage is one. However, when we move that unit of fertilizer into sorghum (Y_2), we increase sorghum production by eight units because, at that input level the marginal physical product (MPP_{Y_2}) is eight. From the ratio of MPP_{Y_1}/MPP_{Y_2} (or $\Delta Y_1/\Delta Y_2$), we can see that the increasing rate of substitution is based on the decreasing returns in the production function. Thus, if the production functions of the two products exhibit decreasing returns, the PPF is bowed out (concave) and exhibits increasing MRPS, as in Figure 6.4.

A second competitive possibility occurs when the products substitute for each other at a constant rate. This situation, indicated by the straight line in Figure 6.5, reflects the constant rate of competition between the two products. Under this substitution situation the manager would find that only one product, the one whose relative price is greater than the constant physical substitution rate,

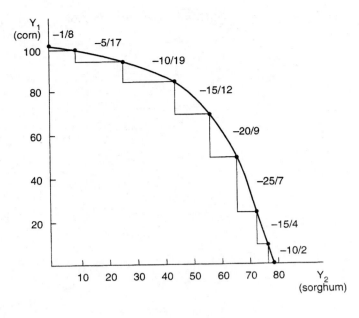

FIGURE 6.4
Production possibilities frontier and marginal rate of product substitution of sorghum (Y_2) for corn (Y_1).

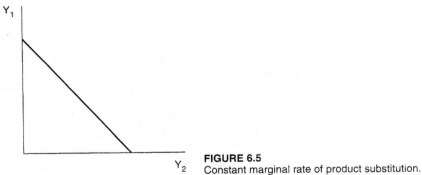

FIGURE 6.5
Constant marginal rate of product substitution.

should be produced. If sorghum substituted for corn at a constant two-to-one rate then as long as the price of sorghum is at least one-half the price of corn, only sorghum would be produced to achieve revenue maximization. Obviously, if the manager feels the prices may change in the future or if the production efficiency of either product might change, he or she might choose to produce some of each.

The third competitive possibility is when the two products substitute for each other at a decreasing rate. This would be expected when the production functions exhibit increasing returns so that less and less has to be given up of one product to increase production of the alternative product. This relationship, shown as a convex curve in Figure 6.6, occurs usually in the early section of the production func-

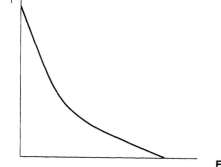

FIGURE 6.6
Decreasing marginal rate of product substitution.

tion where capital or another resource might be limited. (Recall PPF for $X = 4$ in Figure 6.1?) If this relationship holds, only one product, the one with the highest relative price–output ratio, will be produced.

In summary, since the production possibilities frontier is derived from the underlying production functions of the two products the firm can produce, its shape and rate of MRPS are dependent on those production functions. All of these rates of substitution are based on the competitive relationship between products, where one product can only be increased by a decrease in the level of production for the other product since the two products compete for the same inputs. The most common relationship is increasing MRPS because the law of diminishing returns is, in fact, a common occurrence. The manager must continually evaluate what is gained versus what is given up when deciding what combination of products to produce.

ECONOMIC RELATIONSHIPS

The goal of the product–product decision is to maximize revenue for a given level of input use, or expenditure. To determine the revenue associated with the possible output combinations, we must add prices to the technical relationship. Output prices are appropriate since we are examining the product–product decision.

Isorevenue Line

A useful tool for describing the value of our two products is an **isorevenue line,** which is defined as *a line depicting all combinations of two products that will generate a given, or the same, level of total revenue.* It is calculated by adding the total revenue received from selling each product, or

$$TR = (P_{y_1} \times Y_1) + (P_{y_2} \times Y_2)$$

TABLE 6.5 Combinations Generating $150 Revenue with Price of Corn (P_{y_1}) = $2.50 and Price of Sorghum (P_{y_2}) = $5.00

Corn Sold (Y_1)	Revenue from Corn ($P_{y_1} \times Y_1$)	Sorghum Sold (Y_2)	Revenue from Sorghum ($P_{y_2} \times Y_2$)	Total Revenue ($P_{y_1} \times Y_1 + P_{y_2} \times Y_2$)
60	$150	0	$ 0	$150
40	$100	10	$ 50	$150
20	$50	20	$100	$150
0	$0	30	$150	$150

where TR is the total revenue, in dollars, received from sales of both products, P_{y_1} is the price of product 1, P_{y_2} is the price of product 2, and Y_1 and Y_2 are the units of product 1 and 2 sold by the firm.

The amount of revenue received depends on the amount of each product produced and the respective prices of the products. Going back to our crop example, if the crop producer can sell corn for $2.50/bu and sorghum for $5/bu, how many units of each crop will she have to sell to earn $150 in total revenue? Several possible product combinations that will generate $150 in total revenue are listed in Table 6.5. Each of the product combinations listed generates the same level of total revenue. For example, $150 would be earned from selling the first combination (60 bu corn, 0 bu sorghum) or the third combination (20 bu corn, 20 bu sorghum) or even the fourth combination (0 bu corn, 30 bu sorghum). Thus, to earn $150 in revenue the crop producer could sell any of the corn/sorghum combinations in Table 6.5. The isorevenue line for $150 is depicted graphically in Figure 6.7.

If total revenue is $150 and the crop producer sells only Y_2, she would have to sell 30 units of Y_2 to earn $150 ($150 ÷ $5 = 30). If she sells only Y_1, she would have to sell 60 units of Y_1 to earn $150 ($150 ÷ $2.50 = 60). These points are the end points, or intercept terms, of the isorevenue line. Connecting the two end points yields the isorevenue line for $150 (TR = $150 in Figure 6.7). Any combination of outputs on this line will generate the same level of total revenue ($150) from sales. Since all the product combinations listed in Table 6.5 yield the same level of total revenue ($150) they all lie on the isorevenue line for $150 (TR = $150) in Figure 6.7.

If total revenue is $375 and the crop producer only sells Y_2, she would have to sell 75 units ($375 ÷ $5 = 75); if she only sells Y_1, she would have to sell 150 units ($375 ÷ $2.50 = 150). The isorevenue line for $375 (TR = $375) is also drawn in Figure 6.7. As illustrated in Figure 6.7, the distance from the origin indicates the level of revenue. The closer to the origin, the lower the level of total revenue; the farther away from the origin, the higher the level of total revenue.

Of course, an infinite number of isorevenue lines could have been drawn in Figure 6.7, each reflecting a different level of revenue. For illustrative purposes we have only drawn two, one for $150 and one for $375.

The slope of the isorevenue line drawn in Figure 6.7 is $-P_{y_2}/P_{y_1}$, also referred to as the inverse price ratio (IPR). It is derived from the isorevenue equation:

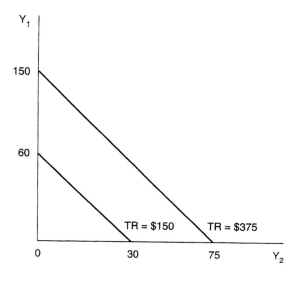

FIGURE 6.7
Isorevenue line for $150 and $375 total revenue.

$$TR = (P_{y_1} \times Y_1) + (P_{y_2} \times Y_2)$$

Since Y_1 is on the vertical axis, we solve for Y_1. First we subtract $(P_{y_2} \times Y_2)$ from both sides:

$$(P_{y_1} \times Y_1) = TR - (P_{y_2} \times Y_2)$$

and then we divide both sides by P_{y_1} to obtain

$$Y_1 = \frac{TR}{P_{y_1}} - \frac{P_{y_2}}{P_{y_1}} \times Y_2$$

This is the familiar formula for a straight line $Y = a + bX$, where the intercept term (a) in this case is TR/P_{y_1} (which identifies the Y_1 intercept on the Y_1 axis) and the slope (b) is $-P_{y_2}/P_{y_1}$. The slope of the isorevenue line is called an inverse price ratio because the price of the product on the horizontal axis is divided by the price of the product on the vertical axis. Thus, the way the graph axes are labeled will dictate which output price is in the numerator and denominator of the price ratio. The slope of the isorevenue line tells us the relative value of the outputs, or how they substitute for one another economically.

Effects of a Price Change

Since the slope of the isorevenue line depends on the relative price ratio of the two products, if the price of one of the products changes, the slope of the isorevenue line also changes.

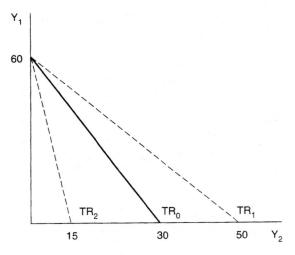

FIGURE 6.8
Effect of price Y_2 change on isorevenue line.

Figure 6.8 illustrates the impacts of a change in P_{y_2}. Recall our previous example where TR = $150, P_{y_1} = $2.50, and P_{y_2} = $5. At these prices, if the crop producer sold only Y_1, she could sell 60 units of Y_1; or if she sold only Y_2 she could sell 30 units of Y_2 (TR_0 in Figure 6.8) to earn $150. If the price of Y_2 changes, the number of units of Y_2 that will earn $150 in revenue also changes. If P_{y_2} increases, Y_2 is now more profitable to produce relative to Y_1, and with the price increase fewer units have to be sold to earn $150. For example, if P_{y_2} increases to $10, it only takes 15 units of Y_2 to earn $150 total revenue instead of the initial 30 units. The isorevenue line for $150, given the new prices (P_{y_1} = $2.50 and P_{y_2} = $10) is TR_2 in Figure 6.8. On the other hand if P_{y_2} decreases, Y_2 is now less profitable to produce relative to Y_1 and with the price decrease additional units must be sold to earn the same level of revenue. For example, if P_{y_2} decreases to $3, it now takes 50 units of Y_2 to earn $150 total revenue instead of the initial 30 units. The isorevenue line for $150 given the new prices (P_{y_1} = $2.50 and P_{y_2} = $3) is TR_1 in Figure 6.8.

As illustrated in Figure 6.8, if the price of product Y_2 changes, the number of units of Y_2 that have to be sold to earn the same level of revenue also changes. If P_{y_2} increases, the isorevenue line becomes steeper, or rotates inward, because the same level of total revenue can now be obtained from selling fewer units of Y_2. If P_{y_2} decreases, the isorevenue line becomes flatter, or rotates outward, because more units of Y_2 must now be sold to earn the same level of total revenue. However, note that the isorevenue line rotates around the axis intercept of the output whose price has not changed (Y_1 in this case). This is because with P_{y_1} = $2.50 the producer still has to sell 60 units of Y_1 to earn $150 in total revenue.

Figure 6.9 illustrates what will happen if the P_{y_1} changes. Recall our previous example where TR = $150, P_{y_1} = $2.50, and P_{y_2} = $5. At these prices, if the crop producer sold only Y_1, she could sell 60 units of Y_1; or if she sold only Y_2 she could sell 30 units of Y_2 (TR_0 in Figure 6.9) to earn $150. If the price of Y_1 changes, the number of units of Y_1 that will earn $150 in revenue also changes. If P_{y_1} increases, Y_1 is

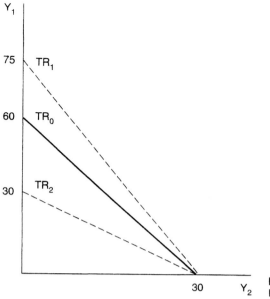

FIGURE 6.9
Effect of price Y_1 change on isorevenue line.

now more profitable to produce relative to Y_2, and with the price increase fewer units have to be sold to earn \$150. For example, if P_{y_1} increases to \$5, it only takes 30 units of Y_1 to earn \$150 total revenue instead of the initial 60 units. The isorevenue line for \$150 given the new prices (P_{y_1} = \$5 and P_{y_2} = \$5) is TR_2 in Figure 6.9. On the other hand if P_{y_1} decreases, Y_1 is now less profitable to produce relative to Y_2, and with the price decrease more units have to be sold to earn the same level of revenue. For example, if P_{y_1} decreases to \$2, it now takes 75 units of Y_1 to earn \$150 total revenue instead of the initial 60 units. The isorevenue line for \$150 given the new prices (P_{y_1} = \$2 and P_{y_2} = \$5) is TR_1 in Figure 6.9.

As illustrated in Figure 6.9, if the price of product Y_1 changes, the number of units of Y_1 that have to be sold to earn the same level of revenue also changes. Therefore, if P_{y_1} decreases, the isorevenue line becomes steeper, or rotates upward, because at the lower price more units of Y_1 must be sold to earn the same level of total revenue. If P_{y_1} increases, the isorevenue line becomes flatter, or rotates downward, because with the higher price we have to sell fewer units of Y_1 to earn the same level of total revenue. However, note that the isorevenue line rotates around the axis intercept of the output whose price has not changed (Y_2 in this case). This is because with P_{y_2} = \$5 the producer still has to sell 30 units of Y_2 to earn \$150 in total revenue.

We can summarize the important points to remember about the isorevenue line as follows:

1. The end points of the isorevenue line show how many units of each output would have to be sold if all the total revenue came from selling only that output.

2. A change in total revenue causes a parallel shift in the isorevenue line.
3. A change in an output price causes the isorevenue line to rotate around the intersection of the isorevenue line and the axis of the other output.
4. The slope of the isorevenue line, termed the inverse price ratio (IPR), is given by the price of the product on the horizontal axis divided by the price of the product on the vertical axis.

In summary, the main items of interest for an isorevenue line are distance from the origin (total revenue) and the slope (inverse price ratio).

The isorevenue line is a concept similar to the isocost line except the isorevenue measures revenue from selling product combinations, whereas the isocost line measures expenditures on purchasing input combinations.

PRODUCT–PRODUCT DECISION

When the manager's goal is to maximize profits, the objective of the producer is to find the combination of products that will maximize revenue for a given level of input use. Three equivalent approaches that can be used to determine the **revenue maximizing combination of products**. These approaches, all of which are based on the same economic trade-offs, are the total revenue, numerical, and graphical approaches.

Total Revenue Approach

Under the total revenue approach the manager simply calculates the total revenue associated with each product combination and chooses the product combination that yields the highest revenue.

We can illustrate this approach using our crop example. Assuming that the crop producer can sell corn for $2.50/bu and sorghum for $5/bu the total revenue generated by each alternative product combination produced using eight units of fertilizer is presented in Table 6.6, where we find that the product combination that maximizes revenue is 70 units of corn (Y_1) and 56 units of sorghum (Y_2) with a total revenue of $455. No other combination will produce this much revenue, as long as the product prices remain the same.

The total revenue approach is the simplest of the three. As indicated in Table 6.6, simply adding the revenue received for each product combination allows us to identify the revenue maximizing combination. However, this approach only allows the firm operator to examine known product combinations.

Numerical (or Mathematical) Approach

Under the numerical approach the manager equates the marginal rate of product substitution (slope of the production possibilities frontier) to the inverse price ratio (slope of the isorevenue line) to determine the output combination that maxi-

TABLE 6.6 Determining Revenue Maximizing Combination of Products for $X = 8$: Total Revenue Approach with Price of Corn (P_{y_1}) = $2.50 and Price of Sorghum (P_{y_2}) = $5.00

Y_1 (Corn)	Y_2 (Sorghum)			Total Revenue $(P_{y_1} \times Y_1 + P_{y_2} \times Y_2)$		
101	0	$252.50	+	$0.00	=	$252.50
100	8	$250.00	+	$40.00	=	$290.00
95	25	$237.50	+	$125.00	=	$362.50
85	44	$212.50	+	$220.00	=	$432.50
70	56	$175.00	+	$280.00	=	$455.00
50	65	$125.00	+	$325.00	=	$450.00
25	72	$62.50	+	$360.00	=	$422.50
10	76	$25.00	+	$380.00	=	$405.00
0	78	$0.00	+	$390.00	=	$390.00

mizes revenue. The marginal rate of product substitution describes the rate of physical (technical) substitution between the two products, while the inverse price ratio (IPR) describes the value (return) substitution between the products. In other words, we are determining the point where the rate of physical substitution (MRPS) is equal to the rate of economic substitution (IPR) for the two products, or for our crop example, where

$$\text{MRPS}_{Y_2 Y_1} = \frac{P_{y_2}}{P_{y_1}}$$

Why? We know that $\text{MRPS}_{Y_2 Y_1} = \Delta Y_1 / \Delta Y_2$, and since the inverse price ratio is negative and MRPS is always negative, the negative signs will cancel so we can write

$$\frac{\Delta Y_1}{\Delta Y_2} = \frac{P_{y_2}}{P_{y_1}}$$

Multiplying both sides by P_{y_1} and ΔY_2 gives

$$P_{y_1} \times \Delta Y_1 = P_{y_2} \times \Delta Y_2$$

which can be described as **value given up (Y_1) = value gained (Y_2).**

So the producer continues to substitute one output (adds Y_2) for the other output (replaces Y_1) until the value gained from the added output is equal to (or greater than) the value given up from the output being replaced. In other words, the producer examines each output combination and:

If $(P_{y_1} \times \Delta Y_1) < (P_{y_2} \times \Delta Y_2)$, then Y_2 should be added because the value given up by replacing Y_1 is less than the value gained by adding Y_2.

If $(P_{y_1} \times \Delta Y_1) > (P_{y_2} \times \Delta Y_2)$, then Y_1 should be added because the value given up by replacing Y_1 is greater than the value gained by adding Y_2.

This comparison is the basis for the product–product decision rule, which tells us that the revenue maximizing combination of products to produce from a given level of input will be found where the physical rate of substitution (MRPS) between the two products is equal to the economic rate of substitution (IPR) between the two products, or where MRPS = IPR.

We can illustrate this approach using our crop example. The appropriate information is presented in Table 6.7. Using our decision rule, $\text{MRPS}_{Y_2Y_1} = -P_{y_2}/P_{y_1}$ and the information in Table 6.7, we find that for an input use level of eight the revenue maximizing product combination is 70 units of corn (Y_1) and 56 units of sorghum (Y_2). Note that this is the same output combination found using the total revenue approach.

Why does this combination of products (70 units of Y_1 and 56 units of Y_2) maximize revenue when $\text{MRPS}_{Y_2Y_1} < -P_{y_2}/P_{y_1}$ at this point? Because each alternative product combination in Table 6.7 is produced using the same input level, we can compare the value gained from adding sorghum (Y_2) to the value given up from replacing corn (Y_1) to illustrate why this combination maximizes revenue for the crop producer. The last two columns in Table 6.7 show the value given up from replacing Y_1 (equal to $\Delta Y_1 \times P_{y_1}$) and the value gained from adding Y_2 (equal to $\Delta Y_2 \times P_{y_2}$). A simple comparison shows that the value gained by adding Y_2 is greater than the value given up, or lost, by replacing Y_1 up to the product combination of 70 units of Y_1 and 56 units of Y_2. Since the producer gains more by adding

TABLE 6.7   Determining Revenue Maximizing Combination of Products for $X = 8$: Numerical Approach with Price of Corn (P_{y_1}) = \$2.50 and Price of Sorghum (P_{y_2}) = \$5.00

Y_1 (Corn)	Y_2 (Sorghum)	ΔY_1	ΔY_2	$\text{MRPS}_{Y_2Y_1}$ ($\Delta Y_1/\Delta Y_2$)	$-\dfrac{P_{y_2}}{P_{y_1}}$ ($-5.00/2.50$)	Value Lost ($\Delta Y_1 \times P_{y_1}$)	Value Gained ($\Delta Y_2 \times P_{y_2}$)
101	0						
		−1	8	−0.125	−2.00	−\$2.50	\$40.00
100	8						
		−5	17	−0.294	−2.00	−\$12.50	\$85.00
95	25						
		−10	19	−0.526	−2.00	−\$25.00	\$95.00
85	44						
		−15	12	−1.250	−2.00	−\$37.50	\$60.00
70	56						
		−20	9	−2.220	−2.00	−\$50.00	\$45.00
50	65						
		−25	7	−3.570	−2.00	−\$62.50	\$35.00
25	72						
		−15	4	−3.750	−2.00	−\$37.50	\$20.00
10	76						
		−10	2	−5.00	−2.00	−\$25.00	\$10.00
0	78						

Y_2 than she loses by replacing Y_1, the producer can increase revenue by replacing corn (Y_1) and adding sorghum (Y_2) until this combination is reached. If the producer were to produce the next output combination (50 units of Y_1 and 65 units of Y_2), the value gained from adding Y_2 would be $45 but the value given up from replacing Y_1 would be $50. Since the producer gains less money ($45) by producing additional Y_2 (from 56 to 65 units) than she gives up ($50) by producing less Y_1 (from 70 to 50 units), the producer would continue to produce the output combination 70 units of Y_1 and 56 units of Y_2.

Graphical Approach

Using the graphical approach, the revenue maximizing combination of products is located at the point of tangency between the isorevenue line and the production possibilities frontier. Recall that the point of tangency between two curves is the point where the slopes of the two curves are equal to each other. Since the slope of the isorevenue line is the inverse price ratio and the slope of the production possibilities is the MRPS, graphically making the two curves tangent to each other is equivalent to equating the two numerically. Thus, the graphical approach is simply a picture of our decision rule $\text{MRPS}_{Y_2 Y_1} = -P_{y_2}/P_{y_1}$ and is illustrated in Figure 6.10.

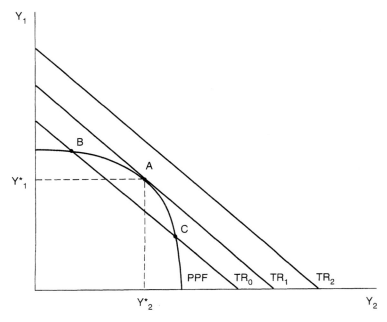

FIGURE 6.10 Product–product decision: determining the revenue maximizing product combination.

Since we assume that the producer is rational and wants to maximize revenue, the producer wants to reach the highest possible level of total revenue for a given level of expenditure. Therefore, the producer wants to find the tangency between the highest obtainable isorevenue line and the production possibilities frontier for his or her given level of input use. The isorevenue lines TR_0, TR_1, and TR_2 in Figure 6.10 represent three of the many possible revenue levels that could have been drawn. The input level for which the producer is trying to maximize revenue is represented by the production possibilities frontier, PPF, in Figure 6.10. The lowest revenue level depicted, the isorevenue line TR_0, intersects the production possibilities frontier at two points, B and C, indicating that either of these product combinations can be produced given the input use level. However, the producer's goal is to produce the output combination that maximizes revenue so the producer would like to reach the highest valued isorevenue line possible. Since they are on the same production possibilities frontier, the product combinations at points B and C can be produced using the same input use level as the product combination at point A in Figure 6.10. However, since the output combinations at points B and C are on a lower isorevenue line, TR_0, they generate less total revenue. Thus, the product combination at point A, which can be produced from the desired input use level, generates a higher revenue than either of the product combinations at points B or C. The revenue maximizing product combination, Y_1^* units of Y_1 and Y_2^* units of Y_2, is at point A in Figure 6.10 where the slope of the highest obtainable isorevenue line (TR_1) is tangent to the production possibilities frontier for the specified input use level (PPF), or where $MRPS_{Y_2Y_1} = -P_{y_2}/P_{y_1}$.

This can be illustrated using our crop example. This product–product decision is illustrated in Figure 6.11. From Table 6.6 we know that the revenue maximizing product combination generates a total revenue of $455. If the producer only

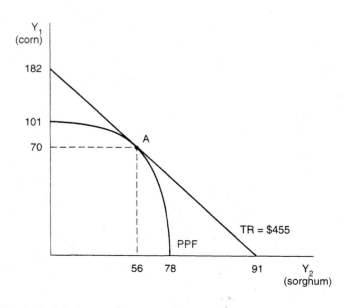

FIGURE 6.11
Revenue maximizing combination of corn and sorghum using eight units of fertilizer.

sold Y_2, she would have to sell 91 units of Y_2 (intercept point of horizontal axis in Figure 6.11) to earn $455 in total revenue. If the producer only sold Y_1 she would have to sell 182 units of Y_1 (intercept on vertical axis in Figure 6.11) to earn $455 in total revenue. At point A, the point of tangency between the production possibilities frontier (PPF) for $X = 8$ and the isorevenue line (TR = $455), the slopes of the two curves are the same. The revenue maximizing product combination is 70 units of corn (Y_1) and 56 units of sorghum (Y_2).

The three methods of determining the combination of products that will maximize revenue are equivalent. In our crop example, the revenue maximizing product combination, as determined under all three approaches, is 70 units of corn (Y_1) and 56 units of sorghum (Y_2) with a total revenue of $455. But what happens to the decision, or determination of the revenue maximizing combination, if the price of one of the products changes? We examine this in the next section.

Impact of Changes in Product Prices

Managers are often faced with changing product prices when making product–product decisions. When the relative prices of products change, the combination of products that maximizes revenue also changes.

Product Price Increase

When the price of one product increases, the profitability of producing that product increases relative to the other product. In this situation the manager must reevaluate the product–product decision to determine the new revenue maximizing product combination. This decision is depicted in Figure 6.12.

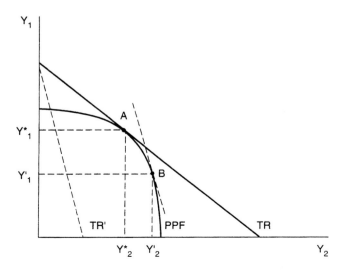

FIGURE 6.12
Product–product decision under changing prices: price of Y_2 increases.

Initially, the producer is at point A in Figure 6.12 where the slope of the production possibilities frontier ($MRPS_{Y_2Y_1}$) is equal to the slope of the isorevenue line ($-P_{y_2}/P_{y_1}$), and the revenue maximizing combination of outputs is Y_1^* units of Y_1 and Y_2^* units of Y_2. If P_{y_2} increases to P'_{y_2}, the isorevenue line becomes steeper, rotating inward from TR to TR′ in Figure 6.12, because the slope is now equal to $-P'_{y_2}/P_{y_1}$. After the price increase, the product Y_2 is more valuable, relative to Y_1, to produce. Since Y_2 is more valuable the producer can increase total revenue by reallocating resources so that more resources are used to produce Y_2 and less are used to produce Y_1. The producer continues to shift resources away from producing Y_1 and into the production of Y_2 until the slope of the new isorevenue line (TR′) is equal to the slope of the production possibilities frontier (PPF). The new revenue maximizing product combination, Y_1' units of Y_1 and Y_2' units of Y_2, is found at point B in Figure 6.12. Point B is found by locating the point of tangency between the production possibilities frontier and a line with the same slope as (parallel to) the new isorevenue line (TR′). After the price increase, the new revenue maximizing product combination incorporates more of Y_2 ($Y_2' > Y_2^*$) and less of Y_1 ($Y_1' < Y_1^*$). Since the new combination (Y_1', Y_2') lies on a higher isorevenue line, it earns a higher level of revenue than the initial combination (Y_1^*, Y_2^*). Due to the price increase, the producer is able to increase total revenue by reallocating input use from the production of Y_1, the relatively less valuable output, to the production of Y_2, the relatively more valuable output, without increasing the level of input use (costs).

We can illustrate the effect of a product price increase on the product–product decision using our crop example. What happens to the revenue maximizing combination of corn and sorghum if the price of sorghum increases to $9? The total revenue received from the output combinations produced using eight units of fertilizer is presented in Table 6.8, which shows that at the initial prices of $P_{y_1} =$ $2.50 and $P_{y_2} = 5 the revenue maximizing product combination is 70 units of corn (Y_1) and 56 units of sorghum (Y_2) with a total revenue of $455. After the price of sorghum increases (from $5 to $9) sorghum is more valuable, relative to corn, as reflected in the higher level of revenue earned by each product combination. Given the new prices, the producer is able to increase total revenue by shifting resources out of the less valuable output (Y_1) and into the more valuable output (Y_2). Using our decision rule, we determine the new revenue maximizing combination by equating the slope of the new isorevenue line (-3.60) to the slope of the production possibilities frontier ($MRPS_{Y_2Y_1}$). With product prices $P_{y_1} = 2.50 and $P'_{y_2} = 9 the new revenue maximizing product combination is 25 units of corn (Y_1) and 72 units of sorghum (Y_2) generating a total revenue of $710.50. This decision is graphically depicted in Figure 6.13.

As Figure 6.13 shows, when sorghum prices increase to $9 the isorevenue line steepens, rotating inward from TR to TR′, showing that sorghum (Y_2) is now more valuable relative to corn (Y_1). Due to the higher price, if the producer sold only Y_2 she would only have to sell 51 units of Y_2 to earn $455 in total revenue instead of the initial 91 units. To maximize revenue at the new prices, the producer reallocates input use between the two products until the slope of the new isorevenue line (TR′) is equal to the slope of the production possibilities frontier (PPF). So the producer

TABLE 6.8 Determining Revenue Maximizing Combination of Products Under Changing Prices: Crop Example, Price of Sorghum Increases

Y_1 (Corn)	Y_2 (Sorghum)	Total Revenue		$MRPS_{Y2Y1}$ $\left(\dfrac{\Delta Y_1}{\Delta Y_2}\right)$	$-\dfrac{P_{y_2}}{P_{y_1}}$ $\left(-\dfrac{5.00}{2.50}\right)$	$-\dfrac{P'_{y_2}}{P_{y_1}}$ $\left(-\dfrac{9.00}{2.50}\right)$
		$P_{y_1} = \$2.50$ $P_{y_2} = \$5.00$	$P_{y_1} = \$2.50$ $P'_{y_2} = \$9.00$			
101	0	\$252.50	\$252.50			
				−0.125	−2.00	−3.60
100	8	\$290.00	\$322.00			
				−0.294	−2.00	−3.60
95	25	\$362.50	\$462.50			
				−0.526	−2.00	−3.60
85	44	\$432.50	\$608.50			
				−1.250	−2.00	−3.60
70	56	\$455.00	\$679.00			
				−2.220	−2.00	−3.60
50	65	\$450.00	\$710.00			
				−3.570	−2.00	−3.60
25	72	\$422.50	\$710.50			
				−3.750	−2.00	−3.60
10	76	\$405.00	\$709.00			
				−5.00	−2.00	−3.60
0	78	\$390.00	\$702.00			

moves from point A in Figure 6.13, producing 70 units of Y_1 and 56 units of Y_2, to point B, producing 25 units of Y_1 and 72 units of Y_2, maintaining input use at eight units. After the price of sorghum increases, the producer has increased production of the relatively more valuable output sorghum (Y_2) and reduced production of the relatively less valuable output corn (Y_1). This substitution of sorghum (adding Y_2 from 56 to 72 units) for corn (replacing Y_1 from 70 to 25 units) represents movement along the production possibilities frontier so input use remains at the same level. As you can see in Table 6.8, at the new price levels our initial revenue maximizing combination (70 units of Y_1 and 56 units of Y_2) earns \$679 in total revenue, which is less than the new output combination (25 units of Y_1 and 72 units of Y_2), which earns \$710.50. Thus, by increasing sorghum production, from 56 to 72 units, the producer is able to increase her total revenue.

As we illustrated earlier, if the price of an output increases, the producer is able to increase total revenue by reallocating input use toward the relatively more valuable product and away from the relatively less valuable product. Thus, when the firm can produce more than one product with its resource base, if the price of an output increases, the firm will produce more of that output and less of the other outputs.

As an exercise to help you understand these concepts better, recalculate the example of Table 6.8 and Figure 6.13 assuming the price of sorghum is \$5/bu and the price of corn increases to \$8.

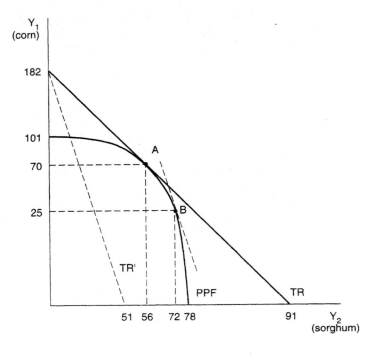

FIGURE 6.13
Revenue maximizing
combination of corn and
sorghum: price of
sorghum increases.

Product Price Decrease

When the price of one product decreases, the profitability of producing that product, relative to the other product, also decreases. In this situation the manager must reevaluate the product–product decision to find the new revenue maximizing product combination. This decision is illustrated in Figure 6.14.

Initially the producer is at point A in Figure 6.14 where the slope of the production possibilities frontier ($MRPS_{Y_2Y_1}$) is equal to the slope of the isorevenue line ($-P_{y_2}/P_{y_1}$), and the revenue maximizing combination of outputs is Y_1^* units of Y_1 and Y_2^* units of Y_2. If P_{y_2} decreases to P'_{y_2}, the isorevenue line becomes flatter, rotating outward from TR to TR′, because the slope is now equal to $-P'_{y_2}/P_{y_1}$. After the price decrease, the product Y_1 is more valuable, relative to the product Y_2, to produce. Since Y_1 is more valuable, the producer can maximize total revenue by reallocating resources so that more resources are used to produce Y_1 and fewer are used to produce Y_2. The producer continues to shift resource use away from producing Y_2 and into the production of Y_1 until the slope of the new isorevenue line (TR′) is equal to the slope of the PPF. The new revenue maximizing product combination, Y_1' units of Y_1 and Y_2' units of Y_2, is located at point B in Figure 6.14. Point B is found by locating the point of tangency between the production possibilities frontier and a line with the same slope as (parallel to) the new isorevenue line (TR′). After P_{y_2} decreases, the new revenue maximizing combination of outputs incorporates more Y_1 ($Y_1' > Y_1^*$) and less Y_2 ($Y_1' < Y_2^*$). However, since the new combination

BOX 6.1   Fancy Roses or Plain Jane Mums?

Managers of commercial greenhouses must make product–product decisions every few weeks. These decisions involve perishable, high-valued products grown on limited greenhouse floor space using advanced climate controls. Commercial greenhouses can pick from hundreds of combinations of plants, different growing periods, and varying levels of chemical and labor inputs to produce flowers for a very competitive market.

A simplified example would be a greenhouse specializing in flowers to be sold in the wholesale market to florists and other retailers. Suppose a nurseryman had 1,000 square meters of usable greenhouse space to be devoted to either roses or mums. Roses are a high-valued product with an average output of 130 stems per square meter (m^2), a wholesale value of $0.35/stem as an annual average, and thus an estimated gross return of $45.50 per square meter. But rose prices can rise and fall with market conditions (seasonal supplies, imports, holidays, etc.). Mums, a flower grown to be sold to landscape companies and consumers for personal home use, have an average output of 9.25 pots/m^2, an average annual price of $1.36 per pot, and a gross return of $12.58 per square meter. Since the same floor space, environmental controls, and inputs are used for both flowers, the nurseryman can produce all roses, all mums, or some combination.

Roses are a perennial crop that produces commercial stems in a year and have a useful economic life of seven years. So when floor space is devoted to roses, it is committed for several years. Roses have the highest gross revenue but are also the product with two seasonal spikes in prices (Valentine's Day and Mother's Day), higher costs, and management problems for perishability, shipping, and chemicals. In addition, roses face stiff competition from foreign sources (e.g., about 70 percent of roses are grown outside the United States) and enormous seasonal price changes. Mums are an annual crop that involves only eight weeks from seed to market. The market for mums is growing, reflecting new uses and consumers, resulting in modestly rising prices. There is much less risk involved in mums.

Few greenhouse managers use microcomputers and mathematical analysis to make product–product decisions. But their decision making follows product–product logic: Compare the MRPS for mums and roses to changing price ratios. However, nurserymen also must recognize risks in production, perishability, and price competition from imports. Relying on experience and trial-and-error, the nurseryman will estimate how much floor space to devote to high-value, high-risk roses ($45.50/$m^2$) and how much to low-risk, low-value mums ($12.58/$m^2$) to find a product combination that will meet minimum income requirements in the face of considerable production and market risk. The nurseryman faces a new product–product decision every eight weeks (the mum growing cycle) so profit maximization is a process of repeated product–product adjustments to floriculture markets, production costs, and import competition.

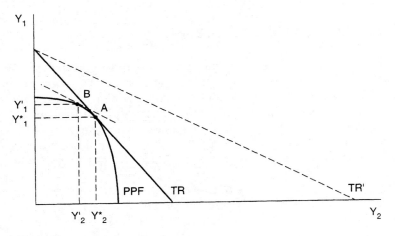

FIGURE 6.14
Product–product decision under changing prices: price of Y_2 decreases.

(Y_1', Y_2') lies on a lower isorevenue line, it earns a lower level of revenue than the initial combination (Y_1^*, Y_2^*). Due to the price decrease, the producer reallocates input use from the production of Y_2, the relatively less valuable output, to producing Y_1, the relatively more valuable output, to generate as much revenue as possible without increasing level of input use (costs).

We can illustrate the effect of a product price decrease on the product–product decision using our crop example. What happens to the revenue maximizing combination of corn and sorghum if the price of sorghum decreases to $1.25? The total revenue generated by the output combinations produced using eight units of fertilizer is presented in Table 6.9, where we see that at the initial prices of $P_{y_1} = \$2.50$ and $P_{y_2} = \$5$ the revenue maximizing combination is 70 units of corn (Y_1) and 56 units of sorghum (Y_2) with a total revenue of $455. After the price of sorghum decreases (from $5 to $1.25), sorghum is less valuable to produce, relative to corn, as reflected in the lower levels of revenue earned by each product combination. Given the new prices, the producer will maximize revenue by shifting resources out of the less valuable output (Y_2) and into the more valuable output (Y_1). Using our decision rule, we locate the point of tangency between the slope of the new isorevenue line (-0.50) and the slope of the production possibilities frontier $(MRPS_{Y_2Y_1})$. With product prices $P_{y_1} = \$2.50$ and $P_{y_2}' = \$1.25$ the new revenue maximizing product combination is 95 units of corn (Y_1) and 25 units of sorghum (Y_2) with a total revenue of $268.75. This decision is depicted graphically in Figure 6.15.

When the price of sorghum decreases to $1.25 the isorevenue line flattens, rotating outward from TR to TR' in Figure 6.15, showing that Y_2 (sorghum) is now less valuable to produce relative to Y_1 (corn). Due to the lower price, if the producer sold only Y_2 she would have to sell 364 units of Y_2 to earn $455 in total revenue instead of the initial 91 units. To maximize revenue at the new prices, the producer

TABLE 6.9 Determining Revenue Maximizing Combination of Products Under Changing Prices: Price of Sorghum Decreases

Y_1 (Corn)	Y_2 (Sorghum)	Total Revenue $P_{y_1} = \$2.50$ $P_{y_2} = \$5.00$	Total Revenue $P_{y_1} = \$2.50$ $P'_{y_2} = \$1.25$	$\text{MRPS}_{Y_2 Y_1}$ $\left(\dfrac{\Delta Y_1}{\Delta Y_2}\right)$	$-\dfrac{P_{y_2}}{P_{y_1}}$ $\left(-\dfrac{5.00}{2.50}\right)$	$-\dfrac{P'_{y_2}}{P_{y_1}}$ $\left(-\dfrac{1.25}{2.50}\right)$
101	0	$252.50	$252.50			
				−0.125	−2.00	−0.50
100	8	$290.00	$260.00			
				−0.294	−2.00	−0.50
95	25	$362.50	$268.75			
				−0.526	−2.00	−0.50
85	44	$432.50	$267.50			
				−1.250	−2.00	−0.50
70	56	$455.00	$245.00			
				−2.220	−2.00	−0.50
50	65	$450.00	$206.25			
				−3.570	−2.00	−0.50
25	72	$422.50	$152.50			
				−3.750	−2.00	−0.50
10	76	$405.00	$120.00			
				−5.00	−2.00	−0.50
0	78	$390.00	$97.50			

reallocates input use between the two products until the slope of the new isorevenue line (TR′) is equal to the slope of the production possibilities frontier (PPF). So the producer moves from point A in Figure 6.15, producing 70 units of Y_1 and 56 units of Y_2, to point B, producing 95 units of Y_1 and 25 units of Y_2, maintaining input use at eight units. After the price of sorghum decreases, the producer is producing more of the relatively more valuable output, corn (Y_1), and less of the relatively less valuable output, sorghum (Y_2). This substitution of corn (adding Y_1 from 70 to 95 units) for sorghum (replacing Y_2 from 56 to 25 units) represents movement along the production possibilities frontier so input use remains at the same level. As you can see in Table 6.9, at the new price levels our initial revenue maximizing combination (70 units of Y_1 and 56 units of Y_2) earns $245 in total revenue, which is less than the new output combination (95 units of Y_1 and 25 units of Y_2), which earns $268.75. Thus, by increasing corn production, from 56 to 95 units, the producer is able to maximize total revenue at the new prices.

As illustrated earlier, if the price of an output decreases the producer will shift input away from that product and toward the other, more valuable products to maximize revenues. Thus, when the firm can produce more than one product with its resource base, if the price of an output decreases, the firm will produce less of that output and more of the other outputs.

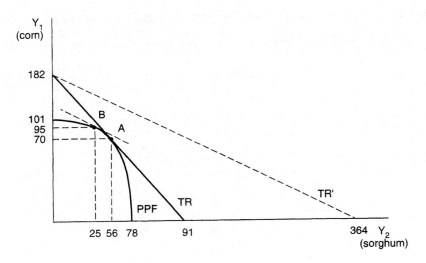

FIGURE 6.15
Revenue maximizing combination of corn and sorghum: price of sorghum decreases.

As an exercise to help you understand these concepts better, recalculate the earlier example of Table 6.9 and Figure 6.15 assuming the price of sorghum is $5/bu and the price of corn decreases to $1.25.

Decision Rule Etiquette

As we discussed earlier, the order of subscripts on MRPS indicates which product is being replaced and which product is being added. Since economic convention places the product being replaced on the vertical axis, the order of subscripts on MRPS also indicates the labeling on the graphs. Therefore, it is very important to have the graphical presentation of the decision rule match the written form, and vice versa.

When the product being replaced is Y_1 and the product being added is Y_2, the decision rule is

$$\text{MRPS}_{Y_2 Y_1} = -\frac{P_{y_2}}{P_{y_1}}$$

since Y_1 is on the vertical axis and Y_2 is on the horizontal axis.

However, if Y_2 is the product being replaced and Y_1 is the product being added, the appropriate decision rule is

$$\text{MRPS}_{Y_1 Y_2} = -\frac{P_{y_1}}{P_{y_2}}$$

because in this case Y_2 will be on the vertical axis and Y_1 will be on the horizontal axis.

Possible Product–Product Relationships

The relationship between products we have been using is a competitive one, depicted by a necessary decrease in the production of one product in order to increase the production of another. This arises because the products require the same inputs at the same time; that is, both corn and sorghum production require the available fertilizer. If fertilizer were unlimited and no other resources were limited, there would not be a competitive relationship and each product could be evaluated as a factor–product question. Competitive products using limited resources can, as discussed earlier, substitute either with increasing, constant, or decreasing MRPS.

Another product relationship faced by managers is the **complementary relationship** in which an increase in one product brings about an increase in the level of production of the other. This relationship is depicted in Figure 6.16 with the range of complementarity being up to point A. Within this area, as Y_1 increases an increase in Y_2 occurs. This type of relationship is extremely attractive to a decision maker because we have "the best of both worlds." In such a situation the need for a manager is not great. However, at some point, the two products will always reach the competitive relationship because some resource will become restricted and competition will start. Examples of such complementary relationships include legumes in a crop rotation in which the nitrogen-fixing properties help the yield of other crops in the rotation or a student in both a math and an econometrics course during the same semester finding that studying math also improves his or her ability in econometrics. But, in both cases, as time and/or space become a constraint, production of one product must compete with the other. A complementary relationship could also arise if, because of a bad management decision, one product

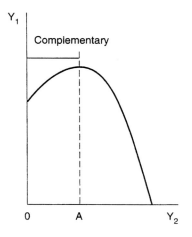

FIGURE 6.16
Complementary product relationship.

was being produced in Stage III. Thus, if we transferred labor from our over-crowded service station to our steel factory, production of both could increase.

Products can also be **supplementary** to each other, a situation in which the amount of one product can be increased without increasing or decreasing the amount of the other product. Figure 6.17 indicates that in the range 0 to A the products Y_1 and Y_2 have a supplementary relationship. This relationship holds if we have excess capacity or underutilized resources. The products then become competitors with each other as any input becomes limited. An example of supplementarity is the small farm flocks of chickens that used to be raised on the wheat farms in the Upper Great Plains. These chickens were tended to by the wife and children in their spare time. This enterprise was not competitive or complementary to the growing of wheat until labor became scarce (school starting in the fall) at which time the chickens were competing with the wheat for the scarce labor resource. Only when a competitive relationship exists are a manager and price information needed.

A fourth product relationship that can exist is that of a **joint product.** Joint products occur when the production of one product actually results in the production of another. Production of flax yields flax straw for linen. Raising sheep generates a given proportion of wool and mutton. In transportation, hauling wheat in one direction (the fronthaul) automatically creates the ability to haul something in the reverse direction (the backhaul). Figure 6.18 indicates that the production possibilities curve in this type of relationship is one single point for each level of resource use. Over the longer run we know we can choose wool breeds of sheep versus those noted for mutton or we can choose a variety of flax with a longer or shorter straw. This type of controlled, joint product relationship is indicated by the short-curved, production possibility curves in Figure 6.18. When no control exists, relative price ratios between the products are not needed. Only when control exists and at least some small competitive relationship sets in is the revenue maximization criterion, MRPS = IPR, needed.

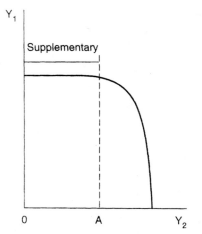

FIGURE 6.17
Supplementary product relationship.

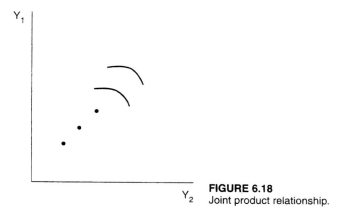

FIGURE 6.18
Joint product relationship.

Profit Maximization

When resources are fixed, as we've assumed in most of our discussion in this chapter, revenue maximization at that input use level is the same as profit maximization. Profit is found by subtracting the costs of any variable and fixed inputs from the revenue obtained at each input use level. But when alternative levels of input use are available, a choice must be made between the different levels of input use so as to maximize profits. In this case, profit maximization will not necessarily result from revenue maximization among output levels. If the added cost of reaching a higher output (revenue) level is greater than the value received for the additional output, then revenue should not be maximized if the highest profit is desired.

The decision rule for maximizing profits is a redefinition of the factor–product case, MRP = MFC. In the unlimited case we use

$$\frac{MRP_{X_1Y_1}}{P_{x_1}} = \frac{MRP_{X_1Y_2}}{P_{x_1}} = \cdots = \frac{MRP_{X_1Y_n}}{P_{x_1}} = 1$$

which states that we will add resources to each individual product until the added value received is just equal to the cost of achieving that value. When resources are limited, the appropriate decision rule is

$$\frac{MRP_{X_1Y_1}}{P_{x_1}} = \frac{MRP_{X_1Y_2}}{P_{x_1}} = \cdots = \frac{MRP_{X_1Y_n}}{P_{x_1}}$$

which states that resources should be used in alternative enterprises in such a manner that their marginal returns are equal. This statement is equivalent to the equimarginal return principle discussed in Chapter 4. As is now evident, the example of opportunity cost and equi-marginal returns discussed earlier is a specific example of profit maximization in a product–product decision.

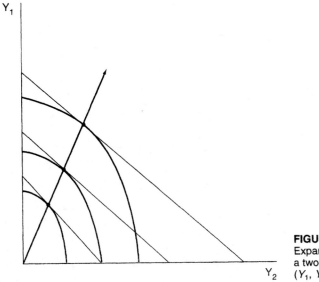

FIGURE 6.19
Expansion path for
a two-product firm
(Y_1, Y_2).

The profit maximization criterion developed here also allows us, as managers, to decide how to expand our farm production, business firm size, or any activity level. An **expansion path** in production is *the line connecting the combination of products that will maximize revenue for any given level of resources, at the price ratio the manager considers relevant and probable.* Obviously, an isocline exists for each price ratio (remember the previous chapter) but the expansion path, as in Figure 6.19, can be used as a planning tool for a manager. It tells the manager what products he or she should be selling, depending on what level of input is available. Again, how far out to go on this expansion path can be determined by evaluating the total cost at each level of input with the total revenue received from selling the products. Or, when the MRP (from $P_{y_1} \times Y_1 + P_{y_2} \times Y_2$) of going to a higher level of input use is less than the cost of increasing input use, increased production is not economically justified.

SUMMARY

In this chapter we developed decision-making tools to answer the third producer question, "What to produce?" These tools now allow us, as managers, to decide what products to produce or what enterprises to include in our production plans. Making this decision requires converting the physical (technical) relationships of substituting one product for another when limited resources exist to an economic (market) value on which to base the substitution decision. This physical substitution is indicated by the slope of the production possibilities curve (MRPS), and the market substitution is indicated by the slope of the isorevenue line, the inverse price ratio (IPR). The combination of products that maximizes revenues is found where the two curves are tangent to each other, or where MRPS = IPR.

The firm faces additional decisions if the price of one of the outputs changes. Being a price taker the firm must take the market price for his or her production. If an output price decreases, the producer will shift production toward the more valuable output (the product whose price has not changed) to maximize revenue. The producer shifts input use away from the product whose price has decreased (production falls) and into the production of the product whose price has not changed (production increases). Thus, if a product price decreases, production of that product also decreases. If an output price increases, the producer will shift production toward the more valuable output (the product whose price has increased) to maximize revenue. Therefore, input use is shifted toward the product whose price has increased (production increases) and away from the production of the product whose price has not changed (production decreases). Thus, if a product price increases, production of that product also increases.

The basic economic logic underlying the product–product decision is the same economic fact underlying much of agricultural and managerial economics: When we minimize opportunity cost (return) by substituting resources so as to achieve equi-marginal returns, we will maximize profits. In other words, we have succeeded as managers if we cannot change our resource use in such a manner that it will increase our returns (equal to further decreasing opportunity costs).

In a more general fashion, there is much that we can learn from the product–product decision analysis we have just worked through. The decision between products can be generalized to a decision, by the manager, as to whether the firm should specialize or diversify its economic activities. When the production function for one enterprise is much more physically productive than other alternative enterprises—due to soil type or weather—we see much specialization by farm and farming area, (wheat in the Great Plains, tobacco in the Southern states, potatoes in Maine and Idaho, and so forth). As we will learn later in this book, the ability to specialize depends heavily on the principle of trade as based on comparative advantage. Another reason why specialization is increasing significantly is the increasing complexity of the production process necessitating a farm or business manager to spend more and more time and resources to achieve the maximum on the production possibilities curve. Twenty years ago, farms in the Dakotas, for example, had many products: small grains, dairy, hogs, chickens, corn, livestock, etc. But the technologies and techniques now necessary to achieve the highest production from the "best enterprise" are so demanding of the manager's scarce resources, time, and expertise that most farms now have only one or two enterprises.

QUESTIONS

1. Why does the production of corn and sorghum display decreasing returns in Table 6.1, when you can clearly see that output increases with each unit of input until eight units are used?

2. What is the difference between the marginal rate of product substitution and the marginal rate of substitution?

3. What can we say about the sign (positive or negative) of the MRPS when our resources are competitive? Why?

4. Looking at Table 6.4 and assuming that sorghum and corn each sell for $1 how would you allocate your scarce resource to maximize your total revenue? What is the MRPS at that allocation level and what do you think it means?

5. Why might the manager of a cattle feed lot find the marginal approach to revenue maximization superior to either the graphical or total revenue approaches?

6. Suppose your major competitor's sorghum crop is stricken by disease, putting him out of business. How might this affect your output decision? Why? What assumptions would you make?

7. Distinguish between complementarity and supplementarity.

8. Joint products are desirable because production of one output results in the same amount of an additional product. Comment.

9. How does the concept of opportunity cost enter into a profit maximization decision in product–product evaluation when resources are limited?

10. Compare and contrast production possibilities curves and isoquants. Now do the same for isocost and isorevenue lines.

CHAPTER 7

Utility Maximization and Consumer Demand Theory

INTRODUCTION

Thus far we have focused on the behavior of the individual firm manager or producer. We have examined, in some detail, decisions concerning input use and output levels that would maximize profits, which combination of inputs would minimize costs, and which combination of outputs would maximize revenue. In answering the producer questions of "How much to produce?," "How to produce?," and "What to produce?," we have left one question unanswered: "Why do firms produce products?"

Firms produce products because those products are in demand by consumers. This demand results in a price being paid to the producer, which, in an economic system, serves as the director of economic activity by directing inputs into those uses desired by consumers. At the firm level the manager can look at price and demand as a signal for his or her activity or can attempt to affect demand in those situations where some control exists. Understanding the consumer's decision-making process and behavior is important for the manager because the consumer initiates the price signals that tell the producer what and how much should be produced.

The economic concept of demand is illustrated by price and quantity relationships. How much of a product is demanded at each possible price? This depends on what the consumer is "willing and able" to purchase at that price. A consumer's willingness, or desire, to purchase a product stems from the satisfaction, called **utility,** derived from consuming the product. However, the consumer's desire for a product is not enough; the consumer must also be able to act on that desire. Hence, we are talking about the **effective demand** for a product. The consumer must be able to translate his or her desire into actual purchases with money in a market economy. Since the consumer has a limited amount of money (income) with which to purchase products, the consumer is faced with the problem of how to allocate those limited funds across the consumption of all available products. The consumer must decide which products to consume, and which products not to consume, given the consumer's limited funds while striving to maximize the satisfaction from consumption. This decision-making process is called **utility maximization** and is similar to the factor–factor decision discussed in Chapter 5.

In This Chapter

The objective of this chapter is to provide the student and manager with an economic view of the consumer's decision-making process. Consumers must decide how to allocate a limited amount of money (budget) among various commodities so as to maximize their satisfaction (economists talk about this satisfaction as utility). The consumers' decisions of what and how much of a product to purchase have direct implications for managers and producers. Therefore, managers need to understand how consumers make their consumption decisions.

Consumer Characteristics

From observing consumer behavior in the marketplace, economists have noted several characteristics exhibited by consumers. First, *consumers spend their entire budget*. That is, consumers spend everything they earn on goods and services. Second, *consumers have unlimited wants*. Consumers never seem to get enough of most things, they prefer more to less, or have insatiable wants. Third, *consumers have limited income*. Consumers cannot buy an infinite amount of goods and services. Finally, *consumers desire variety*. Consumers rarely spend their entire income on a single product, whether out of boredom, nutritional needs, or whatever (e.g., how many snowmobiles does one person want, or how many lattes do we want to drink in a year); instead, they purchase a variety of goods and services.

Since consumers do not have unlimited supplies of money to spend on the products they consume or wish to consume, they must decide how to distribute their available money (or income) across the various products purchased in order to achieve their economic objective, which is the maximization of utility. Thus the consumers' decisions involve interaction between:

1. How much money an individual has to spend;
2. The goods and services available at a price in the market; and
3. The individual consumer's tastes and preferences.

Consumers are usually price takers in the market; thus with a given amount of money, they purchase a variety of goods and services in combinations that will maximize the utility or satisfaction derived from them. The fact that consumers will buy a certain combination of products, and not some other combination, with their available income implies that the combination purchased is more satisfying than some alternative combination of products.

Assumptions Used

The utility maximization decision involves finding the combination of goods, from the different product combinations possible, that will maximize the consumer's level of satisfaction, given his or her available budget. The market provides numerous products that the consumer can purchase with his or her available income. How do we represent these choices made by consumers? We begin with a simple decision problem using the following assumptions:

1. The consumer wants to maximize utility.
2. Two goods are considered at this time; all others are fixed.

The consumer's economic objective is to choose the combination of goods that will maximize his or her utility (or satisfaction). By using only two goods it is possible to make some generalizations about the amount of utility derived from consuming different quantities and combinations of those goods.

So the problem facing the consumer, given our assumptions, is to allocate the given budget, or available money, across the two goods so that the utility (satisfaction) derived from consuming the two goods is maximized. Because the two goods are combined, to a certain extent it is possible to substitute one product for another, both physically and economically, in the consumption process.

To examine the utility maximization decision, we begin by developing a measurement or tool to identify how one good substitutes for another in the consumer's consumption process. Then we will determine which combination of the two goods will maximize the consumer's utility (or satisfaction) given his or her budget constraint or available income. We then use this knowledge to illustrate how to derive an individual's demand curve for a specific commodity. Finally, we examine the similarities between the producer's factor–factor decision and the consumer's utility maximization decision. As we will see, the decision rules developed in this chapter can be used in the same conceptual manner as for the more complex situations with more than two goods.

PHYSICAL RELATIONSHIPS

We begin our discussion of the consumer's utility maximization decision by examining the physical aspects of consumption. To do so we introduce two new concepts. The first concept is utility, a concept that underlies much of consumer demand theory. The second concept is an indifference curve, a theoretical tool used to describe how two commodities substitute for one another in the consumption process.

Utility

As we mentioned earlier, consumers purchase and consume products because they derive satisfaction, or enjoyment, from consuming those products. In economics the word **utility** is used to describe *the satisfaction derived from consuming a product, good, or service.*

TABLE 7.1   Craig's Utility Schedule for Doughnuts

Doughnuts Consumed at Breakfast (Consumption)	Total Utility (TU)	Marginal Utility (MU) (ΔTU/Δconsumption)	
0	0		
		>	24
1	24		
		>	18
2	42		
		>	10
3	52		
		>	4
4	56		
		>	0
5	56		
		>	−1
6	55		
		>	−10
7	45		

Since utility is derived from the inherent characteristics or qualities that make a product desirable, utility may be objective or subjective. Hence, it is unlikely that two individuals would obtain the same level of utility or satisfaction from the same amount of a product. The concept of utility also implies a specified time period. That is, the consumption of and utility derived from a product take place within a specified time frame (i.e., tacos consumed at lunch, the purchase of concert tickets, camping).

Since there is no concrete, quantitative way to measure a person's level of satisfaction, economists use the **util**, which is *a hypothetical numerical measurement of utility,* to represent the satisfaction derived from consuming products. For example, Craig really likes chocolate-covered doughnuts for breakfast. If we could measure the satisfaction Craig obtains from eating doughnuts, his utility schedule might look something like the one presented in Table 7.1. As shown, each additional doughnut adds to Craig's total utility (TU) up to the fourth doughnut. The fifth doughnut neither adds nor subtracts from his total utility but the sixth and seventh doughnuts decrease his total utility (negative utility), indicating that Craig probably feels ill by this time. (The utility schedule in Table 7.1 should look like a production function for utility to you.)

Using Craig's utility schedule for doughnuts, we can examine the additional utility he obtains from consuming one more unit of the commodity (another doughnut). This additional utility is referred to as **marginal utility** and is defined as *the addition to total utility provided by the last unit of the good consumed (ΔTU/Δconsumption).* Craig's marginal utility indicates that his total utility increases at a decreasing rate. This is consistent with the **Law of Diminishing Marginal Utility,** which states that *as additional units of a good are consumed a point is always reached where the utility de-*

BOX 7.1 *The Diamonds–Water Paradox*

Why is it that diamonds, which are useless ornaments, are so valuable, and water, which is so essential to human life, is so cheap? Economists point to marginal utility as an explanation.

In most societies of the developed, high-income nations, diamonds are luxury goods that are required for important social functions such as wedding engagements, marriage, and gift giving. They are also intentionally scarce. The DeBeers diamond cartel controls world supplies of gem diamonds and has successfully managed to create and maintain an image of luxury, affection, and necessity. Thus, the marginal utility of diamonds is very high. After all, how many diamonds do you buy or receive as gifts in a lifetime?

On the other hand, fresh water is abundant. Although the actual human body requirement for water is less than one gallon per day, Johns Hopkins University measured daily residential consumption at 692 gallons for non-metered water systems and 398 gallons per day for metered systems. So water use may be at least 4,000 times actual "need!" And the price of water is very low. In Lexington, Kentucky, the residential price of water is about *one-tenth of one cent* ($0.001) per gallon. (But you could pay more and buy a bottled water like *Evian*, which incidentally is *n-a-i-v-e* spelled backwards.)

So the answer to the paradox about why the essential resource, water, is cheap and diamonds are expensive lies in consumer utility. Water is abundant and the marginal utility is low because marginal utility for water is determined by its lowest valued use, not the value of that first gallon we need for daily survival. So that last gallon of water you used to wash your car this week cost only $0.001 and provided only a small amount of satisfaction. But the diamond you buy will be esteemed and costly, reflecting the high satisfaction and utility Americans place on gemstones, despite their functional uselessness.

---

rived from each successive unit declines. For Craig the first doughnut consumed generates a given amount of utility. But each successive doughnut generates smaller and smaller amounts of satisfaction. Thus, the marginal (additional) utility of consuming the doughnuts is diminishing. As Craig consumes the sixth and seventh doughnuts (too many doughnuts), negative utility occurs.

The Law of Diminishing Marginal Utility illustrates that as the amount of a good consumed increases, the addition to total utility from consuming one more unit of that good decreases. Remember, this is similar to the Law of Diminishing Returns introduced in Chapter 2. In fact, the relationship between marginal utility and total utility is very similar to the relationship between marginal physical product (MPP) and total physical product (TPP) presented in Chapter 2. When marginal utility is greater than zero, total utility is increasing, which mirrors the fact that when MPP is greater than zero, TPP is increasing. When marginal utility is less than zero, total utility is decreasing, which corresponds to when MPP is less than zero and TPP is decreasing. If we were to graph total utility we would see that, by definition, marginal utility is the slope of the total utility curve, just as MPP is the slope of the TPP curve.

Indifference Curves

A theoretical tool used to represent how two commodities substitute for one another in the consumption process is the **indifference curve,** which is defined as *a line showing all the combinations of two goods (or products) that provide the same level of utility.*

Deriving an Indifference Curve

An indifference curve identifies all the combinations of two products that are equally satisfying or provide the same level of utility. Since each combination on an indifference curve provides the same level of utility, the consumer is indifferent to which combination is consumed, hence, the name **indifference curve.**

To illustrate how an indifference curve is derived, consider the following example. During a given week, a typical college student can purchase either sandwiches or tacos, or some combination of the two, for lunches. The product combinations that provide the same level of satisfaction (utility) are listed in Table 7.2. Each combination of tacos and sandwiches listed in this table generates the same level of utility or satisfaction for this hypothetical individual, thus they are all on the same indifference curve. This means the college student would be just as happy consuming the first combination (25 tacos, 5 sandwiches) as the third (14 tacos, 8 sandwiches) or the sixth (5 tacos, 20 sandwiches) combination. As indicated in Table 7.2, to increase the consumption of one good (sandwiches) the consumption of the other good (tacos) must decrease. That is, sandwiches are substituting for tacos in the student's consumption process.

TABLE 7.2   Product Combinations Providing Same Level of Utility

G_1 (Tacos)	G_2 (Sandwiches)
25	5
19	6
14	8
10	11
7	15
5	20

To derive the indifference curve graphically, we simply plot the product combinations, as illustrated in Figure 7.1. The downward slope of the indifference curve shown in Figure 7.1 indicates that if the consumer gives up one commodity (tacos), the resulting loss in satisfaction must be compensated for by consuming additional units of the other commodity (sandwiches) for utility to remain constant. Economic convention places the product being replaced, tacos in this example, on the vertical axis (labeled G_1) and the product being added, sandwiches in this example, on the horizontal axis (labeled G_2). The designation of which good is to be replaced and added is fairly arbitrary, but is important in terms of specifying the marginal rate of substitution (MRS) and the inverse price ratio (IPR), terms that are discussed later.

The indifference curve concept is commonly used in economic analysis to illustrate the consumer's decision process. The indifference curve is used not only to indicate the different combinations of two goods that produce a certain level of utility, but it can represent many levels of utility as well. Since each indifference curve represents a unique level of utility, and there are, theoretically, infinite levels of satisfaction, the indifference curves are considered to be everywhere dense. That is, an indifference curve exists for each level of utility a consumer is capable of experiencing, as illustrated in Figure 7.2, which shows that the distance away from the origin indicates the level of utility. The closer the indifference curve is to the origin (I_0), the lower the level of utility. The farther away from the origin (I_2), the higher the utility level. Higher indifference curves represent larger combinations of both goods. Since consumers have unlimited wants, higher indifference curves represent higher levels of satisfaction. (Yes, this should remind you of our production surface in Chapter 5.)

Since each indifference curve represents a unique utility level, indifference curves can never intersect. Why? If indifference curves intersected, it would mean

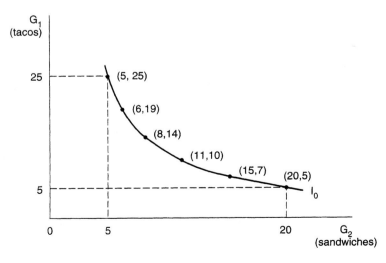

FIGURE 7.1   Indifference curve for combinations of sandwiches and tacos.

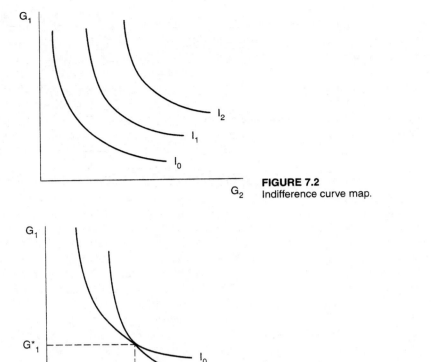

FIGURE 7.2
Indifference curve map.

FIGURE 7.3
Intersecting indifference curves.

that a single combination of goods (or products) would be capable of producing more than one level of utility, or satisfaction. This is depicted graphically in Figure 7.3. If the indifference curves I_0 and I_1 intersected, as in this figure, it would mean that the product combination G_1^* and G_2^* would produce two different levels of utility. Given the unique relationship between the product combination and the consumption process (most goods can only be consumed once), this is not possible.

Marginal Rate of Substitution

As you can see from the derivation just discussed, as we move along the indifference curve the utility level remains constant but the quantities of the goods consumed are changing as one good replaces, or substitutes, for the other. As managers, we are interested in how the two goods substitute for each other. The rate at which one good can physically substitute for another good in the consumption process of the consumer is measured by the **Marginal Rate of Substitution (MRS),** which describes *the rate one good must be or can be decreased as consumption of the other*

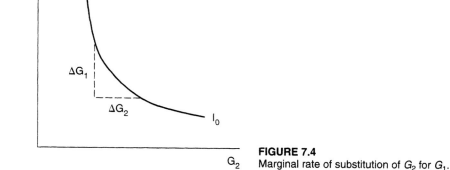

FIGURE 7.4
Marginal rate of substitution of G_2 for G_1.

good increases and, by definition, is the slope of the indifference curve. Since economic convention places the good being replaced on the vertical axis and the good being added on the horizontal axis, MRS is calculated by dividing the change in the replaced good (Δ replaced) by the change in the added good (Δ added). Thus the computational formula and the interpretation of MRS depends on which good is being replaced and which is being added.

For example, if G_1 is the good being replaced and G_2 is the good being added, then the **marginal rate of substitution of G_2 for G_1**, written $\mathrm{MRS}_{G_2 G_1}$ is calculated by dividing the ΔG_1 (good replaced) by the ΔG_2 (good added), or

$$\mathrm{MRS}_{G_2 G_1} = \frac{\Delta G_1}{\Delta G_2} = \frac{\text{replaced}}{\text{added}}$$

Graphically, as shown in Figure 7.4, the good being replaced, G_1 in this case, is on the vertical axis and the good being added, G_2 in this case, is on the horizontal axis.

The slope of the indifference curve in Figure 7.4 is the change in G_1 divided by the change in G_2 ($\Delta G_1 \div \Delta G_2$), which corresponds exactly to the definition of $\mathrm{MRS}_{G_2 G_1}$.

Conversely, if G_2 is the good being replaced and G_1 is the good being added, then the **marginal rate of substitution of G_1 for G_2**, written $\mathrm{MRS}_{G_1 G_2}$, is calculated by dividing the ΔG_2 (good replaced) by the ΔG_1 (good added), or

$$\mathrm{MRS}_{G_1 G_2} = \frac{\Delta G_2}{\Delta G_1} = \frac{\text{replaced}}{\text{added}}$$

Graphically, as shown in Figure 7.5, G_2, the good being replaced, is on the vertical axis, and G_1, the good being added, is on the horizontal axis. The slope of the

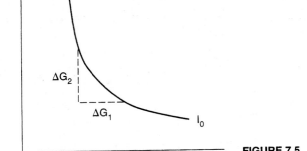

FIGURE 7.5
Marginal rate of substitution of G_1 for G_2.

indifference curve in Figure 7.5 is equal to the change in G_2 divided by the change in G_1 ($\Delta G_2 \div \Delta G_1$) which corresponds exactly to $MRS_{G_1 G_2}$.

Note that the subscripts on the definition of MRS match the axes labels on the graphs in Figures 7.4 and 7.5. Since the order of the subscripts on MRS tells us which good is being added and which good is being replaced, it is very important that the graphical presentation be consistent with the stated MRS, and vice versa.

We indicated and also illustrated earlier that the designation of which good is to be replaced and which good is to be added is an arbitrary designation. This is because

$$MRS_{G_2 G_1} = \frac{1}{MRS_{G_1 G_2}}$$

Why?

$$MRS_{G_2 G_1} = \frac{\Delta G_1}{\Delta G_2} = \frac{1}{\frac{\Delta G_2}{\Delta G_1}} = \frac{1}{MRS_{G_1 G_2}}$$

As illustrated, $MRS_{G_2 G_1}$ is the inverse of $MRS_{G_1 G_2}$ so although they represent the same information, it is important to be aware of which good is being replaced and added for computation and for correct graphical presentation.

Using our college student example, we can now illustrate how to calculate the marginal rate of substitution numerically. As shown in Table 7.2, sandwiches (G_2) are being added and tacos (G_1) are being replaced. Therefore, the correct formula to calculate the marginal rate of substitution between these goods is $MRS_{G_2 G_1} = \Delta G_1 / \Delta G_2$ (or Δ in tacos/Δ in sandwiches) as shown in Table 7.3.

The formula, which states the rate that G_2 substitutes (is added) for G_1 (is replaced), is equal to the change in G_1 divided by the change in G_2, and, as graphically depicted in Figure 7.6, is the slope of the indifference curve. Since consump-

TABLE 7.3 Marginal Rate of Substitution of Sandwiches (G_2) for Tacos (G_1)

G_1 (Tacos)	G_2 (Sandwiches)	ΔG_1 (Replaced)	ΔG_2 (Added)	$MRS_{G_2 G_1}$	$= \Delta G_1 \div \Delta G_2$
25	5				
		-6	1	-6.00	$= \quad -6/1$
19	6				
		-5	2	-2.50	$= \quad -5/2$
14	8				
		-4	3	-1.33	$= \quad -4/3$
10	11				
		-3	4	-0.75	$= \quad -3/4$
7	15				
		-2	5	-0.40	$= \quad -2/5$
5	20				

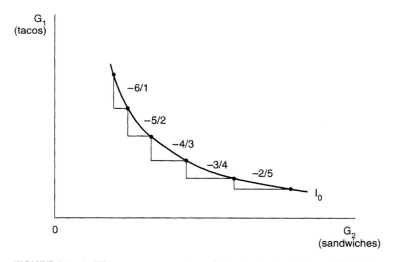

FIGURE 7.6 Indifference curve and marginal rate of substitution of G_2 for G_1.

tion of one product must be decreased to increase consumption of the other, MRS is always negative, as indicated in Table 7.3 and Figure 7.6. The MRS of -6 ($-6/1$) as we go from the first combination (25, 5) to the second combination (19, 6) indicates that 6 tacos (G_1) are replaced, or given up, in order to add 1 sandwich (G_2). This shows that at this point one sandwich substitutes nicely for 6 tacos. Remember, because of diminishing marginal utility this makes sense because the student is consuming several tacos (low MU at this point) and not many sandwiches (high MU).

This -6 also describes the slope of the indifference curve at this point, since by definition, MRS measures the slope of the indifference curve. Again, as shown in Table 7.3, initially sandwiches (G_2) substitute very well for tacos (G_1) because a one-unit increase in G_2 allows a greater than one-unit decrease in G_1; that is, our college student is willing to give up more than one taco to get one additional sandwich. Later, as diminishing marginal utility affects the trade-offs, the two goods substitute less freely. As we substitute sandwiches for tacos in our example, the MRS decreases in magnitude (in terms of absolute value). Thus, this indifference curve exhibits diminishing MRS, indicating that the goods are imperfect substitutes, due to changes in marginal utility. Graphically, as in Figure 7.6, we can state that when the indifference curve has a steep slope (MRS $>$ |1|, where |1| is the absolute value of 1) the good on the horizontal axis, usually G_2, is a good substitute for the other good, usually G_1. As the slope of the indifference curve decreases (MRS $<$ |1|), G_2 becomes a less efficient substitute for G_1.

Recognizing that MRS is the ratio of the marginal utilities of the two products, we can now see the interrelationships between utility and the indifference curve. We know that

$$MU_{G_1} = \frac{\Delta TU}{\Delta G_1}; \text{ or } MU_{G_1} \times \Delta G_1 = \Delta TU$$

$$MU_{G_2} = \frac{\Delta TU}{\Delta G_2}; \text{ or } MU_{G_2} \times \Delta G_2 = \Delta TU$$

where MU is marginal utility and TU is total utility. But along an indifference curve $\Delta TU = 0$ so we can write

$$MU_{G_1} \times \Delta G_1 = 0 \text{ and } MU_{G_2} \times \Delta G_2 = 0$$

or

$$MU_{G_1} \times \Delta G_1 = MU_{G_2} \times \Delta G_2$$

Dividing both sides by ΔG_2 and MU_{G_1} gives the following result:

$$\frac{\Delta G_1}{\Delta G_2} = \frac{MU_{G_2}}{MU_{G_1}}, \text{ or } MRS_{G_2 G_1} = \frac{MU_{G_2}}{MU_{G_1}}$$

The slope of the indifference curve is the MRS between the two goods, which is given by the ratio of the MUs of the two goods. Thus, the substitutability between the two goods is based on the marginal utility the consumer derives from consuming each good.

Possible MRS Relationships

The example developed for the college student had products that were imperfect substitutes for each other, or the indifference curve exhibited diminishing marginal rate of substitution. Diminishing MRS occurs when one good can be exchanged for another, but at a decreasing rate. As illustrated in Figure 7.6, initially G_2 substitutes rather freely for G_1 (a large decrease in tacos for a small increase in sandwiches), but as diminishing marginal utility begins to affect the trade-offs between G_1 and G_2, the two products substitute less freely for one another (a small decrease in tacos for a large increase in sandwiches). Two other possibilities for substitutability between products exist, as discussed next.

Perfect (or constant) substitutability occurs when one unit of a good can be exchanged for another on a consistent basis of one-to-one or some other unchanging ratio. For example, 2% milk from Safeway or from Albertson's, or a chocolate bar from Hershey's or from Nestle. In this constant case, the goods are considered perfect substitutes, diminishing MRS does not occur, and the indifference curve is linear, as illustrated in Figure 7.7(a).

Fixed proportions (or perfect complementarity) occur when consumption goods must be used in a fixed ratio. For example, one hot dog per bun, or one left shoe per left foot and one right shoe per right foot. In such a case, the goods are considered to be perfect complements, substitutability is zero, and the indifference curve is a single point as illustrated in Figure 7.7(b).

The most commonly seen relationship between products is one of diminishing MRS. Why? Because, as we demonstrated earlier, MRS is the ratio of marginal utilities of the two goods, and diminishing marginal utility is evident in most consumption processes.

When it comes to American foods, consumers are always making substitutions among all the products available. In 1994, more than 15,000 new items appeared in grocery stores. Many of them only remained on the shelves a few weeks but many of them survived as successful new products. Advertising, price inducements (e.g., coupons, samples), and changes in display profile are methods used by retailers to encourage consumers to test substitution of these new products.

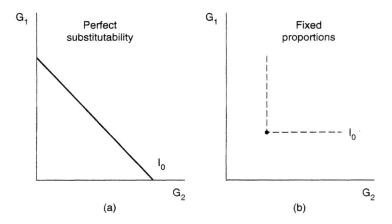

FIGURE 7.7   Perfect substitutability and fixed proportions.

Of course, the appearance of hundreds of new food items is also influenced by social and demographic factors. For example, immigration into the United States is now dominated by Asians and Latin Americans. These new households are influencing the mix of products in grocery stores and food service establishments, leading to substitutions away from the traditional food items favored by the European immigrants of the nineteenth century. Another demographic trend influencing the substitution of items is convenience. Since nearly two-thirds of women work full time, convenience foods have become a major share of grocery store products—replacing the more traditional fresh and canned foods, which require more cooking time and preparation. This trend in food substitution will continue as women continue to join the labor force and remain employed for longer time spans.

ECONOMIC RELATIONSHIPS

Up to this point we have only been comparing physical units of two different commodities and how they substitute for each other in a technical, noneconomic relationship. To achieve our goal of utility maximization, we must introduce prices into the technical relationship. Product prices are appropriate since we are examining the consumer's utility maximization decision.

Budget Line

Consumers do not have unlimited funds with which to purchase the various products they wish to consume. *The amount of money available for purchases in a given time period,* referred to as the consumer's **budget,** limits how much the consumer can purchase. *Price and availability of the goods in the market, along with the size of the budget, place a constraint on consumption,* often referred to as the **budget constraint.** Since the consumer cannot spend more than he or she has, this forces him or her to make trade-offs in consumption.

The budget and budget constraint are represented by the **budget line,** which is defined as *a line indicating all the combinations of two products (goods) that can be purchased using all of the consumer's budget.* It is calculated by adding up the total money spent on each good, or

$$\text{TB} = (P_{g_1} \times G_1) + (P_{g_2} \times G_2)$$

where TB is the total budget available for the consumer to spend on the two goods, P_{g_1} and P_{g_2} are the prices of the goods, and G_1 and G_2 are the amounts of the goods purchased.

The amount of each product that can be purchased and consumed depends on the amount of money to be spent (budget) and the respective prices of the products. Going back to our college student example, if she only has $30 per week to spend on lunches and she can purchase tacos for $1 and sandwiches for $2 how many units of each good could she purchase? Several possible product combinations that could be purchased for $30 are listed in Table 7.4.

Each of the combinations listed in Table 7.4 requires the same budget level. The college student could purchase the first combination (30 tacos, 0 sandwiches) or the third (18 tacos, 6 sandwiches) or even the sixth (0 tacos, 15 sandwiches) with a $30 budget. Thus, with $30 to spend on lunches, the college student could purchase any of the taco/sandwich (G_1, G_2) combinations in Table 7.4. The budget line for $30 is depicted graphically in Figure 7.8.

If the student spent the entire $30 on tacos ($G_1$) she could purchase 30 units of G_1 ($30 ÷ $1 = 30$). If she spent the entire $30 on sandwiches ($G_2$) she could purchase 15 units of G_2 ($30 ÷ $2 = 15$). These points are the end points, or intercept terms, of the budget line. Connecting the two end points yields the budget line for $30 (TB = $30 in Figure 7.8). Any combination of goods on this line requires the

TABLE 7.4 Affordable Product Combinations with Total Budget of $30 and Price of Tacos ($P_{g_1}$) = $1.00 and Price of Sandwiches (P_{g_2}) = $2.00

Tacos Purchased (G_1)	Expenditure on Tacos ($P_{g_1} \times G_1$)	Sandwiches Purchased (G_2)	Expenditure on Sandwiches ($P_{g_2} \times G_2$)	Total Expenditure ($P_{g_1} \times G_1 + P_{g_2} \times G_2$)
30	$30	0	$0	$30
24	$24	3	$6	$30
18	$18	6	$12	$30
12	$12	9	$18	$30
0	$0	15	$30	$30

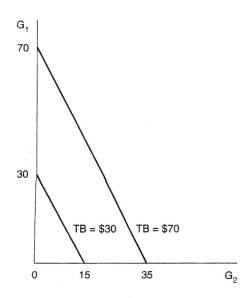

FIGURE 7.8
Budget line for $30 and $70.

same level of total expenditure ($30) to purchase. Thus all of the combinations listed in Table 7.4 lie on the budget line for $30 (TB = $30) in Figure 7.8.

If the student has $70 and purchases only G_1, she could purchase 70 units ($70 ÷ $1 = 70); if she purchases only G_2, she could purchase 35 units ($70 ÷ $2 = 35). The budget line for $70 (TB = $70) is also drawn in Figure 7.8. As illustrated in Figure 7.8, the distance from the origin is an indication of the size of the budget available for the consumer to spend. The closer to the origin, the lower the budget (less money); the farther away from the origin, the larger the budget (more money).

As you are probably aware, an infinite number of budget lines could have been drawn in Figure 7.8, each depicting a different budget amount. For illustrative purposes we have only drawn two, one for $30 and one for $70.

Because consumers spend their entire budget, consumers will consume somewhere on their budget line. Therefore, the optimum combination of goods, that is, the combination of goods that maximizes utility, will lie somewhere on the budget line as shown in Figure 7.9. Every point on the budget line (e.g., point B) in Figure 7.9 represents a combination of the two goods that can be purchased given the consumer's available budget. Combinations to the left of the budget line (e.g., point A) are affordable but are not as desirable as those on the budget line. Combinations to the right of the budget line (e.g., point C) are desirable but are not affordable.

The slope of the budget line drawn in Figure 7.8 (and 7.9) is $-P_{g_2}/P_{g_1}$, and is referred to as the **inverse price ratio (IPR).** It is derived from the budget line equation:

$$TB = (P_{g_1} \times G_1) + (P_{g_2} \times G_2)$$

Since G_1 is on the vertical axis, we solve the equation for G_1. First we subtract $(P_{g_2} \times G_2)$ from both sides:

$$(P_{g_1} \times G_1) = TB - (P_{g_2} \times G_2)$$

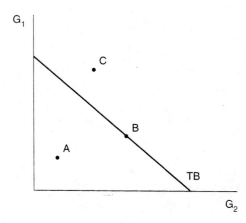

FIGURE 7.9
Affordable and unaffordable combinations for the consumer.

and then we divide both sides by P_{g_1} to obtain

$$G_1 = \frac{TB}{P_{g_1}} - \frac{P_{g_2}}{P_{g_1}} \times G_2$$

This is the familiar formula for a straight line $Y = a + bX$, where the intercept term (a) in this case is TB/P_{g_1} (which identifies the intercept on the G_1 axis) and the slope (b) is $-P_{g_2}/P_{g_1}$. The slope of the budget line is called an **inverse price ratio** because it is *the price of the good on the horizontal axis divided by the price of the good on the vertical axis.* Thus, the way the graph axes are labeled will dictate which product price is in the numerator and denominator of the price ratio. The slope of the budget line tells us the relative value of the two goods, or the rate of economic substitution (how they substitute for one another economically); information like this will allow us to identify the utility maximizing combination of products.

Effects of a Price Change

Since the slope of the budget line depends on the relative price ratio of the two goods, if the price of one of the goods changes, the slope of the budget line also changes.

Figure 7.10 illustrates the impact of a change in P_{g_2}. Recall our previous example where TB = $30, P_{g_1} = $1, and P_{g_2} = $2. At these prices and the student's budget, if she purchased only G_2, she could purchase 15 units of G_2; or if she purchased only G_1, she could purchase 30 units of G_1 (TB$_0$ in Figure 7.10). If the price of G_2 changes, the number of units of G_2 that the student can purchase with $30 also changes. If P_{g_2} increases, G_2 is now more expensive, relative to G_1, to consume

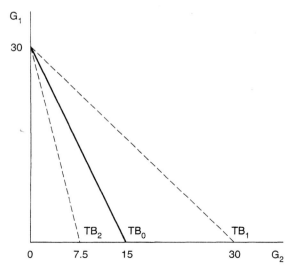

FIGURE 7.10
Effect of a price change on the budget line: price of G_2 changes.

and with the price increase \$30 purchases fewer units of G_2. For example, if P_{g_2} increases to \$4 she can only purchase 7.5 units of G_2 instead of the initial 15 units. The budget line for \$30 given the new prices (P_{g_1} = \$1, P_{g_2} = \$4) is TB_2 in Figure 7.10. On the other hand, if P_{g_2} decreases and P_{g_1} remains the same, G_2 is now less expensive to consume relative to G_1 and with the price decrease the \$30 budget purchases more units of G_2. For example, if P_{g_2} decreases to \$1 she is able to purchase 30 units of G_2 instead of the initial 15 units. The budget line for \$30 given the new prices (P_{g_1} = \$1, P_{g_2} = \$1) is TB_1 in Figure 7.10.

As Figure 7.10 shows, if the price of product G_2 changes the number of units of G_2 that can be purchased for the same budget level also changes. If P_{g_2} increases, the budget line becomes steeper (rotates inward), because the same budget amount now buys fewer units of G_2. If P_{g_2} decreases, the budget line becomes flatter (rotates outward) because the same budget now buys more units of G_2. However, note that the budget line rotates around the axis intercept of G_1, the good whose price has not changed. This is because with the given budget level (\$30) the student can still purchase 30 units of G_1.

Figure 7.11 illustrates the impacts of a change in P_{g_1}. Recall our previous example where TB = \$30, P_{g_1} = \$1, and P_{g_2} = \$2. At these prices and the student's budget, if she purchased only G_2, she could purchase 15 units of G_2; or, if she purchased only G_1, she could purchase 30 units of G_1 (TB_0 in Figure 7.11). If the price of G_1 changes, the number of units of G_1 that the student can purchase with \$30 also changes. If P_{g_1} increases and P_{g_2} remains the same, G_1 is now more expensive, relative to G_2, to consume and with the price increase the \$30 budget purchases fewer units of G_1. For example, if P_{g_1} increases to \$2 she can only purchase 15 units of G_1 instead of the initial 30 units. The budget line for \$30 given the new prices (P_{g_1} = \$2, P_{g_2} = \$2) is TB_2 in Figure 7.11. On the other hand, if P_{g_1} decreases and

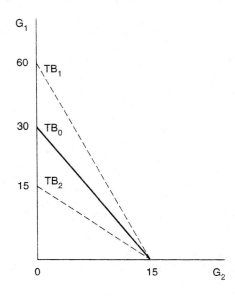

FIGURE 7.11
Effect of a price change on the budget line: price of G_1 changes.

P_{g_2} remains the same, G_1 is now less expensive to consume relative to G_2 and with the price decrease the $30 budget purchases more units of G_1. For example, if P_{g_1} decreases to $0.50 she is able to purchase 60 units of G_1 instead of the initial 30 units. The budget line for $30 given the new prices (P_{g_1} = $0.50, P_{g_2} = $2) is TB_1 in Figure 7.11.

As Figure 7.11 shows, if the price of product G_1 changes, the number of units of G_1 that can be purchased for the same budget level also changes. If P_{g_1} increases, the budget line becomes flatter (rotates downward) because the same budget level now buys fewer units of G_1. If P_{g_1} decreases, the budget line becomes steeper (rotates upward) because the same budget now buys more units of G_1. However, note that once again the budget line rotates around the axis intercept of G_2, the good whose price has not changed. This is because with the given budget level ($30) the student can still purchase 15 units of G_2.

We can summarize the important points to remember about the budget line as follows:

1. The end points of the budget line show how many units of the good could be purchased if all the budget were spent on that good.
2. A change in the budget causes a parallel shift in the budget line.
3. A change in a product price causes the budget line to rotate around the intersection of the budget line and the axis of the other product.
4. With a fixed budget amount, a reduction in the consumption of one good is necessary to increase consumption of the other good. That is, in order to purchase more of one good the consumer must purchase less of the other good.
5. The slope of the budget line, termed the inverse price ratio (IPR), is equal to the price of the good on the horizontal axis divided by the price of the good on the vertical axis.

The problem facing the consumer is how to allocate his or her limited budget between the various goods and services purchased so as to maximize utility. The combination of goods that maximizes a consumer's utility can be found using the indifference curve, which represents the level of satisfaction, and the budget line, which represents the ability to purchase.

UTILITY MAXIMIZATION DECISION

The objective of the consumer is to find the combination of consumption goods that provides the maximum amount of utility for his or her given budget. Since we assume that the consumer is rational and wants to maximize utility, the consumer wants to reach the highest possible level of utility, given his or her budget constraint. Therefore, the consumer wants to find the tangency between the highest obtainable indifference curve, which represents the consumer's utility level, and the budget line, which represents the consumer's budget constraint. At this point of tangency, the rate of physical substitution between the two consumption goods (slope of the indifference curve) will be equal to the rate of economic substitution between the two consumption goods (slope of budget line) and the consumer's utility will be maximized. This decision is depicted graphically in Figure 7.12.

The consumer's budget line, TB in Figure 7.12, represents the consumer's available funds to purchase the goods G_1 and G_2. Indifference curves I_0, I_1, and I_2, represent three of the many possible utility levels for the consumer. As shown in Figure 7.12, all three of the combinations at points A, B, and C, are affordable since they all lie on the budget line. Because the combinations at points B and C both provide the same utility level, they lie on the same indifference curve I_0, and the consumer is indifferent to consuming either of these combinations. Since the combination at point A lies on a higher indifference curve ($I_1 > I_0$), this combination gives a higher level of utility than either of the combinations at points B and C. The combinations on indifference curve I_2, while desirable, are not affordable (to the right of the budget line) and thus not obtainable. The highest indifference curve the consumer can attain, given the budget constraint, is I_1. Thus the combination at point A yields the highest (maximum) utility level for the consumer's budget. The utility maximizing combination of G_1 and G_2 is at point A, where the slope of the indifference curve is equal to the slope of the budget line, and the consumer is consuming G_1^* units of G_1 and G_2^* units of G_2.

To determine the combination of goods (or products) that will maximize utility the consumer equates the marginal rate of substitution (the slope of the indifference curve) to the inverse price ratio (the slope of the budget line). The marginal rate of substitution describes the physical substitution between the two goods,

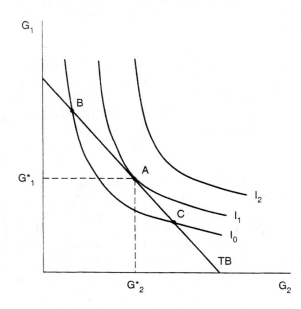

FIGURE 7.12
Utility maximization decision.

while the inverse price ratio (IPR) describes the value (cost) substitution between the two goods. In other words, we are determining where the rate of physical substitution (MRS) is equal to the rate of economic substitution (IPR) for the two goods or, for our college student example, where

$$MRS_{G_2 G_1} = -\frac{P_{g_2}}{P_{g_1}}$$

We can also illustrate this decision criterion using our college student example. Assuming that the college student has a budget of $30, and that tacos can be purchased for $1 and sandwiches for $2, which combination will maximize utility? Keep in mind that the highest possible consumption is found somewhere on the budget line, and that total utility increases as the consumer moves toward higher indifference curves.

With $30 to spend, if the student spends all of it on tacos she could purchase 30 tacos, or if the student spends all of it on sandwiches she could purchase 15 sandwiches. The utility maximization decision for the student with a budget of $30 is presented in Figure 7.13, which illustrates that the highest obtainable indifference curve, given the budget line of $30 (TB = $30), is I_0. Utility maximization occurs where MRS = IPR or at the combination of 14 tacos and 8 sandwiches. At this point the rate of physical substitution between the goods is equal to the economic substitution or the point at which $MRS_{G_2 G_1} = -P_{g_2}/P_{g_1}$.

The point where the indifference curve is tangent to the budget line in Figure 7.13 identifies the product combination (14 tacos, 8 sandwiches) that maximizes the consumer's utility. We can illustrate why this combination maximizes utility using our decision criterion $MRS_{G_2 G_1} = -P_{g_2}/P_{g_1}$ and the information in Table 7.5.

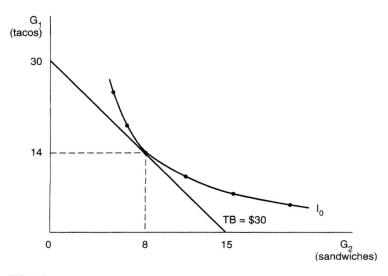

FIGURE 7.13 Utility maximizing combination of tacos and sandwiches.

As shown in this table, with a budget of $30 the only combination affordable is 14 tacos and 8 sandwiches. Thus the indifference curve represented by the taco/sandwich combinations in Table 7.5 is the highest obtainable indifference curve, given the $30 budget, and utility is maximized at this combination.

Alternatively, we can compare the MRS to the IPR to further illustrate why this combination maximizes utility. Recall that $MRS_{G_2 G_1}$ tells us how many tacos the student is willing to give up to obtain an additional sandwich (physical substitution) and that $-P_{g_2}/P_{g_1}$ tells us how valuable tacos are relative to sandwiches (economic substitution). With a limited budget, the student must reduce budget expenditures (consume fewer units) of one good in order to increase consumption (increase budget expenditures) of the other good. Thus, we can compare the budget savings (value of tacos given up) from replacing tacos (equal to $\Delta G_1 \times P_{g_1}$) to the additional budget expenditures (value of sandwiches gained) from adding sandwiches (equal to $\Delta G_2 \times P_{g_2}$) to illustrate why this combination will maximize utility. By reducing taco consumption from 25 to 19, the student saves $6, with which she can finance an additional sandwich purchase from 5 to 6. Since budget savings ($6) from moving to the second combination (19, 6) from the first (25, 5) is greater than the added cost of buying the sandwich ($2), the student would move to the second combination. To move from the second combination to the third (14, 8) will yield cost savings of $5 from replacing tacos and $4 in added costs for additional sandwiches. Again the cost savings are greater than the added costs so the student will move to the third combination. However, to move to the fourth combination (10, 11) would entail $6 in added costs for additional sandwiches but would only provide $4 in cost savings from replacing tacos. Since the added costs are greater than the cost savings the student would remain at the third combination (14, 8) to maximize utility. So not only is this combination (14 tacos, 8 sandwiches) the only one affordable given the $30 bud-

TABLE 7.5 Determining Utility Maximizing Combination of Products for Total Budget of $30 with Price of Tacos ($P_{g_1}$) = $1.00 and Price of Sandwiches (P_{g_2}) = $2.00

G_1 (Tacos)	G_2 (Sandwiches)	Total Expenditure (TB = $P_{g_1} \times G_1 + P_{g_2} \times G_1$)	$MRS_{G_2 G_1}$ $\Delta G_1/\Delta G_2$	$-\dfrac{P_{g_2}}{P_{g_1}}$ (−2.00/1.00)
25	5	$25 + $10 = $35		
			−6.00	−2.00
19	6	$19 + $12 = $31		
			−2.50	−2.00
14	8	$14 + $16 = $30		
			−1.33	−2.00
10	11	$10 + $22 = $32		
			−0.75	−2.00
7	15	$7 + $30 = $37		
			−0.40	−2.00
5	20	$5 + $40 = $45		

get, but the student has no incentive to move from this taco/sandwich combination since the cost savings from replacing tacos is (almost) equal to the cost of adding sandwiches.

The decision criterion for utility maximization (MRS = IPR) also tells us that the optimum combination of goods is found where the marginal utility per dollar spent on each good is equal. Why? We know that utility is maximized where $MRS_{G_2G_1} = -P_{g_2}/P_{g_1}$ (since MRS is always negative, the negative signs will cancel), and we demonstrated earlier that $MRS_{G_2G_1} = MU_{G_2}/MU_{G_1}$. Using this result we can now write our decision criterion as

$$\frac{MU_{G_2}}{MU_{G_1}} = \frac{P_{g_2}}{P_{g_1}}$$

Multiplying both sides by MU_{G_1} and dividing both sides by P_{g_2} gives

$$\frac{MU_{G_2}}{P_{g_1}} = \frac{MU_{G_2}}{P_{g_2}}$$

which shows that utility maximization occurs where the marginal utility per dollar spent on the two goods is equal. When this purchasing pattern is reached, the consumer is said to be at an equilibrium point, as there is no incentive for the consumer to change consumption. Thus, for the stated price level this point is an equilibrium point.

As we illustrated earlier, $MU_{G_2}/P_{g_2} = MU_{G_1}/P_{g_1}$ is a condition for equilibrium. However, when a product price changes the ratio MU/P will also change. This results in a search for a new optimal (utility maximizing) combination of products as the consumer seeks to reestablish a purchase pattern that equalizes the marginal utility per dollar spent on each of the two goods. That is, as the price of one good changes, the combination of goods that will maximize utility also changes. We examine the effects of a price change in the next section.

Impacts of Changes in Product Prices

Consumers are often faced with changing product prices when making consumption decisions. When the relative prices of products change, the combination of products that maximizes utility will also change.

Product Price Increase

When the price of one product changes, consumption of that product becomes more expensive relative to the other product. The consumer must then reevaluate his or her budget allocation across the two goods to find the new utility maximizing product combination. This decision is depicted in Figure 7.14.

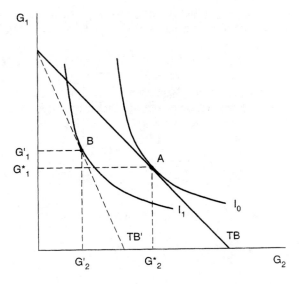

FIGURE 7.14
Utility maximizing decision under changing prices: price of G_2 increases.

Initially the consumer is at point A in Figure 7.14 where the slope of the indifference curve ($MRS_{G_2 G_1}$) is equal to the slope of the budget line ($-P_{g_2}/P_{g_1}$), and the utility maximizing combination of goods is G_1^* units of G_1 and G_2^* units of G_2. If P_{g_2} increases to P'_{g_2}, the budget line becomes steeper, rotating inward from TB to TB' in Figure 7.14, as the slope is now equal to P'_{g_2}/P_{g_1}. When P_{g_2} increases, the good G_2 becomes relatively more expensive to consume compared to the good G_1. Since the consumer is limited by the available budget, when P_{g_2} increases the consumer can no longer afford the product combination at point A (in Figure 7.14) because it now lies on a higher budget line (TB > TB'). Therefore, the consumer must reduce consumption and move to a lower indifference curve (I_1) and a lower level of utility. The consumer moves from the product combination at point A (G_1^*, G_2^*), to point B consuming G_1' units of G_1 and G_2' units of G_2. The new utility maximizing combination of products, point B in Figure 7.14, is the point of tangency between the lower indifference curve (I_1) and the new budget line (TB'), or where the slope of the new budget line (TB') is equal to the slope of the lower indifference curve (I_1). After the price increase the consumer is consuming more of the relatively less expensive good ($G_1' > G_1^*$) and less of the relatively more expensive good ($G_2' < G_2^*$).

We can illustrate the effect of a product price increase on the utility maximization decision using our college student example. What happens to the utility maximizing combination of tacos and sandwiches if the price of sandwiches increases to $3? This decision is depicted graphically in Figure 7.15.

With a budget of $30, and initial prices of P_{g_1} = $1 and P_{g_2} = $2, the utility maximizing combination is 8 sandwiches and 14 tacos (point A in Figure 7.15). If the price of sandwiches increases to $3 the budget line steepens, rotating inward, from TB to TB'. At the new price, if the student bought only sandwiches she could only afford 10 with a $30 budget instead of the initial 15. Due to the price increase,

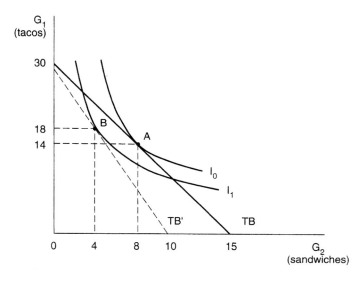

FIGURE 7.15
Utility maximizing combination of tacos and sandwiches when price of sandwiches increases.

the student can no longer afford the product combination at point A (8 sandwiches and 14 tacos, which now costs $38). So the student is forced to move to a lower indifference curve, thus reducing utility. To maximize utility at the new prices (P_{g_1} = $1, P_{g_2} = $3), the student finds the point of tangency between the highest obtainable indifference curve (I_1) and the new budget line for $30 (TB′). The student moves from point A, consuming 8 sandwiches and 14 tacos (which now costs $38), to point B, consuming 18 tacos and 4 sandwiches (which costs $30). When the price of sandwiches increases, the student substitutes the now relatively less expensive product, tacos, for the relatively more expensive product, sandwiches. Thus the student is consuming more tacos and fewer sandwiches, with the same total budget, to maximize utility.

As an exercise to help you understand these concepts better, rework the preceding example assuming the price of sandwiches is $2 and the price of tacos increases to $2.

Product Price Decrease

When the price of one product decreases it becomes less expensive to consume, relative to the other product. The consumer must then reevaluate the budget allocation across the two products to find the new utility maximizing product combination. This decision is illustrated in Figure 7.16.

Initially the consumer is at point A in Figure 7.16 where the slope of the indifference curve ($MRS_{G_2 G_1}$) is equal to the slope of the budget line ($-P_{g_2}/P_{g_1}$), and the utility maximizing product combination is G_1^* units of G_1 and G_2^* units of G_2. If P_{g_2} decreases to P'_{g_2}, the budget line becomes flatter, rotating outward from TB to TB′, as the slope is now equal to $-P'_{g_2}/P_{g_1}$. When P_{g_2} decreases, the good G_2 becomes relatively

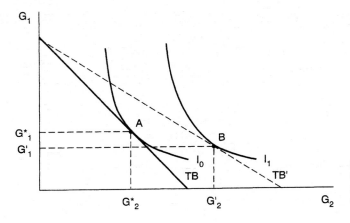

FIGURE 7.16
Utility maximizing decision under changing prices: price of G_2 decreases.

less expensive to consume compared to the good G_1. Since the consumer is limited by the available budget, when P_{g_2} decreases the consumer can afford the combination at point A in Figure 7.16 and have money left over because it now lies on a lower budget line (TB < TB'). Therefore, the consumer is able to increase consumption and move to a higher indifference curve (I_1) and a higher level of utility. The consumer moves from the product combination at point A (G_1^*, G_2^*) in Figure 7.16, to point B, consuming G'_1 units of G_1 and G'_2 units of G_2. The new utility maximizing combination of products, point B in Figure 7.16, is the point of tangency between the higher indifference curve (I_1) and the new budget line (TB'), or where their slopes are equal. After the price decrease, the consumer is consuming more of the relatively less expensive good ($G'_2 > G_2^*$) and less of the relatively more expensive good ($G'_1 < G_1^*$).

We can illustrate the effect of a product price decrease on the utility maximization decision using our college student example. What happens to the utility maximizing combination of tacos and sandwiches if the price of sandwiches decreases to $1? This decision is depicted graphically in Figure 7.17.

With a budget of $30, and initial prices of P_{g_1} = $1 and P_{g_2} = $2, the utility maximizing combination is 8 sandwiches and 14 tacos (point A in Figure 7.17). If the price of sandwiches decreases to $1 the budget line flattens, rotating outward, from TB to TB'. At the new price if the student spent the entire $30 budget on sandwiches she could afford 30 sandwiches instead of the initial 15. Due to the price decrease, the student can afford the product combination at point A (8 sandwiches and 14 tacos, which now costs $22) with money left over. So the student is able to move to a higher indifference curve, thus increasing utility. To maximize utility at the new price level, the student finds the tangency between the highest obtainable indifference curve (I_1) and the new budget line for $30 (TB'). The student moves from point A, consuming 8 sandwiches and 14 tacos (which now costs $22), to point B, consuming 18 sandwiches and 12 tacos (which costs $30). When the price of sandwiches decreases, the student is able to increase consumption and move to a higher indifference curve, because her budget now purchases more. After the

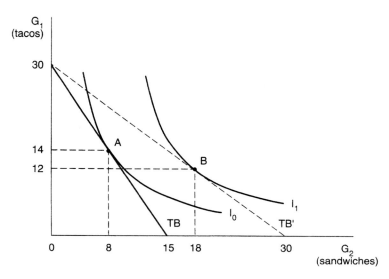

FIGURE 7.17 Utility maximizing combination of tacos and sandwiches when price of sandwiches decreases.

price decrease the student is consuming more sandwiches and fewer tacos, with the same total budget, to maximize utility.

As an exercise to help you understand these concepts better, rework the preceding example assuming the price of sandwiches is $2 and the price of tacos decreases to $0.50.

As we illustrated earlier, as the price of a product changes the combination of products that maximizes utility also changes as the consumer reestablishes a purchase pattern where the marginal utility per dollar spent on the two goods is equalized. More specifically we demonstrated that as the price of a good increases, the consumer purchases less of that good; as the price of a good decreases, the consumer purchases more of that good. Using this information, we can use the budget line and the indifference curve to derive an individual consumer's demand for a product. That is, the quantity of a product that the consumer is **willing** and **able** to purchase at each price.

Deriving a Demand Curve Using Indifference Curves

Managers and economists are interested in consumer demand for the products they consume. But how can we translate what we know about product combinations and utility maximization into a representation of consumer demand? We begin by defining two new concepts.

The first concept is the **demand schedule,** defined as *information on price and quantity (consumption) combinations that give the consumer maximum utility, ceteris*

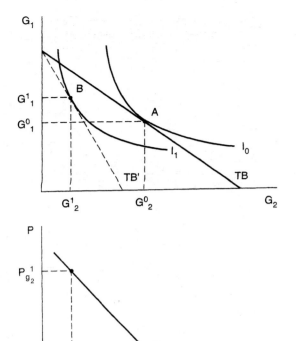

FIGURE 7.18
Demand curve derivation for individual consumer.

paribus (all else constant). The demand schedule is derived assuming that nothing else is changing except the price of the good being examined. Thus, we are assuming that income, tastes and preferences, prices of other goods, etc., are all held at a fixed level. These assumptions are often referred to as the *ceteris paribus conditions.*

The second concept is the **demand curve,** defined as *a line connecting all combinations of price and quantities consumed for a particular good, ceteris paribus.* Each point on a demand curve gives the price and quantity combination of a particular good that a consumer will buy, given his or her budget constraint and the prices of other goods.

We can use the budget line and the indifference curve to derive a consumer's demand for a particular good. Remember that the point of tangency between the indifference curve (MRS) and the budget line (IPR) gives the combination of goods that maximizes utility. Figure 7.18 graphically depicts the derivation of an individual consumer's demand curve.

With a given budget, represented by TB in Figure 7.18, and the initial prices of $P_{g_1}^0$ for G_1 and $P_{g_2}^0$ for G_2, the consumer maximizes utility at point A, where $MRS_{G_2G_1} = -P_{g_2}^0 / P_{g_1}^0$ consuming G_1^0 units of G_1 and G_2^0 units of G_2. So when the price of G_2 is $P_{g_2}^0$ the consumer purchases G_2^0 units of G_2.

If the price of G_2 increases from $P_{g_2}^0$ to $P_{g_2}^1$ the budget line rotates inward toward the origin, from TB to TB' in Figure 7.18, and the slope of the budget line is now equal to $-P_{g_2}^1/P_{g_1}^0$. Since the consumer is limited by the budget, as the price of G_2 increases the consumer is forced to move to a lower indifference curve. To maximize utility given the new prices, the consumer finds the point of tangency between the new budget line (TB') and the highest obtainable indifference curve (I_1). This point of tangency is at point B in Figure 7.18, where $MRS_{G_2G_1} = -P_{g_2}^1/P_{g_1}^0$ and the utility maximizing product combination is G_1^1 units of G_1 and G_2^1 of G_2. Thus, when the price of G_2 is $P_{g_2}^1$ the consumer purchases G_2^1 units of G_2.

We now have two price and quantity combinations ($P_{g_2}^0$, G_2^0 and $P_{g_2}^1$, G_2^1) for the consumer; thus, we can derive the consumer's demand curve for the good G_2. Since demand represents a relationship between the price of a good and the quantity of that good the consumer is **willing** and **able** to purchase, demand curves are graphed in price–quantity space. [Economists generally place price (P) on the vertical axis and quantity (Q) on the horizontal axis.] Plotting the two price and quantity combinations, as shown in the lower portion of Figure 7.18, and connecting the points with a line yields this consumer's demand curve for the product G_2. This relationship shows that at price $P_{g_2}^0$ the quantity demanded is G_2^0; at price $P_{g_2}^1$ quantity demanded is G_2^1. The demand curve slopes downward to the right, indicating an inverse relationship between price and quantity demanded of a good. That is, as the price of a good increases, the quantity demanded of that good decreases (and if price decreases, quantity demanded increases).

Using this knowledge, we can now derive our college student's demand curve for sandwiches. With a budget of \$30, and prices of P_{g_1} (tacos) = \$1 and P_{g_2} (sandwiches) = \$2 the student purchased 8 sandwiches and 14 tacos to maximize utility (see Figure 7.13). When the price of sandwiches increased to \$3, holding the budget and price of tacos constant, the student purchased 4 sandwiches and 18 tacos to maximize utility (see Figure 7.15). When the price of sandwiches decreased to \$1, holding the budget and price of tacos constant, the student purchased 18 sandwiches and 12 tacos to maximize utility (see Figure 7.17). We now have three price and quantity combinations for sandwiches, which are summarized in the demand schedule in Table 7.6 and graphically depicted in Figure 7.19. One of our assumptions used in deriving the demand curve for sandwiches (Figure 7.19) is the holding of all economic factors constant (i.e., budget, price of tacos, etc.) except for the good's own price. Because the changes in the consumption of tacos are a direct result of a change in the price of sandwiches, it reflects a change in one of the *ceteris paribus* conditions, and hence does not represent a demand curve for tacos.

Decision Rule Etiquette

As we discussed earlier, the order of the subscripts on MRS indicates which good is being replaced and which good is being added. Since economic convention places the good being replaced on the vertical axis, the order of subscripts

TABLE 7.6   Demand Schedule for Sandwiches

Price per Sandwich	Quantity Demanded
$3	4
$2	8
$1	18

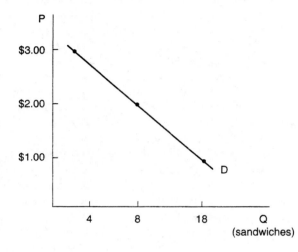

FIGURE 7.19
Demand curve for sandwiches.

on MRS also indicates the labeling on the graphs. Thus, it is very important to have the graphical presentation of the decision rule match the written form, and vice versa.

When the good being replaced is G_1 and the good being added is G_2, the decision rule is

$$MRS_{G_2 G_1} = -\frac{P_{g_2}}{P_{g_1}}$$

since G_1 is on the vertical axis and G_2 is on the horizontal axis.

However, if G_2 is the good being replaced, and G_1 is being added, the decision rule would be

$$MRS_{G_1 G_2} = -\frac{P_{g_1}}{P_{g_2}}$$

because in this case G_2 will be on the vertical axis and G_1 will be on the horizontal axis.

COMPARING THE UTILITY MAXIMIZATION DECISION AND THE FACTOR–FACTOR DECISION

We stated earlier that the consumer's utility maximization decision is very similar to the factor–factor decision faced by the producer. We examine the similarity in this section.

Utility maximization involves finding the combination of products that provides the greatest level of satisfaction for a given budget (available money). The level of satisfaction is represented by the indifference curve, which is a line indicating all combinations of two goods that provide the same level of utility or satisfaction. Consumers consume products for the satisfaction derived from them; therefore, these products can be thought of as inputs in the consumer's production of utility.

The factor–factor decision involves finding the least-cost combination of inputs to produce a given level of output. The level of output is represented by the isoquant, which is a line indicating all combinations of two variable inputs that produce the same level of output. Firms use inputs in the production of output.

Indifference curves and isoquants exhibit very similar characteristics, as summarized here:

1. Indifference curves (isoquants) slope downward to the right, indicating that the consumer (producer) must or can consume (use) less of one product (input) to consume (use) more of the other.
2. Indifference curves (isoquants) can never intersect.
3. The distance from the origin indicates the utility (production) level.
4. Each indifference curve (isoquant) represents a unique utility (production) level. There is an indifference curve (isoquant) for each possible utility (production) level.

The products in the utility maximization decision, and the inputs in the factor–factor decision, can substitute for one another to a certain extent. Therefore, economists and managers are very interested in how they substitute for one another, both physically and economically.

For the utility maximization decision, the marginal rate of substitution measures the rate at which one good will substitute for another and maintain the same level of satisfaction, and, by definition, MRS measures the slope of the indifference curve. Thus, the computational formula is the change in the good on the vertical axis (good being replaced) divided by the change in the good on the horizontal axis (the good being added), or $MRS_{G_2G_1} = \Delta G_1/\Delta G_2$ if the good being replaced is G_1 and the good being added is G_2.

For the factor–factor decision, the marginal rate of substitution measures the rate at which one input will substitute for another and maintain the same level of production, and, by definition, MRS measures the slope of the isoquant. Thus, the computational formula is the change in the input on the vertical axis (input being replaced) divided by the change in the input on the horizontal axis (the input being added), or $MRS_{X_2 X_1} = \Delta X_1 / X_2$ if the input X_1 is being replaced and X_2 is being added.

The economic relationship between consumption goods in the utility maximization decision is represented by the budget line, which is a line indicating all combinations of two consumption goods that can be purchased using all the consumer's budget (available funds).

The economic relationship between inputs in the factor–factor decision is represented by the isocost line, which is a line indicating all combinations of two inputs that can be purchased with the same amount of money.

The budget line and the isocost line exhibit similar characteristics, as summarized here:

1. The end points of the budget line (isocost line) show how many units of a good (input) could be purchased if the entire budget (outlay) were spent on that good (input).
2. With a given budget (outlay) a reduction in the consumption (use) of one good (input) is necessary to increase consumption (use) of the other.
3. The distance from the origin indicates the budget (outlay) level.
4. A change in the budget (outlay) causes a parallel shift in the budget line (isocost line). If the budget (outlay) increases, the budget line (isocost line) moves away from the origin; if the budget (outlay) decreases, the budget line (isocost line) moves toward the origin.
5. The slope of the budget line (isocost line) is the inverse price ratio. It is called the inverse price ratio because it is the price of the good (input) on the horizontal axis divided by the price of the good (input) on the vertical axis.
6. A change in the price of a good (input) causes the budget line (isocost line) to rotate around the intersection of the budget line (isocost line) and the axis of the other good (input).

In the utility maximization decision the economic objective of the consumer is to find the combination of products that will maximize utility for a given budget. Because we assume that the consumer is rational, the consumer wants to reach the highest possible level of utility for his or her given budget. Therefore, the consumer wants to find the tangency between the highest obtainable indifference curve and his or her budget constraint, represented by the budget line. The utility maximizing combination of products is found where the slope of the budget line (IPR) is equal to the slope of the indifference curve (MRS). This decision is depicted graphically in Figure 7.20 where utility maximization occurs at point A, where $MRS_{G_2 G_1} = -P_{g_2} / P_{g_1}$, and the utility maximizing combination of products is G_1^* units of G_1 and G_2^* units of G_2.

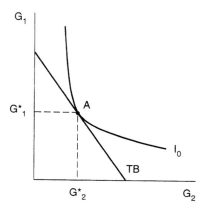

FIGURE 7.20
Utility maximization decision.

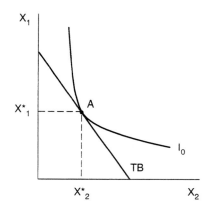

FIGURE 7.21
Factor–factor decision.

In the factor–factor decision the objective of the producer is to find the least-cost combination of inputs to produce a given level of output. Since we assume that the producer is rational, the producer wants to reach the highest possible level of production for the lowest possible expenditure. Therefore, the producer wants to find the tangency between the highest obtainable isoquant and the lowest possible isocost line. The least-cost combination of inputs is found where the slope of the isoquant (MRS) is equal to the slope of the isocost line (IPR). This decision is depicted graphically in Figure 7.21 where the least-cost combination of inputs (X_1^*, X_2^*) to produce the desired output level is found at point A, where $MRS_{X_2X_1} = -P_{x_2}/P_{x_1}$.

The consumer's utility maximization decision is very similar to the producer's factor–factor decision. Consumers use goods to produce utility; producers use inputs to produce output. The consumer's income level is given by the budget line; the producer's expenditure level is given by the isocost line. The consumer's economic objective is to find the combination of products that provides the greatest level of satisfaction for a given amount of money. The producer's economic objective is to find the least-cost combination of inputs that produces the desired output level.

SUMMARY

In this chapter we examined the basic consumer question of how to allocate a limited amount of money (budget) among various commodities so as to maximize satisfaction. To answer this question, we introduced the concept of utility, meaning satisfaction, which underlies much of demand. Consumers desire a product because it gives them satisfaction or enjoyment of some sort. However, the Law of Diminishing Marginal Utility states that as more of a good is consumed, the addition to total utility declines with each successive unit.

The basic consumer question is "How to allocate a limited budget across consumption opportunities so as to maximize utility?" To answer this question, we made the simplifying assumption that there were only two goods available; all others were fixed. In developing the utility maximization criterion, or decision rules, we introduced the concepts of an indifference curve and budget line. The indifference curve describes the physical manner in which the two goods substitute for each other in the consumption process; the budget line describes the economic relationship between relative product prices. Since the consumer's decisions are limited by the available budget, the consumer's objective is to find the combination of goods that maximizes utility subject to the available funds. The decision criteria, or decision rule, equates the slope of the indifference curve (MRS), the rate of physical substitution, to the slope of the budget line (IPR), the rate of economic substitution.

Using this criterion we showed that the consumer maximizes utility by establishing a purchase pattern whereby the marginal utility per dollar of expenditure for each item (MU/P) is equal. When this pattern is reached, the consumer is said to be at an equilibrium point, because there is no incentive to change consumption. Thus, using the indifference curve and budget line, we are able to determine a price and quantity combination for the goods that will maximize utility.

However, when a product price changes, the MU/P ratio also changes, resulting in a search for a new utility maximizing combination of products as the consumer seeks to reestablish a purchase pattern that equalizes the marginal utility per dollar of expenditure across the goods. From this we were able to derive an individual consumer's demand curve for a particular commodity. We showed that an inverse relationship exists between the price of a commodity and the quantity demanded of that commodity. That is, as the price of a product increases, less of that product is demanded; as the price of a product decreases, more of that product is demanded. From the viewpoint of the manager, the consumer dictates much of the manager's activities, particularly in food and agricultural situations. The individual manager must listen to the signals about consumer preferences and react accordingly to maximize profits.

Finally we illustrated the similarities between the utility maximization decision of the consumer and the factor–factor decision of the producer. Utility maximization involves finding the combination of products that provides the greatest level of satisfaction for a given budget; whereas the factor–factor decision involves finding the least-cost combination of inputs to produce a given level of output.

QUESTIONS

1. Utilizing Figure 7.1, what is the difference in utility level achieved by consuming 25 tacos and 5 sandwiches versus consuming 19 tacos and 6 sandwiches?

2. True or false?: Indifference curves are convex to the origin, represent equal utility along any given indifference curve, and increase in utility value as you move farther from the origin. Explain.

3. Suppose $MRS_{G_2G_1}$ is -3. Does this mean that the good G_2 substitutes fairly well for G_1 or vice versa? In general, what can you say about an MRS that is equal to 1.0?

4. Explain the law of diminishing marginal rate of substitution. How is this law related to the steepness of indifference curves? How is it related to the law of diminishing utility?

5. Suppose the price of steak is \$1.50/lb and the price of potatoes is \$0.25/lb. What is the slope of the budget line for steak/potato consumption? If the total budget is \$15 what would the budget line look like? If the budget was \$60? Illustrate with a graph.

6. Using Figure 7.8 as an example, show the effect on the slope of each budget line from a \$2 increase in the price of G_1 (tacos) and then a subsequent \$1.50 decrease in the price of G_2 (sandwiches).

7. From Table 7.5 suppose that the price of G_1 dropped to \$0.50. What combination of tacos and sandwiches would now be optimal? What if the price of G_1 doubled?

8. Using the information presented in Table 7.5, indicate what will happen if the price of sandwiches increases to \$4 and the college student's budget increases to \$46. In this situation, would the student be able to stay on the initial indifference curve? Explain.

9. Define the terms *indifference curve* and *budget line*. What concepts do we use to define the slopes of each?

10. Using the information presented in Table 7.5, calculate $MRS_{G_1G_2}$. (That is, assume sandwiches are being replaced and tacos are being added.) If the price of tacos is \$1 and the price of sandwiches is \$2, what is the utility maximizing product combination? Graphically depict this decision.

CHAPTER 8
Supply and Demand
(As Decision-Making Tools)

INTRODUCTION

In previous chapters we developed decision rules to help managers answer the basic questions faced in deciding what activity to undertake: "What to produce?," "How to produce?," and "How much to produce?" During these discussions we emphasized the decisions that are made at the individual or firm level and we did not examine how these individual decisions affect others or are affected by other's decisions.

In this chapter we examine the two separate economic concepts of supply and demand and learn how a thorough understanding of these concepts can be an important managerial tool. An understanding of supply will allow managers to handle costs of production and other internal firm decisions, comprehend how competitors might be expected to act, and even predict how market prices might vary. Demand, when used as a management tool, can identify potential markets, estimate market prices, aid in product differentiation from competitors, and help in planning future investment decisions.

In This Chapter

In this chapter we move from the individual producer and consumer to the aggregate, or market level, representing many producers and consumers. Thus, we are changing our perspective from that of examining decisions made by the individual to examining the behavior and responsiveness of the consumers and producers that constitute the market for a commodity or product. An important aspect of market supply and demand is the responsiveness of the demand and supply of the commodity or product to changes in the product's price and other economic factors. This responsiveness is measured by elasticity. We begin by examining the concept of supply, its derivation, and the managerial implications of elasticity of supply. Then we explore the concept of demand, moving from the individual's demand, as presented in Chapter 7, to the market demand curve, examining the determinants of demand and its implications for managers.

SUPPLY

The economic concept of supply is a bit different from the layman's common idea of supply. In lay terms, supply means the amount of an item available in an area or time frame; the economic concept of supply deals specifically with what amount of an item will be available in an area or time frame at a specific price. Supply is a direct price and quantity relationship detailing the functional relationship of how suppliers (producers, sellers, and managers) of a product respond to differing price levels. It states what suppliers are **willing** and **able** to supply at a given price.

Derivation of Market Supply Curves

The supply curve for an individual firm is based on the cost structure of the firm and how managers react to alternative product prices as they attempt to maximize profits. As a manager equates MC to MR, the corresponding level of production is determined, as illustrated in Figure 8.1. At P_1 the manager will find Q_1 to be the quantity that maximizes profits because, as we remember from Chapter 4, any additional output greater than Q_1 will cost more to produce than the value received when it is sold. Any output less than Q_1 will mean some profit is foregone because the revenue received from the additional output is greater than the added cost of producing it. If price decreases, from P_1 to P_3, the corresponding quantity produced also decreases, from Q_1 to Q_3. We know from our previous analysis that the manager will only continue to produce as long as variable costs are covered; thus, when price falls below P_3, this producer will not continue in business and no product will be supplied.

Since, as illustrated in Figure 8.1, producers will stop production if price falls below the minimum of average variable cost, the supply curve for an individual firm is simply the "marginal cost curve above the average variable cost curve." As Figure 8.2 shows, if the price falls below P_3 no product is supplied. As

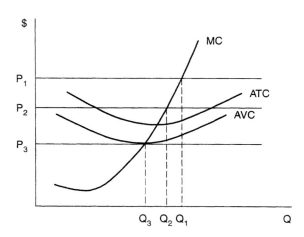

FIGURE 8.1
Profit maximization for the individual producer.

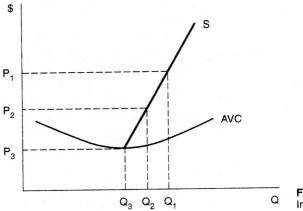

FIGURE 8.2
Individual firm's supply curve.

price increases from P_3 to P_1 the firm responds by increasing production from Q_3 to Q_1. Thus, the MC above the minimum average variable cost traces the firm's response to changing prices in the short-run, which reveals the individual firm's short-run supply curve. In the long-run, the MC above minimum average total cost is the firm's supply curve (in the long run AVC and ATC are the same). Since the MC is the firm's supply curve, whatever factors affect the shape or slope of the marginal cost curve, such as the production function and price of inputs, will affect the individual firm's ability or willingness to supply the product.

The individual firm's supply curve is drawn or derived under the assumption that the only economic variable that is changing is the good's own price. Thus, all other economic factors, such as input prices, technology, and prices of other products, are held at a fixed level as discussed earlier. These other factors are often referred to as the *ceteris paribus* conditions.

To understand how much of a product will be offered in a market at any point in time, it is necessary to derive an industry, or market, supply curve. An industry supply curve is obtained by horizontally summing (adding) all the supply curves for the individual producers in the market. This is illustrated in Figure 8.3, where we have assumed a simplified market with only three producers of calculators. Each individual firm will produce calculators based on its marginal cost curve above average variable cost. To find out what the industry supply curve looks like, we add the potential production at each price. At price $60 the three firms would individually produce 70, 60, and 100 calculators, for a total industry supply in response to a $60 price level of 230 calculators. At the lower price of $30 the three individual firms would, as they attempt to maximize profits, produce 50, 30, and 20 calculators, respectively, yielding a total industry supply of 100 calculators at $30.

A real-world example would be the regional supply curves for wheat (dryland, irrigated, rainfed) in the United States, which have different cost functions but add up to a total supply.

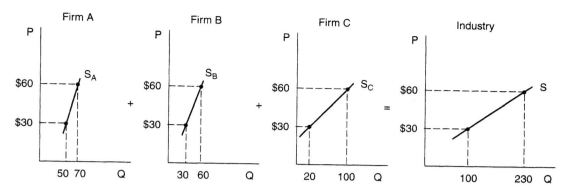

FIGURE 8.3 Derivation of market supply curves.

TABLE 8.1 Supply Schedule for Calculators in Moscow, Idaho

Price per Calculator	Quantity Supplied
$10	55
$20	75
$30	100
$40	140
$50	180
$60	230

Law of Supply

The price and quantity relationship in the economic concept of supply is indicated by a supply schedule as shown in Table 8.1. This mathematical relationship forms the foundation of the supply curve indicated in Figure 8.4. Note that price (P) is always shown on the vertical axis and quantity (Q) is indicated on the horizontal axis. Also note that no information is available nor can a statement be made about prices less than $10 or greater than $60. This supply curve indicates that, with a market price for calculators of $10 in Moscow, Idaho, 55 calculators will be supplied to the market; a market price of $60 will bring 230 calculators to this market.

The relationship between price and quantity, as shown in Table 8.1 and Figure 8.4, is more succinctly stated as the **law of supply**, which says that *the quantity of goods and/or services offered to a market will vary **directly** with the price.* As price increases, a corresponding increase in quantity supplied will be seen; as price decreases, a corresponding decrease in quantity supplied will be seen. The amount of the increase or decrease in the quantity supplied depends on the concept of elasticity of supply, which we discuss next.

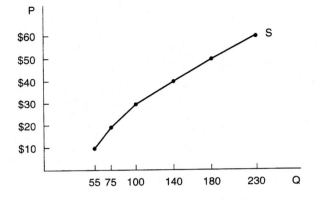

FIGURE 8.4
Supply curve for calculators in Moscow, Idaho.

Elasticity of Supply

Determining how responsive the quantity supplied of a product is to changes in the product's price and other economic factors is a must for today's managers. Thus, elasticity measurements are an important tool for managers and economists. **Elasticity** is defined as *the percentage change in the quantity supplied relative to the percentage change in price (or other economic variable) as we move from one point to another on a supply curve.* It is expressed as

$$E = \frac{\% \text{ change in quantity}}{\% \text{ change in price}} = \frac{\%\Delta Q}{\%\Delta P}$$

where E is the elasticity coefficient.

Elasticity is a measure of responsiveness of quantity to changes in price (or another economic variable). This represents movement along the supply curve and thus elasticity is also a measure of the degree of slope of the curve. This responsiveness can be classified as either inelastic, unitary elastic, or elastic.

An **inelastic** supply response is one in which *a change in price brings about a relatively smaller change in quantity*, or the $\%\Delta Q < \%\Delta P$ so $E < 1.00$. Thus, an inelastic supply curve will have a steep slope. Note that $\%\Delta Q < \%\Delta P$ is equivalent to $\%\Delta P > \%\Delta Q$.

A **unitary elastic** supply response is one in which *a change in price brings about an equivalent change in quantity*, or the $\%\Delta Q = \%\Delta P$ so $E = 1.00$.

An **elastic** supply response is one in which *a change in price brings about a relatively larger change in quantity*, or the $\%\Delta Q > \%\Delta P$ so $E > 1.00$. Hence, an elastic supply curve will have a flat slope. Note that $\%\Delta Q > \%\Delta P$ is equivalent to $\%\Delta P < \%\Delta Q$.

Managers and economists are interested in two types of elasticity measures. The first is **own-price elasticity of supply,** which measures *the responsiveness of the quantity supplied of a good to changes in the price of that good.* That is, how does the quantity supplied of a good respond when its own price changes? The other is **cross-price elasticity of supply,** which measures *the responsiveness of the quantity* *supplied of a good to changes in the price of a related good.* That is, how does the quantity supplied of a good respond when the price of a related good changes? Throughout this section we will be identifying and illustrating how managers can use elasticity measurements in their everyday decisions.

A manager's ability to respond to a price change is affected by many factors. A tool used to evaluate the degree to which managers, and an industry, respond to a given price change is called the elasticity of supply (E_S). Own-price elasticity (meaning response of a product to a change in its own price) of supply (or price elasticity) is calculated as

$$E_S = \frac{\%\Delta Q_S}{\%\Delta P} = \frac{\left[\dfrac{Q_2 - Q_1}{Q_2 + Q_1}\right]}{\left[\dfrac{P_2 - P_1}{P_2 + P_1}\right]}$$

where P_1, Q_1 and P_2, Q_2 are two price and quantity points from the supply curve. Since supply is upward sloping, E_S is always positive. This formula is referred to as the **arc elasticity** since it measures responsiveness along a specific section (or arc) of a supply curve and measures the "average" price elasticity between two points on the curve.

Own-price elasticity of supply, which represents movement along the supply curve, is also a measure of the degree of slope of the supply curve. To illustrate this, consider our calculator example where the supply curves for the individual producers are presented in Table 8.2. When the price of calculators increases from $30 to $60, the quantity supplied by the individual firms changes, but how responsive is each of the three firms to changes in the price of calculators? To calculate the own-price elasticity of supply for each of the firms, we simply plug the appropriate numbers into the formula.

When the price increases from $30 ($P_1$) to $60 ($P_2$), firm A increases its quantity supplied from 50 (Q_1) to 70 (Q_2). Plugging these numbers into the E_S formula, we can calculate firm A's own-price elasticity of supply (E_{S_A}):

$$E_{S_A} = \frac{\left(\dfrac{70 - 50}{70 + 50}\right)}{\left(\dfrac{\$60 - 30}{\$60 + 30}\right)} = \frac{\left(\dfrac{20}{120}\right)}{\left(\dfrac{30}{90}\right)} = \frac{0.1667}{0.333} = 0.501$$

TABLE 8.2 Supply Schedules for Firms Producing Calculators

Price	Firm A	Firm B	Firm C	Industry
$30	50	30	20	100
$60	70	60	100	230
E_S	0.50	1.00	2.00	1.18

Firm A's supply is inelastic since $E_S < 1.00$, which means this firm would respond to an increase (or decrease) in price with a relatively smaller increase (or decrease) in quantity supplied. That is, for a 1 percent increase in price, the quantity supplied by firm A will increase 0.5 percent. Thus, firm A is considered to be fairly unresponsive to price changes.

Similarly, in response to the price increase from \$30 (P_1) to \$60 (P_2), firm B increases quantity supplied from 30 (Q_1) to 60 (Q_2). Using these numbers, firm B's own-price elasticity of supply (E_{S_B}) is calculated as

$$E_{S_B} = \frac{\left(\dfrac{60 - 30}{60 + 30}\right)}{\left(\dfrac{\$60 - 30}{\$60 + 30}\right)} = \frac{\left(\dfrac{30}{90}\right)}{\left(\dfrac{30}{90}\right)} = \frac{0.333}{0.333} = 1.00$$

Since $E_S = 1.00$, firm B's supply is unitary elastic, which means this firm would respond to an increase (or decrease) in price with an equivalent increase (or decrease) in quantity supplied. In other words, if price increases 1 percent, firm B will increase quantity supplied by 1 percent.

Firm C increases quantity supplied from 20 (Q_1) to 100 (Q_2) when price increases from \$30 (P_1) to \$60 (P_2). Thus, the own-price elasticity of supply for firm C (E_{S_C}) is calculated as

$$E_{S_C} = \frac{\left(\dfrac{100 - 20}{100 + 20}\right)}{\left(\dfrac{\$60 - 30}{\$60 + 30}\right)} = \frac{\left(\dfrac{80}{120}\right)}{\left(\dfrac{30}{90}\right)} = \frac{0.667}{0.333} = 2.00$$

Firm C's supply is elastic since $E_S > 1.00$, which means this firm would respond to an increase (decrease) in price with a relatively larger increase (or decrease) in quantity supplied. For a 1 percent increase in price, firm C will increase quantity supplied by 2 percent. Thus, firm C is considered to be very responsive to price changes.

Since elasticity of supply is a measure of responsiveness of the quantity supplied it also reflects the slope of the supply curve. Therefore, the slope of the supply curve also gives an indication of the own-price elasticity of supply. This was illustrated back in Figure 8.3, which graphically represents the individual firm's supply curves. Notice that firm A's (inelastic response) supply curve has a steeper slope than does firm B's or C's, and that firm C (elastic response) has a flatter slope than either firm B or A.

This relationship between elasticity and slope can be seen more clearly in Figure 8.5 where for a given price change a flatter (more elastic) curve, S_2, will generate a larger response than a higher (more inelastic) sloped curve, S_1. For the same change in price, from P_1 to P_2, the increase in quantity supplied, from Q_{21} to Q_{22} is

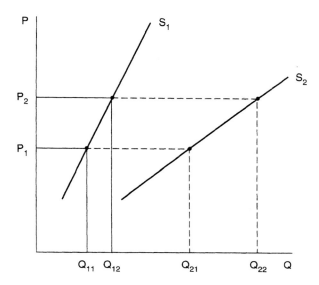

FIGURE 8.5
Supply curves with different elasticities.

significantly larger for the more elastic supply curve, S_2, than the increase in quantity supplied, from Q_{11} to Q_{12}, for the more inelastic supply curve, S_1. Thus, the more elastic (flatter) the supply curve, the more responsive quantity supplied is to changes in price.

Although we normally draw supply curves as straight lines for simplicity's sake, we know the supply curves are based on the marginal cost curve above average variable cost and therefore can be expected to be curved lines. As a result, the elasticity will change along the line, and a manager's ability to respond to a price change is often dependent on where in Stage II of production (remember that?) he or she is operating.

Another critical factor affecting the ability of a producer to respond is the time span under consideration. In the short-run, the elasticity of most products, and certainly farm products, is fairly inelastic because it is difficult to change production plans quickly. Managers may be hesitant to make a large capital investment, for example, or capital funds may not be available, or crops may already have been seeded. As the planning time becomes longer, managers can more easily respond to product price changes by attracting more labor, adding more fertilizer, and securing more plant and equipment.

Another characteristic of supply curves in agricultural production is that they are more elastic for a price increase than for a price decrease. When prices are strong (increasing) producers will purchase inputs, capital items, which then become fixed costs when the next decision has to be made. These inputs may become fixed in their use because of the difference between acquisition value (purchase price) and salvage value (what the input is worth at the end of its useful life) because of their specialized uses for which there are no alternatives (i.e., a combine does not mow lawns very well). Thus, when prices begin to fall, production levels

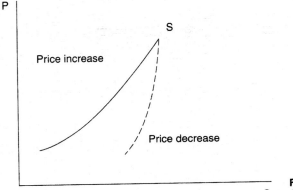

FIGURE 8.6
Supply curves for price changes.

are less responsive because of the inability to pull these inputs out of production. For example, during periods of high prices dairy farmers are encouraged to expand their milking herd, update their milking parlors, and feeding systems. When prices subsequently fall, the inputs remain in production (and hence milk production does not decline rapidly) since milking parlors have few alternative uses and dairy cows are typically sold at a discount (low salvage value). The result is a supply curve situation like that depicted in Figure 8.6 where two curves reflect the reaction to a price decrease or increase.

In summary, a knowledge of the concepts of supply and elasticity of supply aids managers in understanding their firm, their industry, and their competitors. Knowledge of these concepts by public or governmental decision makers aids them in developing programs and/or policies to achieve their goals.

Change in Supply Versus Change in Quantity Supplied

The quantity of a product supplied in a market by an individual firm, and the industry as well, will change in direct response to a change in product price. This change is a movement along an existing supply curve and should be contrasted to a shift in the supply curve itself. The difference between a change in supply versus a change in quantity supplied is graphically illustrated in Figure 8.7.

Changes in the quantity supplied result from changes in the price of the product being examined and are movements along a supply curve. In Figure 8.7, if the price of the product increases from P_1 to P_2, the quantity supplied increases from Q_1 to Q_2. This is movement along the supply curve.

A change in supply results from changes in the quantity sold due to factors other than a change in the product price. That is, if something other than the good's own price changes (i.e., one of the *ceteris paribus* conditions) the position of the curve changes. In Figure 8.7, if there is an increase in supply (one of the *ce-*

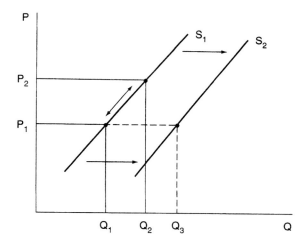

FIGURE 8.7
Change in supply versus change in quantity supplied.

teris paribus conditions changes), the supply curve shifts to the right (from S_1 to S_2) and at the same price level (P_1) the quantity supplied increases (from Q_1 to Q_3). A shift to the right is an increase in supply, whereas a shift to the left is a decrease in supply.

Determinants of Industry Supply

An industry supply curve can be expressed in the following function:

$Q_S = f$ (own price, input prices, technology, prices of alternative products, number of sellers, other factors)

This states that for each supply curve, quantity supplied (Q_S) will vary with price as long as the supply shifters, or supply determinants, are held constant. (Note that the individual supply curve is influenced by all of the factors listed, except the number of sellers.) The factors that will shift the supply curve are those factors earlier assumed to be held constant (the *ceteris paribus* conditions) when deriving and drawing the supply curve. Let's examine these supply shifters (called this because they shift the supply curve left or right) from the manager's perspective.

Input prices. Changes in input prices directly impact the cost structure of the firm. As we saw in Chapter 3 (see Figure 3.7), if input prices increase, the cost of producing each unit of output increases, resulting in an upward shift (inward to the left) of the cost curves. Conversely, if input prices decrease, the cost of producing each unit of output decreases, resulting in a downward shift (outward to the right) of the cost curves (see Figure 3.8). Since the individual firm's supply curve is its marginal cost curve above minimum average variable cost, and the industry supply is the horizontal summation of the individual supply curves, then any factor that affects the position of the firm's marginal cost curve affects the position of both the firm's and the industry's supply curve. If input prices increase, the supply curve will shift inward to the left because each unit of product now costs more to produce. Hence, less is supplied at each price level (a decrease in supply). Conversely, if input prices decrease, costs of production also decrease, shifting supply outward to the right. Hence, more is supplied at each price level (an increase in supply).

Technology.   Technological improvements allow producers to produce more of a good with the same amount of inputs or the same amount of a good with fewer inputs. As we saw in Chapter 2 (see Figure 2.10), a technological improvement shifts the production function (TPP) up, which results in lower production costs (cost curves shift downward to the right). Technological improvements result in an increase in supply (the supply curve shifts outward to the right).

Prices of alternative products.   Most firms can produce more than one product from their available resource bases. Changes in prices of other goods that a firm's inputs could be used to produce affects the position of the supply curve. Recall the "What to produce?" question in Chapter 6. When the price of an alternative product changes, the inverse price ratio changes; thus, the decision MRPS = IPR indicates a new product–product combination (see Figures 6.12 and 6.14). The direction of the shift in supply depends on the relationship between the two products and is reflected in the cross-price elasticity of supply coefficient. Cross-price elasticity of supply is calculated as

$$E_{S_{Y_1Y_2}} = \frac{\%\Delta Q_{Y_1}}{\%\Delta P_{y_2}} = \frac{\left[\dfrac{Q_{Y_12} - Q_{Y_11}}{Q_{Y_12} + Q_{Y_11}}\right]}{\left[\dfrac{P_{y_22} - P_{y_21}}{P_{y_22} + P_{y_21}}\right]}$$

where Y_1 and Y_2 are the products produced by the firm, P_{y_1} and P_{y_2} are the prices of those products, and $E_{s_{Y_1Y_2}}$ is read as the cross-price elasticity of supply for product Y_1 with respect to product Y_2. This cross-price elasticity measures the effect of a change in the price of good Y_2 on the quantity supplied of Y_1. We can use the cross-price elasticity to quantify the relationship between the products produced.

An $E_{s_{Y_1Y_2}} > 0$ implies that as the price of Y_2 increases, the quantity of Y_1 supplied by the firm will also increase (or, conversely, as the price of Y_2 decreases, the quantity of Y_1 produced will decrease). If this relationship holds, the products are complementary, or are produced in combination (recall Figure 6.16). For example, if when the price of wool (P_W) increases from \$20/cwt to \$45/cwt, the quantity of lambs (S_L) produced increases from 200 to 300, then the cross-price elasticity of lambs with respect to the price of wool ($E_{s_{LW}}$) would be (calculated by plugging the appropriate numbers into the formula)

$$E_{S_{LW}} = \frac{\left(\dfrac{300 - 200}{300 + 200}\right)}{\left(\dfrac{\$45 - 20}{\$45 + 20}\right)} = \frac{\left(\dfrac{100}{500}\right)}{\left(\dfrac{25}{65}\right)} = \frac{0.200}{0.385} = 0.52$$

The positive cross-price elasticity coefficient ($E_{s_{Y_1 Y_2}} > 0$) suggests that lambs and wool are **complements in production** for this firm, or they are produced at the same time. In this case, not only will wool production increase in response to an increase in the price of wool, but lamb production will also increase (supply curve for lambs shifts to the right).

An $E_{s_{Y_1 Y_2}} < 0$ implies that as the price of Y_1 increases, the quantity of Y_2 supplied by the firm decreases (or as the price of Y_2 decreases, the quantity of Y_1 supplied increases). If this relationship holds, the products are **substitutes in production** because they compete for the same resources (remember Figure 6.4). For example, if the quantity of acreage planted to barley (S_B) and, hence, barley production decreases from 1,500 to 600 acres in response to an increase in the price of wheat (P_W) from \$4 to \$8 per bushel, then the cross-price elasticity of barley with respect to wheat ($E_{s_{BW}}$) would be

$$E_{S_{BW}} = \frac{\left(\dfrac{600 - 1500}{600 + 1500}\right)}{\left(\dfrac{\$8 - 4}{\$8 + 4}\right)} = \frac{\left(\dfrac{-900}{2100}\right)}{\left(\dfrac{4}{12}\right)} = \frac{-0.429}{0.333} = -1.29$$

The negative cross-price elasticity coefficient ($E_{s_{Y_1 Y_2}} < 0$) indicates that barley and wheat are considered substitutes in production for this firm. Thus, as the price of wheat increases, the producer will remove productive resources out of barley and reallocate them into wheat production (which will increase supply of wheat and decrease supply of barley).

Number of sellers. The number of sellers (firms) in an industry affects the position of the supply curve. As firms enter an industry, supply shifts to the right; as firms exit an industry, supply shifts to the left.

Other factors. Another factor that affects the location of the supply curve, especially in agriculture, is weather (rainfall, drought, flood, etc.). Favorable weather conditions tend to shift the supply to the right; unfavorable weather shifts supply to the left. Actually, this is simply a shift in the production function, similar to a change in technology.

Institutional factors such as taxes, subsidies, and government policies also affect the position of the supply curve. Government programs such as acreage allotments, acreage restrictions (set-asides), and bans on pesticides, herbicides, or other chemicals can affect the position of the supply curve. If the government program results in resources removed from production or increases in the costs of production, then there will be a decrease in supply. Taxes and subsidies also impact the firm's costs of production. If a tax is imposed (raising costs of production), supply will shift inward to the left. If a subsidy is granted, decreasing costs of production, supply will shift outward to the right.

In summary, a change in quantity supplied is considered to have occurred only in response to a product price change, whereas a change in supply is considered to have occurred, if at the same price, more (or less) product is supplied. It is useful to a firm manager or public decision maker to be able to differentiate between these two changes so a better understanding of the market can be developed.

DEMAND

Obviously, the reason for the firm's production is because the item produced is in demand and, most importantly, because that demand results in a price being paid to the producer. In an economic system, price serves as the director of activity by channeling inputs into those uses desired by the consumers. At the firm level the manager can look at price and demand as a signal for his or her activity or can attempt to affect demand in those situations where some control exists. In this section we develop the concept of demand and examine it from the manager's viewpoint.

Derivation of Market Demand Curves

As we discussed in Chapter 7 the economic concept of demand is based on a price and quantity relationship. The amount that consumers are **willing** and **able** to purchase at each price level determines how much of a product is demanded at that price. Consumers desire a product because it gives them satisfaction, or utility. Since the marginal utility of each item purchased by a consumer varies, the consumer, who has a given budget to spend, will maximize total utility by establishing a purchase pattern whereby the marginal utility per dollar of expenditure for each item is equal. Thus, as we derived in Chapter 7, each point on an individual's demand curve represents a consumption level where utility is maximized.

To understand how much of a product will be demanded in a market at any point in time, it is useful to derive a market demand curve. A market demand curve is obtained by horizontally summing (adding) all the demand curves for the individual consumers in the market. This is illustrated in Figure 8.8 where we have assumed a simplified market of three individuals who have quite a different intensity of demand for peanuts. To find out what the market demand curve looks like, we add the potential consumption at each price level. At price $2 the three con-

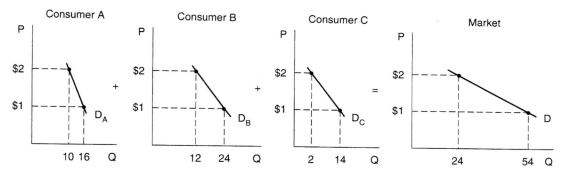

FIGURE 8.8 Market demand derivation.

TABLE 8.3 Demand Schedule for Peanuts in Pullman, Washington

Price per Pound	Quantity Demanded (lbs)
$2.00	24
$1.75	31
$1.50	39
$1.25	46
$1.00	54
$0.75	59

sumers would individually consume 10, 12, and 2 pounds of peanuts for a total market demand of 24 pounds in response to a price of $2. At the lower price of $1 the three individual consumers would, as they strive to maximize utility, demand 16, 24, and 14 pounds of peanuts, respectively, generating a total market demand of 54 at $1.

A real-world example would be markets for soybean products (oil, meal, protein, etc.), which add up to the total domestic demand for soybeans.

Law of Demand

The **law of demand** states that *the quantity of a product demanded will vary **inversely** to the price of that product.* As the price of a commodity increases, the quantity demanded of that product decreases; conversely, a decrease in price will be accompanied by an increase in quantity demanded. This relationship is shown in the demand schedule listed in Table 8.3 and the corresponding demand curve in Figure 8.9. The downward slope of the demand curve in Figure 8.9 reflects the inverse relationship between price and quantity embodied in the law of demand. This relationship occurs for three reasons.

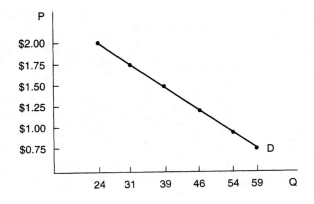

FIGURE 8.9
Demand curve for peanuts in Pullman, Washington.

Income Variation

A characteristic of most societies is a broad and uneven distribution of incomes. As a result, even if we all loved peanuts equally, those with lower incomes might not be able to purchase peanuts when the price rose to $2 per pound. As a result, less quantity would be demanded in the market at that higher price.

Diminishing Marginal Utility

As discussed earlier, the amount of additional satisfaction we receive from consuming more and more of one commodity decreases. As a result, because we might get a great deal of satisfaction from the first pound of peanuts and are willing to pay a high price for it, we will only buy another pound if the price decreases somewhat. Again, the lower the price, the more of a commodity a consumer with diminishing marginal utility will buy.

Personal Preference

Your authors love peanuts and receive great utility from them and, as a result, would be willing to pay $2 or more per pound. However, other consumers might not be "peanut freaks," and some consumers might even dislike the taste and would only use peanuts to feed the birds. Those consumers would only purchase peanuts if the price were very low; so once more, if price increases, these people would no longer be willing to buy, so the quantity demanded will decrease.

A more formal statement of the inverse relationship between price and quantity demanded is often presented in the form of the *income-substitution effect*. The income effect derives from the fixed budget of a consumer. When the price of a commodity decreases, more of that commodity can be purchased for a fixed amount; when the price increases, less of that commodity can be purchased with the fixed budget. Both of these contribute to the downward-sloping demand curve.

The substitution effect is based on the marginal utility per dollar of expenditure on all products in a consumer's purchase pattern. When the price of one good (peanuts) decreases, the marginal utility per dollar of expenditure increases relative to the other goods so the consumer shifts (substitutes) toward more purchases of the now cheaper good.

Elasticity of Demand

Determining how responsive the quantity demanded of a product is to changes in the product's price and other economic factors is a must for today's managers. Earlier we discussed the concept of elasticity and how it related to supply. In this section we discuss elasticity and how it relates to demand. In this case, **elasticity** is defined as *the percentage change in the quantity demanded relative to the percentage change in price as we move from one point to another on a demand curve.* It is expressed as

$$E = \frac{\% \text{ change in quantity}}{\% \text{ change in price}} = \frac{\%\Delta Q}{\%\Delta P}$$

where E is the elasticity coefficient.

The law of demand states the relationship between price and quantity, but elasticity of demand (E_D) can tell the manager the strength of that relationship. Elasticity of demand indicates how strongly or weakly buyers in a market will react to a price change or how responsive quantity demanded is to a price change. Own-price elasticity of demand (or price elasticity) is calculated as

$$E_D = \frac{\%\Delta Q_D}{\%\Delta P} = \frac{\left[\dfrac{Q_2 - Q_1}{Q_2 + Q_1}\right]}{\left[\dfrac{P_2 - P_1}{P_2 + P_1}\right]}$$

where P_1, Q_1 and P_2, Q_2 are price and quantity combinations on the demand curve. This is the familiar arc elasticity formula. Since demand is downward sloping, E_D is always negative. However for convenience and to avoid confusion in interpretation, the absolute value of the coefficient ($|E_D|$) is often used to describe the amount of elasticity.

Elasticity represents movement along the demand curve; thus, elasticity is also a measure of the degree of slope of the curve. Elasticity of demand is similar to the elasticity of supply and as such the responsiveness is classified as either inelastic, unitary elastic, or elastic.

An **inelastic** demand response is one in which a change in price brings about a relatively smaller change in quantity, or the $\%\Delta Q < \%\Delta P$ so $|E| < 1.00$ (where $|E|$ is the absolute value of the elasticity coefficient). Thus, an inelastic demand curve will have a steep slope. Note that $\%\Delta Q < \%\Delta P$ is equivalent to $\%\Delta P > \%\Delta Q$.

A **unitary elastic** demand response is one in which a change in price brings about an equivalent change in quantity, or the $\%\Delta Q = \%\Delta P$ so $|E| = 1.00$.

An **elastic** demand response is one in which a change in price brings about a relatively larger change in quantity, or the $\%\Delta Q > \%\Delta P$ so $|E| > 1.00$. Hence, an elastic demand curve will have a flat slope. Note that $\%\Delta Q > \%\Delta P$ is equivalent to $\%\Delta P < \%\Delta Q$.

Managers and economists are interested in two types of demand elasticity measures. The first is **own-price elasticity of demand,** which *measures the responsiveness of the quantity demanded of a good to changes in the price of that good.* That is, how does the quantity demanded of a good respond when its own price changes? The other is **cross-price elasticity of demand,** which *measures the responsiveness of the quantity demanded of a good to changes in the price of a related good.* That is, how does the quantity demanded of a good respond when the price of a related good changes? Throughout this section we will be identifying and illustrating how managers can use elasticity measurements in their everyday decisions.

Own-price elasticity of demand, which represents movement along the demand curve, is a measure of the degree of slope of the demand curve. To illustrate this, consider our peanut example where the demand curves for the individual consumers are presented in Table 8.4. From this table, we can see that when the price of peanuts increases from \$1 to \$2, the quantity demanded by each individual changes, but how responsive is each of the three consumers to changes in the price of peanuts? To calculate the own-price elasticity of demand for each of the consumers, we simply plug the appropriate numbers into the E_D formula.

When the price increases from \$1 (P_1) to \$2 (P_2) the quantity demanded by consumer A decreases from 16 (Q_1) to 10 (Q_2). Plugging these numbers into the formula for E_D we can calculate consumer A's own-price elasticity of demand (E_{D_A}) for peanuts:

$$E_{D_A} = \frac{\left(\dfrac{10 - 16}{10 + 16}\right)}{\left(\dfrac{\$2 - 1}{\$2 + 1}\right)} = \frac{\left(\dfrac{-6}{26}\right)}{\left(\dfrac{1}{3}\right)} = \frac{-0.231}{0.333} = -0.694$$

Consumer A is considered fairly unresponsive to the price of peanuts since her demand for peanuts is inelastic ($|E_D| < 1.00$). If the price of peanuts increases (de-

TABLE 8.4 Demand Schedules for Individuals Consuming Peanuts in Pullman, Washington

Price	Consumer A	Consumer B	Consumer C	Market
\$2.00	10	12	2	24
\$1.00	16	24	14	54
E_D	−0.694	−1.000	−2.252	−1.154

creases), consumer A will reduce (increase) her consumption of peanuts by a relatively smaller amount. In other words, if the price of peanuts increases by 1 percent we would expect, in this example, to see the quantity demanded by consumer A decrease by 0.69 percent.

In response to the price increase from $1 ($P_1$) to $2 ($P_2$) consumer B's quantity demanded decreases from 24 (Q_1) to 12 (Q_2). Using these numbers, consumer B's own-price elasticity of demand (E_{D_B}) is calculated as

$$E_{D_B} = \frac{\left(\dfrac{12 - 24}{12 + 24}\right)}{\left(\dfrac{\$2 - 1}{\$2 + 1}\right)} = \frac{\left(\dfrac{-12}{36}\right)}{\left(\dfrac{1}{3}\right)} = \frac{-0.333}{0.333} = -1.00$$

Consumer B's demand for peanuts is unitary elastic since $|E_D| = 1.00$. As the price of peanuts increases (decreases), consumer B's consumption of peanuts will decrease (increase) by an equivalent amount. For a 1 percent increase in the price of peanuts, we would expect consumer B to decrease quantity demanded by 1 percent.

The quantity demanded by consumer C decreases from 14 (Q_1) to 2 (Q_2) when the price increases from $1 ($P_1$) to $2 ($P_2$). Thus, the own-price elasticity of demand for consumer C (E_{D_C}) is calculated as

$$E_{D_C} = \frac{\left(\dfrac{2 - 14}{2 + 14}\right)}{\left(\dfrac{\$2 - 1}{\$2 + 1}\right)} = \frac{\left(\dfrac{-12}{16}\right)}{\left(\dfrac{1}{3}\right)} = \frac{-0.75}{0.333} = -2.252$$

Consumer C's demand for peanuts is elastic ($|E_D| > 1.00$), which implies that as the price of peanuts increases (decreases), his consumption of peanuts will decrease (increase) by a relatively larger amount. That is, for a 1 percent increase in the price of peanuts, we would expect the quantity demanded by consumer C to decrease by 2.25 percent. Thus, consumer C is considered very responsive to the price of peanuts.

Note that in all three examples given the E_D coefficient is negative. We simply used the absolute value to classify the responsiveness as either elastic, inelastic, or unitary elastic.

Because elasticity of demand is a measure of responsiveness of the quantity demanded it also reflects the slope of the demand curve. Therefore, the slope of the demand curve gives an indication of the own-price elasticity of demand. This was illustrated earlier in Figure 8.8, which graphically represents the individual consumer's demand curves. Notice that the demand curve for consumer A (inelastic response) has a steeper slope than the demand curves for consumers B and C, and that the demand curve for consumer C (elastic response) has a flatter slope than either of the demand curves for consumer B or A.

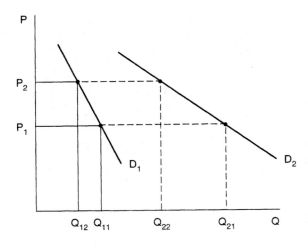

FIGURE 8.10
Demand curves with different elasticities.

Elasticity, which measures responsiveness, indicates just how much a consumer is **willing** and **able** to respond to a price change at a given time and place, and is a numerical description of the degree of slope of the demand curve. This relationship between elasticity and slope can be seen more clearly in Figure 8.10 where for a given price change a flatter (more elastic) curve, D_2, will generate a larger response than a higher (more inelastic) sloped curve, D_1. For the same increase in price, from P_1 to P_2, the resulting decrease in quantity demanded, from Q_{11} to Q_{12}, is significantly smaller for the more inelastic demand curve, D_1, than the decrease in quantity demanded, from Q_{21} to Q_{22}, for the more elastic demand curve, D_2.

Elasticity of Demand and Total Revenue

Elasticity of demand offers other information that is useful to a manager. We know that as price increases, less will be purchased (law of demand), but we can now see how much less and, more importantly, we can see what happens to the total revenue received by a firm. Since total revenue is price times quantity, the additional revenue gained by increasing the price is only realized on those units still sold under the higher price. Thus, the responsiveness of quantity demanded to changes in price will determine the impact on revenue. Understanding this relationship allows the manager who has some price-setting discretion to set prices that will achieve the total revenue minus total cost result desired. The difference in response can be shown by looking at demand curves with differing elasticities as in Figure 8.8 and Table 8.4.

An inelastic demand response occurs when the percentage change in quantity is less than the percentage change in price. Thus, an increase in price will result in a smaller decrease in quantity yielding higher revenues for the firm. Consider consumer A, who, as the price of peanuts increases from $1 to $2, decreases consumption from 16 to 10. Due to the price increase, the total revenue from sales to consumer A increases from $16 ($1 $\times$ 16 = $16) to $20 ($2 $\times$ 10 = $20).

When demand is inelastic, a price increase will increase total revenue for the firm, while in this case, the firm can produce less product and hence, incur fewer costs.

An elastic demand response occurs when the percentage change in quantity is greater than the percentage change in price. In this case, a price increase means that the firm will give up a greater number of sales than that added to total revenue, so total revenue decreases. Consider consumer C, who as the price of peanuts increases from $1 to $2, decreases consumption from $14 to $2. Due to the price increase the total revenue from sales to consumer C decreases from $14 ($1 × 14 = $14) to $4 ($2 × 2 = $4).

On the other hand, a price increase or decrease would have no impact on revenue when demand is unitary elastic since the percentage change in price is met with an equivalent change in quantity. Thus, the sales given up will exactly equal the addition to total revenue. Consider consumer B, who, as the price of peanuts increases from $1 to $2, decreases consumption from 24 to 12 lbs. The total revenue from sales to consumer B remains at $24 at both price levels ($1 × 24 = $24; $2 × 12 = $24).

The relationship between a price change and total revenue can be summarized as follows:

1. Inelastic demand: As $P\uparrow$ TR$\uparrow$; or as $P\downarrow$ TR$\downarrow$.
2. Unitary elastic demand: As $P\uparrow$ TR$\rightarrow$ (no change); or as $P\downarrow$ TR$\leftarrow$ (no change).
3. Elastic demand: As $P\uparrow$ TR$\downarrow$; or as $P\downarrow$ TR$\uparrow$.

The elasticity of a demand curve is important to the individual manager because, if inelastic, the price can be raised and TR increased, even though less of the product is sold. This is doubly important because under this situation, TR is increasing while total cost is decreasing, two positive effects on profit.

Understanding Elasticity of Demand

Elasticity of demand is a measure of responsiveness and as we illustrated earlier differs across consumers and goods. What determines if the demand for a good is inelastic or elastic?

The demand for salt, or for insulin by diabetics, are two reasonable examples of inelastic demand. A small amount of salt is needed in our diet, and no matter what the price is, individuals will consume roughly the same amount. People whose medical problems have required them to use insulin will also be nonresponsive, or inelastic, to the price of this medical aid. An example of an almost perfectly elastic demand is that of the individual demand curve faced by the agricultural producer. If the producer tries to raise his price above the market price, quantity demanded responds so strongly that nothing may be sold. (As we will see in Chapter 10 this is the direct result of producing a homogeneous product in a competitive market.) In such a situation an agricultural producer finds he or she has no control over price and so accepts the market price. But the producer then finds it useful to produce as much as possible, subject to costs, because total revenue directly increases with volume of production sold.

Products that are considered necessities are often inelastic in demand. Grains, milk, and potatoes have lower elasticities than meats and vegetables. Also, products that have good substitutes in the minds of the consumer have more elastic demand curves because, faced with a price increase, consumers can shift to substitute goods. A good that represents a large percentage of a consumer's budget will also be very responsive to price changes, particularly if that item has characteristics of being a luxury and not a necessity.

Research economists have spent considerable time estimating demand elasticities for food and other products. Using sophisticated mathematical techniques, the price elasticities for food products have been estimated several times during the last few years. Table 8.5 summarizes some of the most recent and comprehensive estimates of actual food groups. Of course, all of these price elasticities of demand are negative, reflecting the basic nature of the demand curve. In addition, since all of these actual elasticities are less than 1.0, they reflect the **inelastic demand** for most food items. For beef and veal products, a 1% increase in price will result in a 0.62% reduction in quantity demanded. Notice that for staple foods (e.g., milk and flour) there is almost no change in quantity demanded to small increases (1%) in price. In other words, the demand is very inelastic because American families and food businesses buy these products almost every day or at least every week as an important component of the everyday diet.

All of these factors that affect elasticity are part of the *ceteris paribus* conditions mentioned earlier. If a manager can affect the elasticity of the demand for her or his product, the manager's merchandising position is improved. As we will see, many of the determinants of demand (demand shifters) offer the opportunity to affect both elasticity (slope of the demand curve) and the level of demand (distance from the origin). When consumers are convinced that our product is better than our competitors' product or when consumer income increases, the demand for our product is positively affected.

TABLE 8.5 Demand Elasticities for Selected Foods

Food Items	For a 1% Change in Price, Quantity Demanded Will Change by:
Beef and veal	−0.62%
Pork	−0.73%
Chicken	−0.37%
Turkey	−0.53%
Milk	−0.04%
Flour	−0.08%
Fresh fruits	−0.19%
Fresh vegetables	−0.13%
Coffee and tea	−0.18%

Source: Kuo S. Huang, USDA Economic Research Service Technical Bulletin 1821, 1993.

Elasticities often become part of the public policy debates. In 1997 Republican Senator Orrin Hatch proposed raising the excise tax on cigarettes by $0.43 per pack in order to finance health insurance for poor children not covered by Medicaid (*New York Times News Service*, April 12, 1997). Senator Hatch cited research conducted at the University of Illinois, which estimated that for every 10 percent increase in cigarette prices, teenage smoking would decline by 7 percent. This reflected the general nature of the inelastic demand for tobacco products. But Senator Hatch went on to point out that based on this research, a $0.43 tax increase would lead to a 15.7 percent decline in teenage smoking. Thus, price elasticity of demand was a key element in the public policy debate over smoking.

Elasticity and Linear Demand Curves

We have been using linear (straight line) demand curves. It is important to note that the elasticity of demand changes as we move along a linear demand curve. This is illustrated in Figure 8.11 where the demand curve is elastic from point A to B, unitary elastic from point B to C, and inelastic from point C to D. Thus, along the linear demand curve the elasticity of demand is changing. Why? Because $E_D = (\Delta Q / \Delta P) \times (P/Q)$, and for a linear line the slope $(\Delta Q / \Delta P)$ is constant, but the ratio (P/Q) is changing as we move along the demand curve because quantity is changing. Hence, the elasticity changes as we move along a linear demand curve.

For your own understanding, use the formula for arc elasticity to prove that elasticity does change along the demand curve in Figure 8.11.

Change in Demand Versus Change in Quantity Demanded

As in our discussion of supply, it is useful to differentiate between a change in quantity demanded versus a change in demand. Figure 8.12 shows the difference between the two changes. A change in the quantity demanded results from changes in the price of the product being examined and are movements along a demand curve.

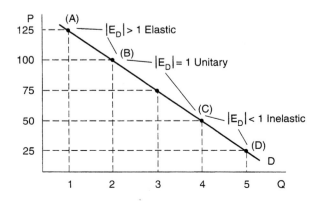

FIGURE 8.11
Elasticity and linear demand curves.

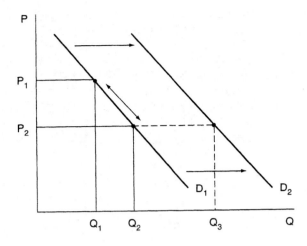

FIGURE 8.12
Change in demand versus change in quantity demanded.

In Figure 8.12, if the price of the product decreases from P_1 to P_2 the quantity demanded increases from Q_1 to Q_2. This is movement along the demand curve (D_1).

A change in demand results from changes in the quantity purchased due to factors other than a change in the product price. That is, if something other than price changes (i.e., one of the *ceteris paribus* conditions) the position of the curve changes. In Figure 8.12, if there is an increase in demand the demand curve shifts to the right (from D_1 to D_2) and at the same price level (P_1) the quantity demanded increases (from Q_1 to Q_3). A shift to the right is an increase in demand, whereas a shift to the left is a decrease in demand.

Determinants of Demand

So, what determines the location and slope of the market demand curve? A market demand curve can be expressed in the following function:

$$Q_D = f \text{(own price} \mid \text{price and availability of substitute goods, income, population, tastes and preferences)}$$

This equation states that for each demand curve, quantity demanded (Q_D) will vary with the price as long as the four demand shifters or demand determinants are held constant or fixed. Since as managers we would normally like to see an inelastic and increasing demand for our product, let's examine these demand shifters (called this because they "shift" the demand curve left or right or change slope) from the manager's perspective. The factors that will shift the demand curve are those factors assumed to be held constant (the *ceteris paribus* conditions) when deriving and drawing the demand curve. As we discuss them, you should be able to identify other ways in which you, as a manager, can use these demand determinants to affect the demand curve for your products.

Price and availability of substitutes.   Changes in the prices or availability of other goods will cause the demand curve for a product to shift; the direction of the shift depends on whether the other good is a substitute or a complement. A tool used to quantify the relationship between goods is the **cross-price elasticity of demand,** which is calculated as

$$E_{D_{AB}} = \frac{\%\Delta Q_A}{\%\Delta P_B} = \frac{\left[\dfrac{Q_{2A} - Q_{1A}}{Q_{2A} + Q_{1A}}\right]}{\left[\dfrac{P_{2B} - P_{1B}}{P_{2B} + P_{1B}}\right]}$$

where Q_{1A}, Q_{2A} are quantities of good A; P_{1B}, P_{2B} are prices of good B; and $E_{D_{AB}}$ is read as the cross-price elasticity of demand for commodity A with respect to commodity B. The cross-price elasticity of demand shows the percentage change in the quantity demanded of a good (good A) in response to a given percentage change in the price of another good (good B), *ceteris paribus*. We can use cross-price elasticity measurements to quantify the relationship between commodities.

If $E_{D_{AB}} > 0$ then A and B are **substitutes in consumption,** which implies that as the price of good B (P_B) increases, the quantity demanded of good A (Q_A) by the consumer also increases (or as P_B decreases, Q_A decreases). The larger the elasticity coefficient the greater the degree of substitutability. For example, if when the price of apples (P_A) increases from \$1.50 to \$2/lb the quantity demanded of bananas (D_B) increases from 3 to 5 lbs, then the cross-price elasticity of bananas with respect to apples ($E_{D_{BA}}$) would be

$$E_{D_{BA}} = \frac{\left(\dfrac{5 - 3}{5 + 3}\right)}{\left(\dfrac{\$2 - 1.50}{\$2 + 1.50}\right)} = \frac{\left(\dfrac{2}{8}\right)}{\left(\dfrac{0.50}{3.50}\right)} = \frac{0.25}{0.143} = 1.748$$

Since the cross-price elasticity coefficient is positive ($E_{D_{AB}} > 0$), bananas and apples are considered substitutes. As the price of apples increases, consumers will switch consumption away from apples to the relatively less expensive good, bananas, which implies that an increase in the price of apples will shift the demand curve for bananas to the right, an increase in demand for bananas. (Remember MRS = IPR from Chapter 7?)

Generally, the more close substitutes for a good that exist, the more elastic is the demand curve for that good and the less control a manager has over price. When the coefficient indicates strong elasticity (a large number), this suggests that the two goods are close substitutes and as a manager we should consider and use the price of the other good in our pricing and production decisions. For example, beef and pork are good substitutes for each other but pork and light bulbs, except in a circus act, really are not close substitutes for each other.

If $E_{D_{AB}} < 0$ then A and B are considered **complements,** or the goods are consumed together, since it implies that as the price of good B (P_B) increases the quantity demanded of good A (Q_A) will decrease (or as P_B decreases, Q_A increases). The larger the negative coefficient, the greater the degree of complementarity. For example, if when the price of bread (P_B) increases from \$1 to \$1.50/loaf the quantity demanded of margarine (D_M) decreases from 3 to 1 sticks, then the cross-price elasticity of margarine with respect to bread ($E_{D_{MB}}$) is

$$E_{D_{MB}} = \frac{\left(\dfrac{1-3}{1+3}\right)}{\left(\dfrac{\$1.50 - 1}{\$1.50 + 1}\right)} = \frac{\left(\dfrac{-2}{4}\right)}{\left(\dfrac{0.50}{2.50}\right)} = \frac{-0.50}{0.20} = -2.50$$

The negative cross-price elasticity coefficient ($E_{D_{AB}} < 0$) indicates that margarine and bread are considered complements in consumption. The increase in the price of bread may decrease purchases of bread, which could also decrease sales of margarine (a decrease in demand for margarine). Again, the manager producing one product must be broad enough in approach to include more products than just his or her own in the decision process. For example, a firm producing nondairy coffee creamer should pay attention to coffee prices when making production decisions.

If $E_{D_{AB}} = 0$ the two commodities (A and B) are considered to be independent; that is, a change in the price of good B has no effect on the quantity demanded of good A. For example, when the price of sugar (P_S) increases from \$3 to \$4/lb the quantity demanded of Ford trucks (D_F) remains unchanged at 2. The cross-price elasticity of demand for Ford trucks with respect to sugar ($E_{D_{FS}}$) is

$$E_{D_{FS}} = \frac{\left(\dfrac{2-2}{2+2}\right)}{\left(\dfrac{\$4 - 3}{\$4 + 3}\right)} = \frac{\left(\dfrac{0}{4}\right)}{\left(\dfrac{1.00}{7.00}\right)} = \frac{0}{0.142} = 0$$

Since the cross-price elasticity coefficient is zero ($E_{D_{AB}} = 0$) Ford trucks and sugar are considered independent commodities. That is, a change in the price of sugar does not affect the consumption and, hence, the demand for Ford trucks.

Note that while own-price elasticity of demand is a measure arising from movement along the demand curve, cross-price elasticity arises from a change in the price of another commodity. Thus, cross-price elasticity reflects a shift in the demand curve, and cross-price elasticity coefficients allow us to quantify the relationship between different commodities. Since the sign of the cross-price elasticity coefficient tells us the relationship between the two goods, we do not use the absolute value of the coefficient as we did for own-price elasticity of demand.

Income. A demand curve represents the amount at each price that consumers are **willing** and **able** to purchase. Thus, the amount of income available to consumers has a direct effect on their effective demand for food, education, automobiles, and so forth. If the consumer's income increases (decreases) the position of the demand curve will also change, the direction of the shift depends on if the good is a normal (demand increases with income) or an inferior good (demand decreases with increases in income). Managers and economists are interested in how responsive consumer's purchases are to changes in income levels. A measurement of this is the **income elasticity of demand,** which is calculated as

$$E_{D_I} = \frac{\%\Delta Q_D}{\%\Delta I} = \frac{\left[\dfrac{Q_2 - Q_1}{Q_2 + Q_1}\right]}{\left[\dfrac{I_2 - I_1}{I_2 + I_1}\right]}$$

where I is income level. Income elasticity of demand measures the percentage change in the quantity demanded of a good resulting from a percentage change in income, *ceteris paribus.* So we can use the income elasticity coefficient to quantify the relationship between the demand for a commodity and changes in the consumer's income.

An $E_{D_I} > 0$ implies that as income increases the quantity demanded also increases (or as income decreases, quantity demanded decreases). If this relationship holds the good is considered to be a **normal** or **superior good.** For example, if income increases from \$100 to \$150 per week and the quantity demanded of lobster (D_L) increases from 2 pounds to 6 pounds per week, then the income elasticity of demand for lobsters $(E_{D_{IL}})$ is

$$E_{D_{IL}} = \frac{\left(\dfrac{6 - 2}{6 + 2}\right)}{\left(\dfrac{\$150 - 100}{\$150 + 100}\right)} = \frac{\left(\dfrac{4}{8}\right)}{\left(\dfrac{50}{250}\right)} = \frac{0.50}{0.20} = 2.50$$

Since the income elasticity coefficient is positive $(E_{D_I} > 0)$ lobster is considered a normal or superior good. Thus, the demand curve for lobster will shift outward to the right (increase in demand) as income increases. Positive income elasticity coefficients vary widely among goods and services, especially food commodities, because there is such a difference in desirability.

An $E_{D_I} < 0$ implies that as income increases the quantity demanded decreases (or as income decreases, quantity demanded increases). If this relationship holds the good is considered to be an **inferior good.** For example, if income increases from \$100 to \$150 per week, the quantity demanded of canned meat (D_M) decreases from 4 cans to 2 cans per week, reflecting the undesirability by the consumer. The income elasticity of demand for canned meat $(E_{D_{IM}})$ is

$$E_{D_{IM}} = \frac{\left(\dfrac{2 - 4}{2 + 4}\right)}{\left(\dfrac{\$150 - 100}{\$150 + 100}\right)} = \frac{\left(\dfrac{-2}{6}\right)}{\left(\dfrac{50}{250}\right)} = \frac{-0.33}{0.20} = -1.65$$

The income elasticity coefficient is negative ($E_{D_I} < 0$); therefore, canned meat is considered an inferior good. This implies that as income increases the demand for canned meat will decrease (the demand curve will shift inward to the left). Note that it may be very difficult to find a true inferior good since the level of income has a great deal to do with which goods are considered to be inferior.

In summary, when consumption of a product increases as income rises, the good is considered to be a superior good, such as T-bone steak, a second calculator, movie tickets, or a sports car. When an income increase is associated with a decrease in consumption of a good (a negative income elasticity coefficient) the good is considered to be an inferior good, by that consumer, as might be the case for black-and-white televisions.

A manager should become aware of the income elasticity of demand for the product produced in his or her firm and, just as importantly, must broaden his or her investigation to include the characteristics of income in the firm's potential market. A trend toward stable or decreasing disposable income should be a warning signal to the manager producing a luxury or superior product such as snowmobiles. The demand for food products is fairly income inelastic, but the demand for services (packaging, premixing, and so forth) is income elastic; so, in periods of rising incomes, a manager may decide to add more services to the basic product.

Population.   The number of buyers in the market has an obvious effect on the demand for products. Generally, if population increases, the demand curve will shift outward to the right (increase in demand). Conversely, if population decreases the demand curve will shift inward to the left (decrease in demand). Although the manager can seldom directly affect this demand shifter, he or she must monitor the change in this variable. We do know it would not be prudent to start a new newspaper that is heavily dependent on readership and advertising in the rural areas of states such as North Dakota that continue to register a population decrease. Finally, the use of satellite stores, chain stores, or the installation of gasoline service stations at crossroads of major highways has been successfully used to increase the population (number of potential buyers) exposed to a given product line.

Tastes and preferences.   The final major determinant of demand is consumer tastes and preferences. Generally, if the manager can influence consumers' tastes and preferences in a positive manner, demand for the product will increase (demand shifts outward to the right). If consumers find our product acceptable to their tastes and superior to our competitors' products or other alternative products, then they will actively enter into the market for our product (remember the indifference curves from Chapter 7?). However, occasionally a negative impact on tastes and preferences will occur. An example of negative impacts include the perceived link between cholesterol (in red meat such as beef) and heart disease, and the link between cigarettes and cancer. In both of these instances, the impact of the news on consumers' tastes and preferences has been negative, resulting in a decrease in per-capita demand for both beef and cigarettes (demand curves shifted inward to the left).

Customs and habits are strong indicators of tastes and preferences (turkey at Thanksgiving, ham at Easter, or flowers on Mother's Day). Changes occurring as a society evolves also affect demand for certain products. Lean bacon is now preferred over fat bacon; fresh foods are increasingly demanded over heavily processed ones; outdoor recreation and physical fitness have greatly affected the demand for athletic supplies. Clothing trends toward jeans and "relaxed styles" have exploded the demand for these product types.

BOX 8.1   DeBeers: The Company That Tells College Students How Much to Spend on Their Product

One of the most interesting examples of how corporations affect consumer tastes and preferences in order to shift the demand for their product is that of diamonds as jewelry. Diamonds are neither scarce nor rare in their natural state, and artificial diamonds are readily available that compare favorably with gem diamonds in color, clarity, and cut. However, it is through the widespread and very influential media campaigns of DeBeers Consolidated Mines, Ltd., that diamonds remain a highly superior good with a carefully protected image.

DeBeers really created the modern market for diamonds after the Depression Era of the 1930s. They gained worldwide control over the supply of gem diamonds through a marketing cartel, the Central Selling Organization. Then they undertook a massive advertising campaign, including providing jewels to actresses in movies and other public events. DeBeers created in the minds of Americans the concept that "diamonds = love & commitment." This is reflected in one of the most successful and enduring advertising slogans: "A Diamond Is Forever."

DeBeers still advertises heavily, including in *Sports Illustrated* and other magazines popular with college students. In these ads, DeBeers gives young brides and grooms advice on picking a diamond engagement ring and instructs them to spend "two months' salary." And just in case the readers don't appreciate their point, the ad continues: "Spend less and the relatives will talk." So despite the availability of substitutes at a fraction of the cost of a mined diamond, DeBeers maintains a consumer preference for the "real thing."

Can you think of any other companies that create family and peer group pressure to spend a certain amount on their product?

It is in the area of tastes and preferences that most of the managers' activities and budgets in our economy are directed. The use of advertising (see Box 8.1 on page 231) as a means of achieving product differentiation continues to increase. Product differentiation is the creation in the minds of the consumer of a positive difference between our product and that of our competitors. It does not matter, economically, if the difference is real as long as we are successful in getting the consumer to perceive a difference. When we have successfully differentiated our product, we have decreased the elasticity of our demand by decreasing the number of acceptable substitutes. Examples of products whose brand names have almost become synonymous with the general product are Kleenex (tissue), Crescent wrenches (adjustable wrenches), Jello (gelatin), Vice Grip (lockjaw wrench), and Scotch tape (adhesive tape). You can easily think of other products that are equally as successful.

A manager or firm should be able to identify tastes and preferences as well as attempt to change them. When Ford Motor Company introduced the Edsel as its new car, it severely misidentified the tastes and preferences of the American consumer. Even with extensive and massive advertising expenditures by Ford Motor Company it was not able to convince the consumer, who had slowly but surely been shifting to smaller, less expensive automobiles, to accept the bulky and relatively expensive Edsel.

SUMMARY

In this chapter we have developed the concepts of supply and demand and indicated how knowledge of each concept is useful to managerial decision making. These concepts are totally separate from each other, but both are price and quantity relationships. Individual supply curves are derived from the cost curve for each firm, and the industry supply curve reflects the composite cost structure of all individual firms. The supply curve is affected by managerial decisions in response to changes in technology, price of inputs, alternative products, and so forth. Elasticity of supply is a concept that allows us to evaluate how quickly producers can and will respond to product price changes. The relationships of supply and supply response are based directly on the internal decision-making process of the firm.

Demand, on the other hand, is derived external to the firm and reflects consumers' willingness and ability to purchase products at given prices. Market demand curves are the summation of individual demand curves, which are based on marginal utility, income variation, and personal preferences. The own-price elasticity of demand can tell a manager what will happen to the total revenue of his firm in response to a price change. This response is a change in quantity demanded and is a movement along an existing demand curve. A change in demand is a shift of the entire demand curve and is brought about by a change in the determinants of demand: tastes and preferences, population, consumer disposable income, and price and availability of substitute goods. Consumer expectations about future prices, product availability, and income can also affect demand.

These demand shifters, particularly tastes and preferences, receive a lot of attention from managers attempting to learn more about the market for their product. Cross-price elasticity and income elasticity of demand are used to identify competitors and/or growth areas for particular products. Advertising is used to achieve product differentiation (or to retaliate against competitors' attempts) in order to shift the level or elasticity of demand.

Previous chapters have taught us the microtheory tools of production economics. These tools increase the efficiency of production and are particularly important when the firm is faced with a highly elastic demand curve, as in agriculture. In the next chapter we will see how the two separate concepts of supply and demand are brought together in the market and how the appropriate market price and quantity are determined. The special role of market price in providing signals to managers as they allocate resources within their firm will receive detailed attention.

QUESTIONS

1. Specify what is meant by the *economic concept of supply.*
2. How would the following affect a pea producer's willingness and ability to supply peas? Discuss and show graphically.
   (a) Availability of new high-yield seed
   (b) A 20% increase in OPEC oil prices
   (c) An invasion of harmful insects
3. Utilizing graphical analysis, show clearly that you understand the distinction between a change in supply and a change in quantity supplied. Be specific; assume your audience understands very little about the concepts involved.
4. Do farm managers have an elastic or inelastic supply function? Why?
5. Define what is meant by the *law of demand.* What does this have to do with utility?
6. Distinguish between elastic, inelastic, and unitary demand functions.
7. Suppose you are a producer facing an inelastic demand function. How would your total revenue be affected if you raised your market price? What additional information would you need to know in order to determine what happened to profit?
8. Explain what we mean by the *cross-price elasticity of demand.* What is the significance of the sign of the coefficient on cross-price elasticity?
9. Consider the four major demand shifters presented in this chapter. Which do you suppose would be most important in determining the quantity demanded of:
   (a) Cadillacs
   (b) New York steak
   (c) Milk
   (d) Tobacco

10. Explain why supply and demand are two separate and distinct concepts.
11. How do we derive the supply curve for an individual producer in the short-run? Show your answer graphically.
12. What would cause a change in supply for wheat versus a change in supply for bread?
13. Why might elasticity of supply be different for farm producers in response to a product price rise as compared to a product price decrease?
14. Show how you would develop a demand curve for calculators by the students in your classroom.

CHAPTER 9
Market Price Determination

INTRODUCTION

In the previous chapter we examined, in some detail, the concepts of supply and demand. These two concepts, which are separate in their origin and derivation, are the two forces that, when brought together in the marketplace, establish the amount of a commodity that will be exchanged and the price at which that commodity will be traded. For the manager to be able to answer the questions of how, what, and how much to produce, information on current market price is critical.

The economic system facing decision makers is not static and market prices are continually changing. Therefore, an understanding of how supply and demand interact and the resultant market price determination enables the manager to forecast and understand prices and make appropriate managerial decisions. Too often we have seen wheat producers who have produced their crop very efficiently and contracted it at $3 per bushel for future delivery, only to miss out on a $6.50 cash price. In a case such as this, improper understanding and use of the market concepts of supply and demand can easily negate the excellent managerial decisions made in production. As an example, think of all the forestry, history, and English majors in college in the 1990s who were unable to find jobs in their fields after graduation and thus received a hard lesson in market supply and demand.

Furthermore, if the manager is operating in an industry where he or she is a price taker (faces an extremely elastic demand curve), it is useful to be able to predict price movements and react accordingly. In the market situation where the manager is searching for or setting prices to maximize profits (imperfect competition, as we'll learn later), the knowledge of how to affect or shift demand and change market price is an attribute of good managerial economics. In truth, knowledge of the market price determination process can be useful to managers of firms and households, managers of the political economy, and students of life.

In This Chapter

In this chapter, we bring together the concepts of supply and demand to examine how the buying and selling decisions interact to determine both the prices and quantities of goods traded in a market. We then use our knowledge of supply and demand curves to predict the effect on market prices and quantities of a change in either supply or demand, or both supply and demand.

Assumptions Used

In every market, buyers and sellers must agree on the terms of exchange, most importantly the price. The principles underlying market price determination can be examined at one point in time (static analysis) or over time (comparative static analysis). As we examine the determination of market price, keep in mind the assumptions we are using. We assume that inputs, commodities, and price information move freely from market to market. That is, there is nothing in the economic system to prevent adjustments to changes. Firms are assumed to be rational and operate as profit maximizers; consumers are assumed to be rational and act as utility maximizers. Finally, there is competition between firms, thus firms are price takers in the market, as are consumers. In essence, we are using the assumptions that we have used at each stage in building our base of economic knowledge. In later chapters, we will relax these assumptions and make our analysis more realistic and complex.

MARKET PRICE DETERMINATION

A **market** is defined as *an institution or arrangement that brings buyers and sellers together.* The market is made up of numerous buyers, or consumers, whose demand for commodities is represented by the market demand curve, which we derived in Chapter 8. The market demand curve represents the aggregate response of all individual consumers, who make choices among available goods so as to maximize utility given a limited budget, within a given market. In other words, market demand is a schedule showing the quantities of a good that consumers are **willing** and **able** to purchase at a series of prices during a specified period in a given market. The market is also made up of numerous individual producers, or firms, whose production of commodities is represented by the market supply curve (also derived in Chapter 8). The market supply curve represents the aggregate response of all the individual producers, who determine how much to produce (and hence supply) by equating their MR to their MC, within a given market. That is, market supply represents the various amounts of a good that producers are **willing** and **able** to produce and make available at each of a series of prices during a specified period in a given market.

The supply and demand curves for a good or service are constructed under the assumption that the good's own price is the only economic variable that changes and that price determines the quantity supplied and demanded of a good. Thus, the curves are drawn under the assumption that many important determinants other than price, such as technology, income, tastes and preferences, and prices of other goods (substitutes and complements), do not change. These conditions or determinants, as we discussed earlier, are referred to as the *ceteris paribus* conditions.

The market system of allocating goods and services brings together the supply and demand curves for a product through the interactions of buyers and sellers. Producers will provide various quantities at different prices and consumers will purchase various quantities at different prices. However, there is one, and only one, price that equates the quantity of a good offered for sale with the quantity buyers are willing and able to buy. This mutually agreeable price at which willing buyers and sellers exchange a good or product is referred to as the **market price.**

Market Equilibrium

The interaction between supply and demand determines the market price. This interaction is depicted graphically in Figure 9.1 where the actions of the consumer are represented by the demand curve (*D*) and the actions of the producer are represented by the supply curve (*S*). *The price that clears the market,* that is, where the quantity supplied exactly equals the quantity demanded, is referred to as the **equilibrium price,** P_e. The quantity traded at that price is referred to as the **equilibrium quantity,** Q_e. Thus, **market equilibrium** occurs when *the quantity of a good offered by sellers at a given price equals the quantity buyers are willing and able to purchase at that same price* and no conditions exist to move the market away from this point.

Consider the market supply and demand curves for ham sandwiches in Kearney, Nebraska, shown in Figure 9.2. The supply curve depicts the combination of marginal cost curves above average variable costs for all ham-sandwich firms in the area and identifies how many ham sandwiches will be supplied to the market at each price. It says that at $4, 225 ham sandwiches will be produced because at

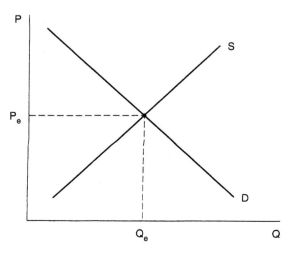

FIGURE 9.1
Market equilibrium.

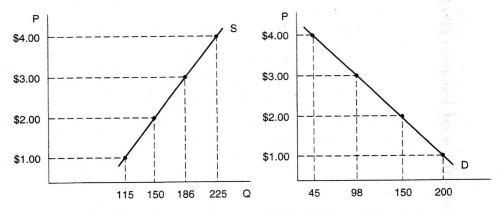

FIGURE 9.2 Market supply and demand for ham sandwiches in Kearney, Nebraska.

that high price, firms will shift from producing other products to producing ham sandwiches and inefficient (high-cost) firms can stay in production. At a market price of $1 only 115 ham sandwiches will be supplied because only firms with low marginal and average variable costs can afford to produce at this price level and other firms will have shifted to production of other commodities (the product–product decision). Each point on the supply curve states that if this price appears, then this quantity of sandwiches will be supplied.

The demand for ham sandwiches in Kearney is also presented in Figure 9.2. At the high price of $4 per sandwich only 45 sandwiches are demanded, reflecting the personal preference of some buyers, the high income of some buyers, and a high marginal utility for that first ham sandwich at any price. If the price drops to $1, a great deal of excitement would occur in the market because our hungry ham-sandwich lovers would be willing and able to purchase 200 sandwiches. Remember, each point on the demand curve represents the utility and satisfaction derived from sandwiches by stating that if the price is at this level, then this quantity of sandwiches will be demanded.

It is truly a marvel how simply market price is determined. Bringing the market supply and demand curves from Figure 9.2 together in Figure 9.3 illustrates how the interaction between supply and demand determines the market price for sandwiches. This is a static analysis in that we are conducting our investigation at one point in time.

When, and only when, supply and demand interact in the market can a price and quantity be identified that will "clear the market" (a situation without surpluses or shortages). As illustrated in Figure 9.3, the equilibrium price and quantity for ham sandwiches in Kearney is $2 and 150 sandwiches, respectively. At this price ($2), and only at this price, the quantity demanded by all consumers in the market (150) exactly equals the amount that all producers are willing to supply (150).

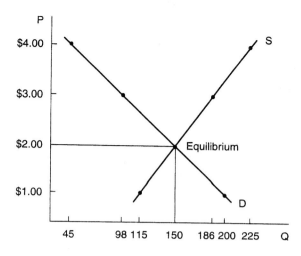

FIGURE 9.3
Market equilibrium for ham sandwiches in Kearney, Nebraska.

Markets in Disequilibrium

As we illustrated earlier, within a market there is only one price, referred to as the *equilibrium price*, at which the quantity supplied by producers exactly equals the quantity demanded. However, if the price observed in the market is not the equilibrium price, shortages or surpluses result and the market is considered to be in **disequilibrium.**

In an unregulated market, if the market price is either too high or too low, referred to as the *disequilibrium price*, economic forces come into play to move price back toward the equilibrium price. The adjustment process of markets moving toward equilibrium provides the signals to producers and consumers to produce and consume more or less. In the process, resources are allocated to the production of alternative competing goods and services. This is graphically depicted in Figure 9.4. In Figure 9.4(a), the current price, P_1, is higher than the equilibrium price (P_e), resulting in a **surplus** in the market. At this price (P_1) producers are willing to supply quantity Q_s but consumers are only willing to consume quantity Q_d. At the current price level, P_1, the quantity supplied (Q_s) in the market exceeds the quantity demanded in the market (Q_d) resulting in a **surplus** of the commodity, equal to $Q_s - Q_d$. Since producers are willing to supply more product than is being consumed, they will begin to lower prices in an effort to persuade consumers to purchase additional quantities. A surplus ($Q_s > Q_d$) occurs in a market when the current price ($P_1 > P_e$) is too high for consumers to purchase all that is supplied. As a result, economic forces will exert downward pressure on prices until the market reestablishes equilibrium.

In Figure 9.4(b), the current price in the market, P_2, is lower than the equilibrium price (P_e), resulting in a **shortage** in the market. At this price (P_2) consumers are willing and able to consume quantity Q_d but producers are only willing to supply quantity Q_s. At the current price level, P_2, a **shortage**, equal to $Q_d - Q_s$, exists

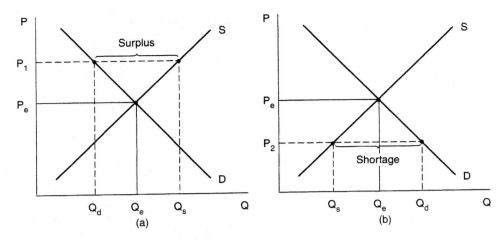

FIGURE 9.4 Market in disequilibrium.

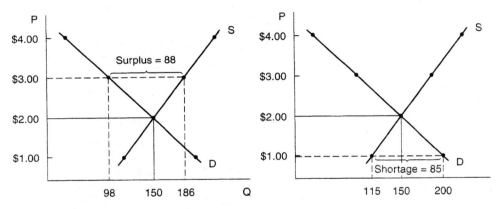

FIGURE 9.5 Surplus and shortage in sandwich market.

in the market since the quantity supplied (Q_s) is less than the quantity demanded (Q_d) by consumers. Since consumers desire more product than producers are willing to supply, consumers will bid up the product's price as they try to obtain the product. A shortage ($Q_d > Q_s$) occurs in a market when the current price ($P_2 < P_e$) is too low for producers to supply all that consumers are willing to purchase. As a result economic forces (consumers expressing their desires) will exert upward pressure on prices until the market reestablishes equilibrium.

Equilibrium price and quantity are reached through a bidding process that we can illustrate using the sandwich market in Kearney. For example, if the price of ham sandwiches started at $3 [Figure 9.5(a)], the market would be in disequilibrium. At $3, buyers would demand only 98 sandwiches, but suppliers would be willing and able to supply 186 sandwiches, causing a surplus of 88 sandwiches. As producers attempt to get rid of this surplus they would do so by lowering their price until the surplus no longer existed, at a price of $2.

If the price of ham sandwiches were $1, the opposite would occur as shown in Figure 9.5(b). At this price consumers would like to have 200 ham sandwiches, but producers would only be willing and able to supply 115; hence, there would be a shortage of 85 sandwiches. Consumers would then bid the price up in an attempt to secure ham sandwiches, until the equilibrium price and quantity, $2 and 150, are reached and the bidding process stops.

The bidding process in real life is a dynamic and never-ending movement of the market toward an elusive price and quantity that will produce equilibrium. Although in this example we ignored the time span of reaction by producers and consumers, obviously the time it takes producers and consumers to react will affect price movements over time. The elasticities of supply and demand at any point in time will also dictate how quickly producers and consumers can and will respond in this search for equilibrium. As we will see, equilibrium price and quantity are based on so many dynamic and interactive forces that seldom does the exact price and quantity exist, but movement toward that point continually exists in our economy.

Mathematical Representations

Market supply and demand curves are relationships that can be expressed in mathematical terms. As such, we can numerically express the equilibrium price and quantity and also determine if a shortage or a surplus exists.

For example, assume supply (Q_S) and demand (Q_D) are expressed as

$$Q_s = -20 + 7P$$

and

$$Q_D = 40 - 3P$$

where P is the price for the commodity. We know that for equilibrium to occur, the condition $Q_S = Q_D$ must hold. Therefore, we can determine the equilibrium price level by setting the two equations equal to each other and solving for price (P). Setting Q_S to Q_D yields

$$-20 + 7P = 40 - 3P$$

Adding $3P$ to each side, and adding 20 to each side gives

$$10P = 60$$

Dividing both sides by 10 to solve for price gives

$$P = 6$$

The equilibrium price in this market is $6. At this price level, the quantity supplied will equal the quantity demanded. To determine the equilibrium quantity, we simply plug the equilibrium price into either the supply or demand function. At a price of $6 the quantity supplied would be

$$Q_S = -20 + (7 \times 6) = -20 + 42 = 22$$

and the quantity demanded would be

$$Q_D = 40 - (3 \times 6) = 40 - 18 = 22$$

As you can see, at the equilibrium price of $6 the quantity supplied equals the quantity demanded. After we have identified the equilibrium price and quantity, we can then determine if a price observed in the market will result in a shortage or a surplus, and can calculate the exact amount of the shortage or a surplus.

For example, if the price in the market is $10, we know there will be a surplus in the market since $P > P_e$ ($10 > $6). At a price of $10, the quantity supplied in the market would be

$$Q_S = -20 + (7 \times 10) = -20 + 70 = 50$$

and the quantity demanded would be

$$Q_D = 40 - (3 \times 10) = 40 - 30 = 10$$

At a price of $10 the quantity supplied is greater than the quantity demanded ($Q_S > Q_D$) and a surplus of 40 units exists ($Q_s - Q_d = 50 - 10 = 40$).

On the other hand, if the price in the market is $4, we know there will be a shortage in the market since $P < P_e$ ($4 < $6). At a price of $4, the quantity supplied in the market would be

$$Q_S = -20 + (7 \times 4) = -20 + 28 = 8$$

and the quantity demanded would be

$$Q_D = 40 - (3 \times 4) = 40 - 12 = 28$$

At a price of $4 the quantity demanded is greater than the quantity supplied ($Q_D > Q_S$) and a shortage of 20 units exists ($Q_d - Q_s = 28 - 8 = 20$).

For practice, plot these supply and demand curves and graphically show the equilibrium price and quantity.

As constructed, the supply and demand curves reflect specific conditions in a specific market during a specific period of time. As the market price changes, with all other conditions remaining the same, the quantity supplied and the quantity demanded will change, which is movement along the curves. However, if something other than the good's own price changes (one of the *ceteris paribus* conditions), a shift in the curves will occur. The resultant change in prices and quantities is discussed below.

COMPARATIVE STATIC ANALYSIS

Comparative static analysis involves comparing two separate market situations (static pictures of the market) before and after a change in the marketplace. These shifts in the demand and/or supply curves are brought about by changes in the *ceteris paribus* conditions mentioned earlier.

Changes in Demand

When one of the factors determining demand changes, the demand curve will shift. The resulting equilibrium price and quantity will depend on the direction of the shift in demand.

Increase in Demand

An increase in demand is shown by an outward shift in the demand curve (shifts to the right) indicating that at every price level more of the commodity is demanded. This is graphically depicted in Figure 9.6, where the initial market equilibrium occurs where demand (D_1) equals supply (S) with an equilibrium price and quantity of P_1 and Q_1, respectively. Given the increase in demand (from D_1 to D_2 in Figure 9.6), the quantity demanded by consumers (Q_3) at the initial price level (P_1) is greater than the quantity producers are willing to supply (Q_1). The resulting shortage ($Q_3 - Q_1$) causes the price to be bid up to the new market clearing price (P_2) and quantity (Q_2) where supply (S) equals the new, higher demand (D_2). The increase in demand, from D_1 to D_2, could be due to an increase in population or income, a change in tastes and preferences, or a change in prices and availability of substitute goods. Possible examples are the impact on demand for bananas in the market of a newly arrived consumer's desire for fresh fruit, a sharp increase in the price of substitute fruits, or a tax decrease that increases consumer disposable income. The net impact of an increase in demand is a higher price ($P_2 > P_1$) and a larger quantity ($Q_2 > Q_1$) in the new static equilibrium.

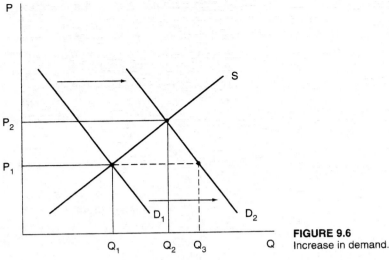

FIGURE 9.6
Increase in demand.

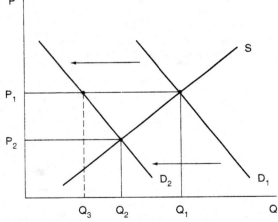

FIGURE 9.7
Decrease in demand.

Decrease in Demand

A decrease in demand is indicated by the demand curve shifting inward to the left, showing that at every price level less of the commodity is demanded. This is graphically depicted in Figure 9.7, where the initial market equilibrium occurs where the demand (D_1) equals the supply (S) with an equilibrium price and quantity of P_1 and Q_1, respectively. Given the decrease in demand (from D_1 to D_2 in Figure 9.7) the quantity demanded (Q_3) at the initial price level (P_1) is less than producers are willing to supply (Q_1). A decrease in the level of demand results in a surplus ($Q_1 - Q_3$) of the product already existing in the market. To "move" the product, producers will lower their asking price until a new market-clearing price (P_2) and quantity (Q_2)

are reached [where supply (S) equals the new lower demand (D_2)]. Examples of such shifts, from D_1 to D_2, might be the effect of high gasoline prices on big "gas hog" automobiles, the embargo of Soviet purchases on corn prices, lower consumer income on color television sales, or lower housing starts on lumber products. Ultimately, a decrease in demand results in an equilibrium price that is lower ($P_2 < P_1$) than before the change and a quantity ($Q_2 < Q_1$) that is smaller.

In summary, if the demand for a product increases, the demand curve shifts outward to the right, resulting in a higher equilibrium price and quantity for the product. If the demand for a product decreases, the demand curve shifts inward to the left, and the equilibrium price and quantity for the product will decrease. Thus, when demand increases (or decreases) the new equilibrium price and quantity are higher (or lower) than the initial equilibrium price and quantity.

Changes in Supply

When one of the factors determining supply changes, the supply curve will shift. The resulting equilibrium price and quantity will depend on the direction of the shift in supply.

Increase in Supply

An increase in supply is shown by an outward shift in the supply curve (shifts to the right), indicating that at every price level more of the commodity is supplied. This is graphically depicted in Figure 9.8, where the initial market equilibrium occurs where the supply (S_1) equals the demand (D) with an equilibrium price and quantity of P_1 and Q_1, respectively. Given the increase in supply (from S_1 to S_2 in Figure 9.8) producers are willing to supply (Q_3) more at the initial price (P_1)

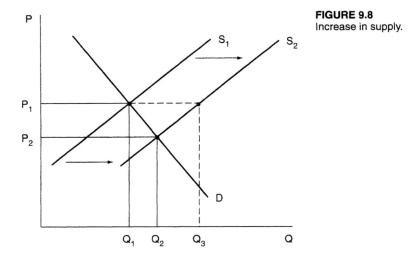

FIGURE 9.8
Increase in supply.

than consumers are willing to purchase (Q_1). The resulting surplus ($Q_3 - Q_1$) exerts downward pressure on price, causing the price to fall to the new market-clearing price (P_2) and quantity (Q_2) where demand (D) equals the new higher supply (S_2). The increase in supply, from S_1 to S_2, could be due to the effect of new producers of electronic calculators entering the original calculator market, an increase in yield of wheat due to new varieties, or the effect a decrease in the price of corn has on the production of beef. An increase in supply results in a lower price ($P_2 < P_1$) and a larger quantity ($Q_2 > Q_1$) in the new static equilibrium.

Decrease in Supply

A decrease in supply is indicated by the supply curve shifting inward to the left, showing that at every price level less of the commodity is supplied. This is graphically depicted in Figure 9.9, where the initial market equilibrium occurs where the supply (S_1) equals the demand (D) with an equilibrium price and quantity of P_1 and Q_1, respectively. Given the decrease in supply (from S_1 to S_2 in Figure 9.9), producers are willing to supply (Q_3) at the initial price level (P_1), which is less than what consumers are willing to buy (Q_1). A decrease in the level of supply results in a shortage ($Q_1 - Q_3$) of the product in the market. This results in bidding by consumers for the available supply, which exerts upward pressure on prices until a new market-clearing price (P_2) and quantity (Q_2) are reached where demand (D) equals the new lower supply (S_2). Examples of such shifts, from S_1 to S_2, are supply of paper products when forest harvest restrictions are imposed, which increases costs; wheat production, when drought or hail conditions lower the crop yield; or greenhouse production under higher energy costs. Ultimately, a decrease in supply results in an equilibrium price that is higher ($P_2 > P_1$) than before the change and a quantity ($Q_2 < Q_1$) that is smaller.

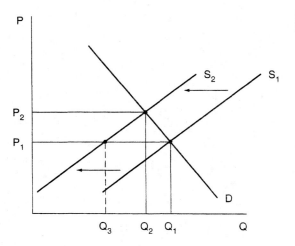

FIGURE 9.9
Decrease in supply.

In summary, if the supply for a product increases, the supply curve shifts outward to the right and the equilibrium price decreases and the quantity increases. If the supply for a product decreases, the supply curve shifts inward to the left and the equilibrium price increases and the quantity decreases. Thus, for an increase (or decrease) in supply, equilibrium price and quantity move in opposite directions.

Change in Supply with Change in Demand

As we suggested earlier, the market system is not static but is continuously changing. Therefore, both supply and demand are often changing in any given market. The resulting equilibrium price and quantity will depend on the direction of movement in both the curves, as well as the relative strength of the shifts in the curves.

If two price-increasing factors, such as a decrease in supply and an increase in demand, occur simultaneously, a strong increase in equilibrium price will be seen, as illustrated in Figure 9.10. Initial equilibrium price (P_1) and quantity (Q_1) occur where initial supply (S_1) and demand (D_1) intersect in Figure 9.10. If a decrease in supply (from S_1 to S_2) occurs simultaneously with an increase in demand (from D_1 to D_2) the resulting equilibrium price (P_2) and quantity (Q_2) are found where the new lower supply (S_2) and higher demand (D_2) intersect. The new equilibrium price (P_2) is much higher than the initial equilibrium price (P_1) because both an increase in demand and a decrease in supply tend to increase prices. The new equilibrium quantity (Q_2) is larger than the initial equilibrium quantity (Q_1) but not by much.

The net effect on equilibrium quantity will depend on the relative strength of the movements in supply and demand. Why? Because although a decrease in supply and an increase in demand both increase price, they have opposite effects on

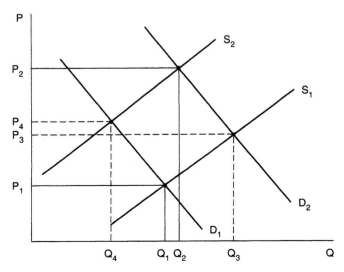

FIGURE 9.10
Decrease in supply with an increase in demand.

quantity. Individually, an increase in demand (from D_1 to D_2) tends to increase both equilibrium price (from P_1 to P_3) and quantity (from Q_1 to Q_3) as shown by the intersection of the initial supply (S_1) and the new higher demand (D_2) in Figure 9.10. By itself a decrease in supply (from S_1 to S_2) tends to increase equilibrium price (from P_1 to P_4) and decrease equilibrium quantity (from Q_1 to Q_4) as shown by the intersection of the initial demand (D_1) and the new lower supply (S_2) in Figure 9.10. Examples of such shifts, from S_1 to S_2 and from D_1 to D_2, are supply of paper products when forest harvest restrictions are imposed (which increases costs) combined with rising newspaper subscription numbers (which increases demand); or wheat production, when drought or hail conditions lower crop yields combined with increased export demand. Thus, when two price increasing factors occur, a strong increase in the equilibrium price will be seen, but the net effect on the equilibrium quantity will depend on the relative strength of the two movements.

When two changes occur simultaneously in the market that have opposite effects on prices, such as a decrease in demand and a decrease in supply, the net effect on price will be determined by the relative change of each, as illustrated in Figure 9.11. Initial equilibrium price (P_1) and quantity (Q_1) occur where initial supply (S_1) and demand (D_1) intersect in Figure 9.11. If a decrease in demand (from D_1 to D_2) occurs simultaneously with a decrease in supply (from S_1 to S_2) the new equilibrium price (P_2) and quantity (Q_2) are found where the new lower supply (S_2) and the lower demand (D_2) intersect. Since both of these changes tend to decrease equilibrium quantity, the final quantity exchanged will be significantly decreased. The new equilibrium price (P_2) is higher than the initial equilibrium price (P_1), but not by much.

The net effect on equilibrium price will depend on the relative strength of the movements of supply and demand. Why? Because although a decrease in supply tends to increase prices, a decrease in demand tends to decrease prices.

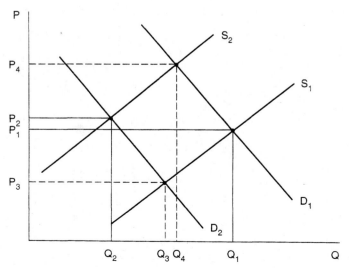

FIGURE 9.11
Decrease in demand with a decrease in supply.

Individually, a decrease in demand (from D_1 to D_2) tends to decrease both equilibrium price (from P_1 to P_3) and quantity (from Q_1 to Q_3) as shown by the intersection of the initial supply (S_1) and the new lower demand (D_2) in Figure 9.11. On the other hand, by itself a decrease in supply (from S_1 to S_2) tends to increase equilibrium price (from P_1 to P_4) and decrease equilibrium quantity (from Q_1 to Q_4) as shown by the intersection of the initial demand (D_1) and the new lower supply (S_2) in Figure 9.11. Examples of such shifts, from D_1 to D_2 and from S_1 to S_2, might be the effect of high gasoline prices on big "gas hog" automobiles combined with the discovery of a new oil reserve (which increases supply); or an embargo of Soviet corn purchases (which decreases demand) in conjunction with favorable weather conditions (which increases yields). Thus, when two changes that have opposite effects on prices occur, a strong decrease in equilibrium quantity will be seen, but the net effect on the equilibrium price will depend on the relative strength of the two movements.

You should be able to use the tools of supply and demand to further identify the effects of many possible change combinations.

Using Comparative Static Analysis

As we indicated earlier, it is very important for managers to be able to understand and predict the impacts of economic changes on the equilibrium price and quantity of their products. In the following discussion we illustrate, using examples, how managers can use this economic knowledge to predict the effects of changes in the market on price and quantities for their business.

When presented with a problem using comparative static analysis, the answer comes in a series of steps. First, identify which product the supply and demand curves represent. Second, identify which curve is affected by the change (remember, the supply curve represents producer response and the demand curve represents consumer response). Finally, identify which direction the curve will shift (increase or decrease). To illustrate how to answer a question using these steps, we will work through a few examples.

For example, if the price of labor used in producing clothing has increased because of the union's successful negotiations, what will happen to the equilibrium price and quantity in the clothing market? In this example the price of labor, which is an input in producing clothing, has increased. Thus, the costs of production are now higher than they were before making it more expensive to produce each unit of clothing. We know that when an input price increases, the marginal cost curve shifts upward to the left (since $MC = P_x \div MPP$). Since the supply curve is the marginal cost curve above the average variable cost curve, the increase in costs is shown by the supply curve shifting inward to the left. This is graphically depicted in Figure 9.12.

In Figure 9.12, the initial equilibrium price (P_1) and quantity (Q_1) are at the intersection of supply (S_1) and demand (D). When the price of labor increases, the supply curve shifts inward to the left from S_1 to S_2, which is a decrease in supply.

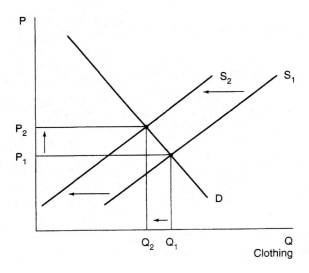

FIGURE 9.12
Cost of labor increase in clothing market.

As a result of the increased cost of producing clothing, less clothing is available at each price level (supply decreases), resulting in upward pressure on prices as consumers bid up the price of available clothing. The new market equilibrium is found at the intersection of the new lower supply (S_2) and demand (D) with a market price of P_2 and quantity Q_2 traded. Due to the increase in costs of production, the quantity of clothing exchanged decreases (from Q_1 to Q_2) and price increases (from P_1 to P_2). The effect of this change on total revenue for the industry will depend on the elasticity of demand for clothing. If the demand for clothing is inelastic, the increase in price will result in higher revenues; if the demand for clothing is elastic, the increase in price will result in lower revenues.

Consider what would happen to the equilibrium price and quantity of wheat if the price of barley (a product that can be produced using the same resources) were to decrease. Since wheat and barley can both be produced using the same resources, as the price of barley decreases, the production of wheat will increase. Why? As the price of barley decreases, wheat becomes relatively more profitable to produce. Thus, the farmer will remove resources from barley production and add them to wheat production (product–product decision). As the price of barley decreases, the supply of wheat will shift outward to the right, as illustrated in Figure 9.13.

The initial equilibrium price (P_1) and quantity (Q_1) is at the intersection of supply (S_1) and demand (D) in Figure 9.13. As resources are added to wheat production, the supply curve for wheat shifts outward to the right (from S_1 to S_2), which causes the equilibrium price of wheat to decrease (from P_1 to P_2) and quantity to increase (from Q_1 to Q_2). Due to the decrease in barley price, wheat production (and, hence, supply) increases and wheat prices decline. The impact of this market change on the revenues of wheat producers will depend on the elasticity of demand for wheat. An inelastic demand implies that revenues will fall as price decreases, and an elastic demand implies that revenues will rise as prices fall.

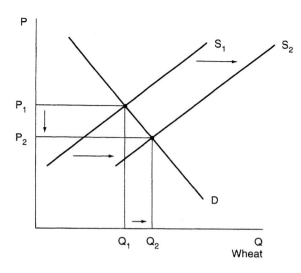

FIGURE 9.13
Impact of price of barley decrease in wheat market.

As another example, assume a bright university agricultural technology and management graduate invented a mechanical harvester that cuts the labor needed to harvest an apple crop. At about the same time, changes to the federal immigration law have decreased the number of laborers allowed into the United States. How does this situation affect the equilibrium price and quantity for the mechanical harvester? A common source of labor to harvest apples comes from immigrant labor. Since the federal immigration law reduces the number of laborers allowed in the country, the size of the available labor pool has been diminished, which results in labor costing more than it previously did. Since the mechanical harvester represents a substitute for labor, the apple producer can use the mechanical harvester to harvest his crop instead of using immigrant labor. Thus, as the price of immigrant labor increases due to decreased supply (changing the price of a substitute good), the apple producer will use less immigrant labor and demand more mechanical harvesters. This situation is depicted graphically in Figure 9.14.

In Figure 9.14, the initial equilibrium price (P_1) and quantity (Q_1) is at the intersection of supply (S) and demand (D_1). As apple producers substitute use of the mechanical harvester for human labor, the demand for mechanical harvesters will increase (from D_1 to D_2), resulting in the equilibrium price and quantity increasing (from P_1 to P_2, and Q_1 to Q_2, respectively). Due to changes in the labor market, both the price and quantity traded of mechanical harvesters increase. The amount of the increase will depend on the responsiveness of both producers (elasticity of supply) and consumers (elasticity of demand). The impact on revenue for the manufacturers of the mechanical harvester will also depend on the elasticity of demand.

Price Control Policies

You can also use comparative static analysis to analyze the impact of price controls on the supply and demand situation. Price controls are not unknown in the

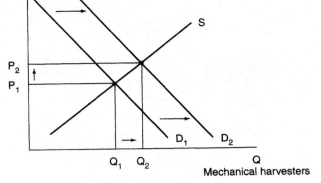

FIGURE 9.14
Impact of cost of labor change in market for mechanical harvesters.

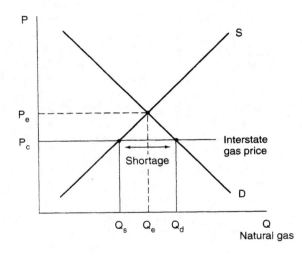

FIGURE 9.15
Impact of price controls on natural gas market.

American economy. President Nixon tried to freeze the price of meat in the 1970s. In 1938 the Federal Power Commission began to regulate the price of natural gas shipped across state borders in pipelines. The commission's mandate was to maintain "just" prices for natural gas, and it developed an elaborate scheme for bureaucratic regulation of price controls on pipelines and producers.

The price ceilings on natural gas were well below equilibrium levels when the energy crisis hit the United States in the 1970s. The impact of the commission's regulation can be illustrated using Figure 9.15. If the market had been in equilibrium, the gas price would have been P_e and the quantity demanded would have been Q_e. But with Federal Power Commission price regulation, the price ceiling was below equilibrium resulting in consumer demand of Q_d and producer willingness to supply of Q_s. There was a serious shortage in the market as producers held gas back in the face of rising consumer demand since "natural gas was cheap!"

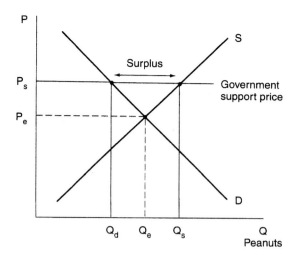

FIGURE 9.16
Impact of government support price for agricultural commodity.

During the cold winter of 1974–75 serious problems arose. The Federal Power Commission was forced to take steps to ration available supplies. Some natural gas industrial customers experienced supply interruptions or "curtailments," as the Federal Power commission described them, in order for home heating to receive all the gas necessary in the face of a serious winter over much of the nation. The shortage amounted to about 2.0 trillion cubic feet of gas or about 10 percent of the marketed production. The cause of the problem was government intervention in market dynamics, which created below-equilibrium prices and eventual shortages. In 1978 Congress passed legislation to phase out natural gas price regulation and today we have an open market for natural gas and prices that substantially reflect market supply and demand conditions.

The government also implements policies that raise prices above equilibrium levels. Take the case of peanuts, a major agricultural commodity in Georgia, the Carolinas, and Virginia. In the midst of low prices for peanut production in the 1930s, Congress passed legislation directing the Department of Agriculture to stabilize peanut prices by making payments to farmers to take land out of peanut production. Eventually this program was changed to a guaranteed minimum price if farmers restricted production to a specified poundage limit for the industry.

Unfortunately, this guaranteed price was often above the market equilibrium levels—a result of political pressure in the Congress. The impact of the USDA peanut program is illustrated in Figure 9.16. With a guaranteed price ("support price") above the market equilibrium level of P_e, the consumers of peanuts (candy companies, households, airlines) demanded only Q_d while the peanut farmers were willing and eager to supply quantity Q_s. The result was a classic case of market surplus brought on by government well-intentioned government policies.

Since 1962 the cost to the taxpayer for the "peanut program" has been about $1.1 billion and of course consumers have been paying higher-than-normal prices for all peanut products due to this government intervention. The USDA peanut program is now operated at no net cost to the taxpayer but production is still restricted so that prices exceed the market equilibrium levels.

Interdependence in the Economy

The previous discussion and examples are quite simplistic in that they do not consider the effects a change in one market will have on other markets. For example, when the demand for Coors beer increases, the price of the beer and the quantity sold in the market will increase. Yet, this change may affect the demand for barley to make the beer, fertilizer to grow the barley, transportation facilities to move the fertilizer, energy to operate the transportation facilities, labor to explore and develop new energy sources, and beer (probably Coors) for the working men and women. Yes, it is a complex world we are trying to analyze . . . but it is an important attempt to make if managers are to be able to react to and plan for changes.

Price plays an important role, as we've suggested, in directing the economy as it performs the basic functions. It guides consumption by serving as the measuring stick by which consumers evaluate utility per dollar of expenditure for alternative products. It guides production by identifying consumer desires in the form of revenue dollars available from alternative enterprises. Producers at the managerial level then allocate resources in response to these dollar signals from consumers. Price also serves to distribute production back to the consumer whose income first allowed an expression of desire to be made. The level of prices is directly related to the purchasing pattern dictated by the existing income distribution in the economy. Low prices allow more consumers to participate in the purchasing, while high prices restrict the purchase to only higher income category consumers who have a strong preference for a commodity.

SUMMARY

In this chapter we've reviewed the process of market price determination. Impacts of changes in supply and/or demand on equilibrium price and quantity were identified. The analysis was developed in a simple fashion so that the information and understanding gained from the discussion will aid all decision makers—wives, husbands, farmers, agribusiness managers, and so forth—in evaluating the economic environment surrounding them. For the manager who is a price taker (perfect competition, as we'll see), the market information will aid him or her in the planning process for the future. For the price searcher or maker (imperfect competition), market knowledge allows pricing, advertising, and merchandising decisions to be more profitably made.

In the following chapters we will examine the socioeconomic environment surrounding the manager. In addition, we will also relax some of the assumptions of perfect knowledge and time that have been made up to this point.

QUESTIONS

1. Why do the authors specify "ability" as well as willingness in their discussion of supply, demand, and market price determination?
2. Does a situation of excess supply exist at a price above or below the market equilibrium price? Illustrate with a graph.
3. Suppose, due to a federal income tax cut, that demand for kiwi fruit increases. Further, assume that the government drops the 20 percent tariff on that fruit, which results in an increased supply. Will the new equilibrium price be higher, lower, or the same as before the changes? How do you know and under what conditions?
4. In your own words describe the role that price plays in directing the smooth functioning of the economy.
5. Although we have not discussed it, how do you suppose that price–quantity decisions are made in centrally planned economies such as Cuba or the People's Republic of China?
6. Define what we mean by *comparative static analysis*.
7. Assume the supply for widgets (Q_s) is given by $Q_s = -10 + 3P$. The demand for widgets (Q_d) is given by $Q_d = 60 - 2P$. What is the equilibrium price and quantity for this widget market? Graphically illustrate these curves and the market equilibrium.
8. Are you sure you remember the difference between change in demand versus a change in quantity demanded? Use an example where the demand curve shifts and then discuss the difference.
9. Graphically illustrate and verbally explain what would happen to the equilibrium price and quantity of chicken if the surgeon general were to put out a report indicating that beef consumption is excellent for one's health.
10. Graphically illustrate and verbally explain what would happen to the equilibrium price and quantity of wheat if the government were to remove from the market a herbicide that controls weeds in barley production.

PART II

MANAGERIAL TOOLS

CHAPTER 10

Competitive Decision Making

INTRODUCTION

The U.S. economy is really a collection of many interdependent subparts or industries. Each industry consists of one or, more usually, many firms or individual economic units. Some of the larger, more important industries make news headlines from time to time (e.g., housing, automobile, and soybean). Most industries, though, are smaller and rarely noticed by the public.

Regardless of the size of the company, economists and managers are interested in how the firms in each industry are organized and how decisions are made. This organization takes many forms in the United States. Automobiles are produced in an industry dominated by three large firms. Soybeans are produced by thousands of farm "firms." Electric service is typically available from only one company in a given area, but medical service is usually available from many different clinics or offices in the same area, but both of these situations are now changing.

Over the years, economists have developed a systematic method for describing and analyzing these many different markets. This method involves four basic structures, or forms, of market competition that characterize the seller side of the market. In each case, we focus on the characteristics of the market's structure and what the implications are for the determination of price and output levels within that market, and how much discretion the manager has in making decisions. Figure 10.1 illustrates how the four market models can be arranged.

As we move from perfect competition toward a monopoly situation in Figure 10.1, the number of firms in an industry decreases, but the barriers to entry and exit increase, as does the individual firm's influence over the prices received for production. Thus, the differences between each market structure lies

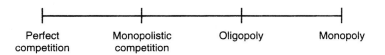

FIGURE 10.1   Forms of market competition.

not in the economic objective of the firm (all are profit maximizers) but in the structure of the market itself.

In This Chapter

We have been studying market level supply and demand and the determination of market price. Economists are also interested in market behavior, or how the market operates to determine prices and quantities. In this chapter, we discuss the basic assumptions that make up the competitive model and outline the characteristics and results of competitive behavior in markets. We also examine the similarities between the purely competitive market model and production agriculture. The remainder of the chapter is devoted to the challenges and strategies of managing in a competitive environment.

THE COMPETITIVE MODEL

The competitive model, also referred to as **pure** or **perfect competition,** provides a benchmark against which other market forms are compared and is especially useful in agricultural economics and agribusiness management. An understanding of the operation of the competitive market also permits us to understand other market organizations that are not competitive (monopolies, oligopolies, and monopolistic competition). The competitive ideal becomes the standard of judgment. Why? We will elaborate on this as we discuss the competitive model.

Characteristics of Competition

When economists use the term **pure competition** they are referring to *a market or industry with four general characteristics: a large number of buyers and sellers, homogeneous products, freedom of entry and exit, and perfect information.*

Large Numbers of Buyers and Sellers

In a competitive industry the firms are so small and so numerous that no single firm can noticeably influence market prices or quantities. That is, the firm is a **price taker.** The only economic variables a competitive firm can influence are its own production level and the combination of inputs used to affect production costs.

Homogeneous Product

The output of a competitive industry is a standardized, **homogeneous product**. Economists refer to this as "undifferentiated output" because each firm produces an identical product. There are no "brand name" products in the competitive industry.

Freedom of Entry and Exit

Managers are free to enter a competitive industry or to remove resources and exit. There are no governmental regulations or artificial restraints on "getting into business."

Perfect Knowledge and Information

All firms in a competitive industry have the same access to new knowledge (technological advances) and market information (prices, quality, and quantity). No firm has an exclusive patent, special information, or restrictive franchise.

These are the general characteristics or assumptions (also referred to as the "conditions of competition") necessary for an industry to be described as purely competitive. Using the conditions of pure competition, economists have developed a model of market behavior for the competitive firm and industry. It is this economic model that enables economists to describe and predict the behavior of real-world businesses in a competitive environment.

Results of Competition

In the competitive model, the demand curve faced by the individual firm is horizontal, as illustrated in Figure 10.2, indicating that at the market price, P_m, the firm can sell all it produces. If the firm demanded a higher price, say, P_n, it would sell nothing. Buyers would simply go elsewhere since the product is homogeneous. This is why competitive firms are described as price takers and as being at the "mercy of the market." The prices they receive are determined by forces beyond the control of the firm.

The competitive firm determines its most profitable, or **optimal,** output level by operating where marginal revenue is equal to marginal cost. Since the

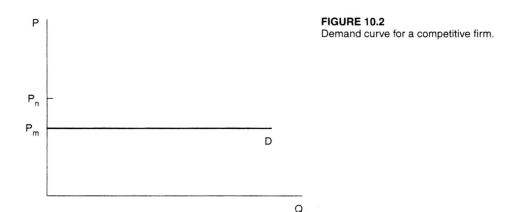

FIGURE 10.2
Demand curve for a competitive firm.

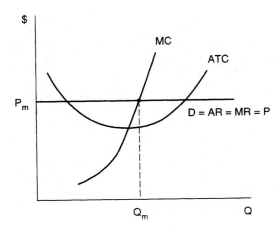

FIGURE 10.3
Optimal level of output for a competitive firm.

firm receives the same market price for each unit sold, the MR curve is horizontal, as shown in Figure 10.3, and is also the demand curve for the individual firm. The MC curve represents the additional cost incurred from producing another unit of output, and its shape reflects the increasing/decreasing returns exhibited by the production function. The point where MC exactly equals MR is the optimal level of output, Q_m in Figure 10.3, for the firm. If the firm produces fewer units than Q_m, the firm can increase profits by expanding production; to produce more units than Q_m would add more to costs than it would add to revenue. Remember that in the short-run the MC above minimum AVC is the firm's supply curve. In the long-run, as market prices vary, the manager of a competitive firm adjusts output along that portion of the MC above minimum ATC. (Note that up to now in our discussion of producer and consumer decision making we have been assuming a competitive environment.) The output level of the individual firm is determined by these production-cost relationships. The firm is a price taker so it must be very cost-conscious.

The actions of many buyers and sellers in a competitive market produce aggregate or market supply and demand relationships. As we learned earlier, the demand curve reflects the inverse relationship between price and quantity purchased (as price goes down, the quantity demanded increases) by consumers. The supply curve reflects the positive relationship between price and quantity supplied (as price decreases, the quantity supplied decreases) by producers. This is represented by a downward-sloping demand curve and an upward-sloping supply curve, as shown in Figure 10.4, where the market is in equilibrium at price P_m with quantity Q_m traded.

In this competitive model, Adam Smith, the father of economics, saw an "invisible hand" at work. The **invisible hand** is *the unregulated operation of the market that results in efficient production.* Efficient always means lowest cost. Competition produces this result impersonally through the direction and discipline of the market. To accomplish this, the "hand" operates in two ways.

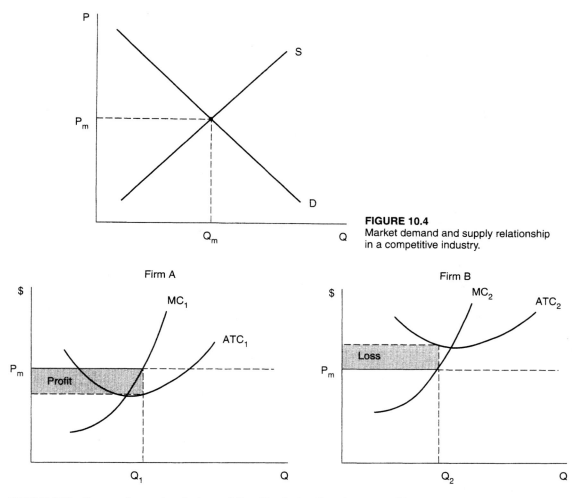

FIGURE 10.4 Market demand and supply relationship in a competitive industry.

FIGURE 10.5 Comparative cost and return relationships for two firms in a competitive market.

First, no firm can charge exorbitant or "rip-off" prices since it would lose all customers to competing firms that sell the same homogeneous product. This point is illustrated in Figure 10.2, where if the firm charges a price higher than the market price, it sells nothing.

Second, firms that have higher cost curves, reflecting higher per-unit costs of production, are forced out of business. This point is illustrated in Figure 10.5, where firm A has costs of production that are reflected by MC_1 and ATC_1 and an optimal level of output at Q_1, and firm B has higher costs of production, as reflected by MC_2 and ATC_2, and an optimal output Q_2. These higher per-unit costs could stem from many reasons, such as poorer land, bad management, and/or obsolete machinery. As price takers, both firms face the same market price, P_m. Firm B, therefore, is in a loss situation at Q_2 and, in the long-run, will be forced out of production (MR < min ATC). Firm B is not as efficient as firm A and will be disciplined (bankrupted) by the "invisible hand" of the competitive market.

Specific Results of Competition

An industry that is competitively organized has specific implications for the determination of prices and output levels, as well as firm behavior. Some important points about the results of a competitively organized industry are summarized next.

Price takers. The buyers and sellers in a competitive market are price takers and not price makers. In a competitive environment the manager reacts to market price and conditions.

Optimal output. The competitive firm determines its most profitable output level by recurrent comparisons of MR and MC. The MC curve above average variable cost is the supply curve for the firm. In the long-run, as market prices vary, the manager in the competitive environment adjusts output along that portion of the MC curve that lies above the ATC curve (in the long-run, ATC and AVC are the same).

No product differentiation. Competitive firms do not advertise or otherwise differentiate their product from the product (output) of other competitive firms. To do so would raise costs but not influence sales, since the competitive firm can sell all its output at the market price or none at a higher price.

Market equilibrium. The competitive market constantly tends toward an equilibrium of price and quantity. The equilibrium market price will be market-clearing and remove all the product from the market. This search for equilibrium occurs naturally and without intervention from governments.

Stimulating new technology. Competitive firms are constantly searching for new technology or other improvements that will increase output (since a firm can sell all it wants at the market price) or reduce production costs (which means more profit even at constant output). This behavior has been referred to as a "treadmill." As soon as a few growers adopt a new hybrid seed and find it successful, most growers follow suit, and the advantage to the early adopters is lost. Thus, the manager in a competitive environment must necessarily be very aware of any technological improvement in production, marketing, management, or transportation that will potentially affect costs or output.

Efficiency. The competitive industry is considered an efficient industry. Resources are attracted to the industry when profits rise, and resources leave the industry when profits decline. The firms that survive over the long-run are the firms that can produce at least cost. The badly managed firms or those firms without good resources (land, labor, and capital) are forced out. Thus it is concluded that **efficiency** governs the operation of the competitive industry and enables products to be supplied to society at the least possible cost. Certainly no money is "wasted" on advertising.

Why Economists Love Competition

Economists talk constantly of the "competitive industry," the "competitive solution," the need for "more competition" in American industry. We have discussed the specialized meanings of the economists' use of "competition." We discussed some of the characteristics and results of competition. Why, though, does it seem that economists are in love with competition?

There are at least two important reasons for this behavior. First, the competitive model is a norm or "measuring stick" for the performance of other types of industries. Because the competitive solution provides products at marginal (or least) costs, economists can look at other industries and conclude that industry Y is pricing products above the level that would prevail under competitive conditions. Thus, the competitive model gives economists a standard by which to judge the pricing performance of all types of industries.

A second reason economists might be in love with competition is the attractiveness of how the competitive market operates. In theory, the competitive market can be efficient without government intervention. The discipline of efficiency operates like an "invisible hand" to keep only the most efficient firms producing and keeping the entire market in equilibrium. The competitive market is assumed to treat everyone the same and does so in an impersonal and silent manner.

AGRICULTURE AND THE COMPETITIVE MARKET

As discussed earlier, there are four general characteristics or assumptions necessary for an industry to be described as purely competitive. Does any industry meet these conditions? No. Agriculture, however, comes as close as any real-world example. This is why the competitive model is so important in agricultural economics.

Agriculture's Competitive Characteristics

In production agriculture today, there are about 2.1 million farms. Although some of these farms may be large in land size (thousands of acres) or total sales (millions of dollars), each farm is small relative to the total market. No single farm producer of a major agricultural commodity will actually sell more than 1 percent of the total market value of corn, hogs, oranges, or whatever. This represents the situation of the competitive firm as "one competing against many" and qualifies agriculture under the first characteristic of perfect competition. Of course, this situation refers to production agriculture, and not to the agribusiness industry serving agriculture.

The products in agriculture are also homogeneous to a large extent. Economists refer to these types of products as undifferentiated products because, for all practical purposes, there are no differences in the products from farms A, B, and C or farms X, Y, and Z. Grade "A" milk from many dairy barns is blended, packaged, and sold as if it came from the same cow. A grain elevator stores wheat from many farms in the same silo. Fruits and vegetables are graded for quality and marketed through cooperative businesses where the source of an individual apple or tomato is unknown. This homogeneity of products qualifies agriculture under the second characteristic.

Freedom of entry and exit in agriculture is also possible. In fact, exit has been the common behavior of the last 40 years. In 1950, there were 5.6 million farms; today about 2.1 million remain. Many corporations and individuals start farming each year and can do so without a government license or a franchise. There is no need to purchase the right to farm from someone else. It is a free and competitive market as required under the third assumption.

Agricultural information is provided to everyone through newspapers, radio, market news services, and government offices such as the Statistical Reporting Service and the Cooperative Extension Service. Likewise, new technology in the form of new farming practices, new machinery, or new seeds is available to all buyers and users. This is the freedom of information and knowledge outlined under the fourth characteristic.

Agriculture's Departures from Competition

As close as agriculture comes to the competitive ideal, there are several important departures. Although large numbers of buyers and sellers remain, some prices have been affected by the influence of one buyer (the Soviet Union in the grain deal of 1973) or collections of sellers (marketing cooperatives). Freedom of entry into agriculture is hampered by the current high land and capital costs. This situation has given prominence to the homily "The only way to get a farm these days is to marry it or inherit it." It is also apparent that not every farmer operates with the same technology or level of information. The computers, chemicals, and machines used on the largest of American farms set these farms apart from the more numerous but smaller farms.

Despite these departures, agriculture is close enough to the competitive ideal for competitive market behavior to be observed. Supply, demand, and price determination occur in a specific manner. What does this imply for managing production agriculture within the competitive market? We address this in the following discussion.

MANAGING IN A COMPETITIVE ENVIRONMENT

A purely competitive industry does not exist outside the pages of economic textbooks. There are some industries, such as agriculture, that resemble a competitive environment. Retailing is another example. In the retail industry we have many buyers and sellers ("a store on every corner") and entry and exit are fairly easy. But retail stores rarely deal in homogeneous, undifferentiated products. In fact, the product differences and price variation (Sale—50% Off) are keys to success in retailing.

So, all we have for real-world examples of pure competition are industries that resemble the competitive model. Because the resemblance is close on many points, discussing how to manage in the competitive environment is important. Good management will focus on minimizing costs, seeking the benefits of bargaining power, and planning for changes in product demand.

Minimizing Costs

The competitive environment makes the manager a price taker. Necessarily then, the manager must focus on that side of the profit equation over which he or she can exert some influence: costs of production. Being cost conscious is a way of life for the firm and the industry.

An obvious method of minimizing costs can be best described as "plain good management." Here we are talking about getting the most from the resources available to the firm or individual. Before considering a new equipment purchase or additional land acquisition, the manager should use "plain good management" and carefully analyze to see if costs can be reduced by better (more efficient) use of the existing equipment, buildings, land, and labor resources. If a manager more efficiently uses existing resources, costs per-unit of production will decline.

Another common method of reducing per-unit costs of production is innovation or new technology in order to affect the production function. In agriculture, new technology does not just mean a new tractor or another hybrid seed. In fact, technology comes in many forms. In livestock, technological innovations occur not only in the form of new breeds or types but in systems of livestock production such as confined hogs, industrial-scale feed lots, and drylot dairy production. In plant production, technological innovations include plant varieties that produce more yield per plant or that are resistant to disease, drought, or pests. New methods of using the land, such as no-till corn production, are also innovations.

New technology can help managers reduce costs in two ways:

1. By increasing production from the same level of resources and thereby reducing per-unit costs; and
2. By producing the same output using fewer resources and thereby reducing per-unit costs.

The effects of new technology are illustrated in Figure 10.6, where if the present cost relationship is expressed by MC_1 and ATC_1, the new technology allows increased production at a lower per-unit cost (MC_2 and ATC_2). At the same market price (P_m) the new, lower costs indicate to the manager that output can be increased profitably, from Q_m to Q_n in Figure 10.6. Given the same market price, with the new technology profit increases from area P_mABC to area P_mDEF.

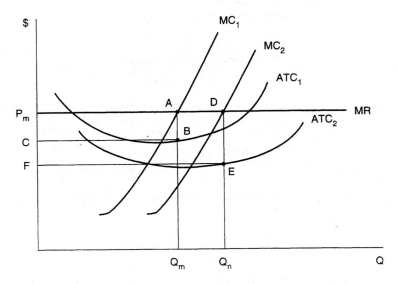

FIGURE 10.6 Effect of new technology on cost relationships for a competitive firm.

The manager must realize that new technology, in whatever form, will in many cases increase total costs of production. Examples can include automated feeding systems, electronically controlled harvesters, or costly pesticides. What is important is the effect of the technological change on per-unit costs of production. Before adopting the new technology, the manager needs to examine carefully whether the results of adopting the new technology (increased production, lower per-unit production costs) will offset the costs of the new technology.

A popular phrase in agricultural circles is the **cost–price squeeze.** This phrase reflects the competitive nature of agriculture. First, agricultural markets are characterized as price takers, that is, having no influence over product prices. Secondly, agriculture as we have discussed is a cost-conscious industry, and, therefore, it is also technology and innovation conscious. Managers are constantly trying to lower per-unit costs of production by being more efficient. But what happens when one or more input prices rise dramatically? Answer: The manager is often caught in a cost–price squeeze. That is, the manager is caught in a situation where the costs of production are increasing while the prices received for his product remain unchanged, or, in a worse situation, prices decrease simultaneously.

In other noncompetitive industries, managers can react to a cost–price squeeze by passing on price increases to the consumer. A firm in a noncompetitive environment can raise its price and still sell most of its product and thereby escape some of the effects of a cost–price squeeze. The competitive firms in agriculture have no influence on price. To raise prices as a method of escaping the cost–price squeeze would be a disaster—the firm would not be able to sell any of its product.

Generally, competitive firms react to cost–price situations in one of four ways:

1. Go out of business altogether, or drop one or more products from production.
2. Take economic losses in the short-run (one season) and then try to innovate and continue in business in the long-run.
3. Seek bargaining power through cooperative marketing arrangements.
4. Seek federal or state government relief in the form of grants, disaster aid, loans, or other special actions.

The first two alternatives are straightforward. If a firm cannot maintain liquidity (adequate cash flow to meet expenses) in the short-run, it is bankrupt. Those firms that are able to survive short-run losses in the face of cost–price squeezes will be able to return for another production season. We discuss the third alternative in the next section. Discussion of the fourth alternative is deferred until Chapter 14's policy review.

The Benefits of Bargaining Power

Firms in a competitive environment can achieve some economic benefits by acting together in their marketing efforts and input purchases. Farmer cooperatives provide this important function for their member-patrons. A cooperative is a business organization that is owned and managed for its member-patrons for the purpose of providing marketing and/or marketing goods and services to its patrons at cost. Farmer cooperatives, by buying in bulk quantities, are able to secure cost savings on agricultural inputs for their members. A number of competitive firms producing a substantial amount of any one commodity may be able to influence price through group bargaining power. In agriculture, marketing cooperatives provide this service to their member-patrons. Individual farmer-members sell their production through the marketing cooperative, which then pools all members' production and negotiates the sales of the commodity in the market. By this method the manager in a competitive firm (farm) hopes to no longer be a price taker at the mercy of the market, but a price maker with some control of the market.

Bargaining power movements are not new to American agriculture. The American Society of Equity, an organization of Midwest farmers, tried to raise wheat prices in 1903 by withholding grain from the market. During the 1920s Aaron Sapiro organized California producers into cooperatives that bargained for strong contracts between growers and processors. Milo Reno, of the Farmer's Holiday Association, urged farmers in 1932 not to deliver products to market at prices lower than production costs.

The modern versions of these bargaining power movements are the National Farmers Organization (NFO) and the American Agricultural Movement (AAM). The NFO is an echo of the Farmer's Holiday Association in its demands that members withhold products until prices exceed production costs. The NFO's influence reached its peak in the early 1970s. The AAM formed as a result of the cost–price squeeze on farmers in the late 1970s and has called for members not to plant grain until the government guarantees a fair price.

The bargaining power movements have not succeeded in their efforts to give agriculture much influence over product prices. There are two important reasons why.

First, since bargaining associations are voluntary, they are unable to force all producers to join. The nonmember producers who continue to market products undermine any withholding action and erode the bargaining position of the association. Failure is almost inevitable.

The second reason is that for the association to be able to influence price it must also be able to control marketing and shipments. There may be a surplus amount of product that cannot be sold at the bargained price. This will have to be destroyed or shipped overseas. If the surplus production enters the market (illegally or through charity), the bargaining position is undermined so the association must be able to spread the burden of surplus production evenly among members. This is a difficult, if not impossible, task.

Competitive managers can seek the benefits of bargaining power through other systematic marketing means. These involve government-sponsored and government-controlled marketing orders and marketing agreements.

The Agricultural Marketing Agreement Act of 1937 is the legislation that permits producers to organize and bargain with buyers by means of marketing orders. The marketing order for any particular commodity (milk, fruit, vegetables, or whatever) must be approved by a majority of the producers. Once approval is complete, the order is binding on all producers. The marketing order legally allows the producers to enhance product price through control over the quantity and quality of the product entering the market, and through advertising or promotion of the product as a group. The Secretary of Agriculture must approve the order (if it is a federal order) and the USDA administers the operation. Thus, the government supervises the process and overcomes the major weaknesses of the bargaining movements (disciplining all producers and disposing of surpluses).

The marketing order provides the competitive manager with many benefits not available to a producer acting alone. Marketing orders provide advertising, promotion, and research to differentiate products. They also provide quantity controls to govern the flow of product to the market and handle surplus production, and also provide quality controls. Marketing orders also provide some bargaining power for price, with all growers acting as a group.

Marketing orders have been most successful for single commodities grown in restricted geographic areas. Examples are the citrus fruits and nuts of California and fruits in Michigan. (Milk is an exception to this generalization.) Not all producers or managers consider the marketing order to be an answer to the problems of the competitive environment. Potato producers in Florida and turkey producers nationally have rejected marketing orders.

A marketing order is mandatory, whereas a marketing agreement is voluntary. A marketing agreement is a contract between the producers and the handlers of a particular commodity. The contract must be supported, or endorsed, by the Secretary of Agriculture, and is binding only on those producers who sign it.

Planning for Changes in Demand

In the short-run, managers are preoccupied with the problems of current season production. Often overlooked is the serious management challenge of changes in product demand over a period of years or in the long-run. Change is a fact of life for managers in agriculture. One subtle, but extremely important, source of change concerns the demand for agricultural products. Tastes and preferences change; population grows; incomes rise; and inflation erodes earning power. All of these changes affect the demand for agricultural products.

Long-run changes in product demand are evidenced by changes in the per-capita consumption of agricultural commodities over time. For example, since the early 1980s consumers have been drinking more low-fat milk (from 82 lbs in 1981 to 103 lbs in 1995) and less whole milk (from 140 lbs in 1981 to 75 lbs in 1995). During this same period, poultry consumption increased from 44 lbs in 1981 to 49 lbs in 1995, whereas beef consumption decreased from 73 lbs in 1981 to 64 lbs in 1995. Consumers are also eating more fruits and vegetables (126 and 174 lbs, respectively, in 1995) than in previous years (84 and 71 lbs, respectively, in 1981).

How are these changes in product demand reflected in the competitive decision-making model? As demand increases, many firms respond to the opportunity for new profits by expanding production of the desired product type. Firm managers, as we have discussed, will constantly compare MC with present and anticipated prices (MR to the competitive firm). Total output will increase in the face of rising demand. However, it has been the common experience that output will increase too rapidly and prices will begin to fall. Managers get caught in a cost–price squeeze, and an adjustment period occurs during which some firms may fail, other firms switch to different products, and finally output and price stabilize as a new competitive equilibrium is reached.

A real-world example of this experience is sunflower production in the upper Midwest. With favorable sunflower-seed prices, thousands of farmers switched away from corn, wheat, and soybeans to sunflowers. Output increased from 353,000 tons in 1973–1974 to 1.8 million tons in 1978–1979. Domestic consumption and foreign exports were on the rise, but the expansion in output came too quickly and prices fell markedly in 1979. An adjustment period ensued in which prices and production fluctuated as the market struggled to reach a new equilibrium. Production levels in 1992–1993 averaged 1.1 million tons.

The lesson for the competitive manager is this: Adjustments must occur to long-run changes in product demand as well as short-run changes in costs and prices. These long-run changes may not be as dramatic as the rise in energy prices in the 1970s or the drop in grain prices after the Russian embargoes of

1979 and 1982, but they are nevertheless important to the success and survival of the firm. Production choices made now should hinge on both long-run and short-run factors.

SUMMARY

Competition is used by economists to describe a certain market condition or environment. Pure competition refers to a market characterized by large numbers of buyers and sellers, homogeneous products, freedom of entry and exit, and perfect knowledge and information. The competitive model is an economic abstraction built on the market conditions of pure competition.

There are several important characteristics, or results, of the competitive model that make it appealing to economists. The individual producer faces a horizontal demand curve, making the competitive manager a price taker. Being a price taker implies that the marginal revenue of the firm is the market price for the product (P is MR). The competitive firm determines optimal output by equating marginal revenue to marginal cost. Through the impersonal operation of the market, the competitive market achieves efficiency (low-cost production) by disciplining high-cost producers, and the market itself tends toward equilibrium price and quantity without outside intervention.

Economists have a great fondness for the competitive model. Not only is it a stepping stone to further economic analysis, but it also represents an economic system where prices equal marginal cost, or the value of resources used in producing that unit. This result ($P = $ MC) has become a standard or norm for measuring the performance of other industries. Although the competitive model has no real-world counterpart, production agriculture comes very close. Competition describes in general terms the market conditions for agriculture and thus is the basic economic tool for analysis and management of farms and related firms.

Managing in a competitive environment is a challenge. Having no influence on price, managers must focus on minimizing costs through "plain good management" and technological innovation. But despite good management, competitive firms are often caught in a cost–price squeeze whenever input prices increase dramatically or product prices decease. The competitive firm is unable to pass on price increases to the consumer and must adjust accordingly.

The nature of competition in agriculture has motivated producers to seek bargaining power in order to enhance their influence on price. Bargaining power movements have not been successful, but government-sponsored marketing orders and agreements have achieved some of the same ends. A marketing order allows the producers to legally enhance price through bargaining, advertising, and promoting products as a group; through quality control; and through the regulation of disposal or destruction of surpluses.

Long-run changes in product demand also affect the management environment in agriculture. Tastes, preferences, population, income, and inflation all contribute to changes in the kinds of agricultural products desired by the American public and foreign purchasers. These sometimes subtle changes should be taken into account by the manager as he or she prepares and plans for the future.

QUESTIONS

1. Briefly outline the major assumptions of pure competition.
2. What does the term *market-clearing price* mean?
3. Why do individual firms face a horizontal demand curve in the competitive model?
4. Using graphical analysis, illustrate how the more efficient farmer A can continue to produce at a lower product price than farmer B.
5. Describe how new technology can help farmers reduce their costs. On a graph show how this affects their profit situation.
6. What options do farmers have to combat the proverbial "cost–price squeeze"?
7. What important factors affecting demand for agricultural products should the manager constantly be considering?
8. As a Michigan apple producer, evaluate the following hypothetical situation: Washington apple producers have been suffering from sluggish demand but have recently succeeded in gaining access to the vast Asian markets. Should you, as a Michigan apple producer, plant more trees in anticipation of future increased demand? Defend your answer.

CHAPTER 11

Noncompetitive Decision Making

INTRODUCTION

Competition is only one type of market structure or business environment. As we described at the beginning of Chapter 10, goods and services in the United States are provided in a variety of ways. For many agricultural products, the market structure is competition, or as close to it as you'll find in the "real" world. However, pure competition does not describe how electric service is traditionally supplied (one utility company per area, but with the deregulation currently under way this will be changing), or automobiles produced (the "Big Three"), or furniture manufactured (hundreds of firms).

In the preceding chapter we defined perfect competition and examined the conditions necessary for a market to be considered perfectly competitive. However, much of what we observe about industry in the United States and around the world is best described as **noncompetitive**, or as **imperfect competition**. That is, the structure of markets for many goods and services does not meet all of the competitive assumptions—many buyers and sellers, homogeneous products, freedom of entry and exit, and perfect knowledge. Recall that economists have developed four models to describe how these markets, both competitive and noncompetitive, behave. These market models allow economists to describe and predict the behavior of real-world businesses in both a competitive and noncompetitive environment. The four market models first seen in Chapter 10 are listed again in Figure 11.1.

As we move from perfect competition toward a monopoly situation in Figure 11.1, the number of firms in the industry decreases, but the barriers to entry and exit increase, as does the individual firm's influence over the price received for its products. Thus, the differences between each market structure lie not in the economic objectives of the firm but in the structure of the market itself. The three models that describe how imperfectly competitive, or noncompetitive (not as competitive), industries behave (monopoly, monopolistic competition, and oligopoly) are all based on the competitive model and use it as a starting point.

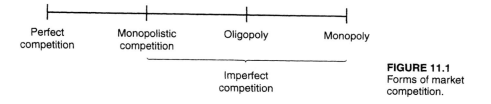

FIGURE 11.1
Forms of market competition.

Monopoly and oligopoly have sinister connotations to many people. Monopoly is "bad," whereas competition is "good." Economists have undoubtedly contributed more than their fair share to the development of these attitudes. This is a mistake. Economists can pronounce judgments about efficiency and welfare but need not comment about "goodness" and "badness." As a matter of fact, exactly the same motive drives both monopolists and competitors: profits, net returns, and the "bottom line." Consumers and society can decide, and they have, whether monopolies are bad, as we'll discuss.

In This Chapter

The differences between competitive and noncompetitive firms lie not in their motives (profit making) but in the nature of the industry. How many firms? Are the products differentiated? Are entry and exit possible? The answers to these questions lead us to want to look more closely at noncompetitive firms to see how they make price and output decisions and how managers can operate in noncompetitive environments.

In this chapter we examine each of the three noncompetitive market models more closely, examining the characteristics of each market structure and what the implications are for price determination in each market. We use the perfectly competitive model as a starting point and for comparison to determine if noncompetitive markets result in higher prices for the consumer. We begin our discussion by briefly reviewing the competitive market model.

PURE COMPETITION

We earlier examined at length the nature of pure competition as a market form. The competitive model is based on four important assumptions:

1. Many buyers and sellers;
2. Homogeneous products;
3. Freedom of entry and exit; and
4. Perfect knowledge and information.

The nature of the competitive business environment means the individual firm is a price taker and faces a horizontal demand curve. At the market price the firm can sell all of its output. Bargaining for higher prices is fruitless for an individual firm.

Without an influence on price, the managers in competitive situations look to cost minimization as the primary method of increasing profits. This accounts for the strong motivation on the part of competitive firms to seek new cost-reducing or output-increasing technology.

Competitive firms often look to group action in order to secure some bargaining power in the marketing of their products or purchasing their inputs. Although bargaining power movements have not been successful in agriculture, government-sponsored group action has worked to stabilize and enhance product prices. Thus, group action is the most effective method for firms in competitive industries to use if they are to have any influence on product price.

We have defined perfect competition and examined the conditions necessary for a market to be considered perfectly competitive. However, much of what we observe about industry in the real world is best described as imperfectly competitive, or as noncompetitive. In the next section we examine the characteristics, and the implication for price and quantity determination, of the noncompetitive market structure.

IMPERFECT COMPETITION

What is meant by imperfect competition? The key to the definition, or what separates a competitive firm from a noncompetitive firm, is whether the firm has some (it doesn't have to be complete) control over the price it receives for its product. **Imperfect competition** *exists whenever the individual firm has some control over prices received for the goods it produces and markets*, which implies that the individual firm faces a downward sloping demand curve. That is, the firm must decrease prices in order to increase sales, or if price is increased, some product will still be sold. A downward-sloping demand curve also implies that the marginal revenue curve will be downward sloping. Thus, in a noncompetitive market, price does not equal marginal revenue as it does in competition. Firms in an imperfectly competitive market can change their prices (within limits) to expand or reduce their sales or attempt to change the demand for their product through advertising.

Although firms in noncompetitive markets can influence price, their control over price is seldom absolute. The individual firm cannot ignore its competition. How much influence on market price the firm has depends on its position in the market. Important variables are the size of the firm and its share of the market, the firm's strategies for buying and selling, the market itself, and the actions of any competing firms. In contrast, the competitive firm cannot influence the market price because its share of the market is small. As we stated earlier, the differences between noncompetitive firms and competitive firms lie not in their motives (profit maximization) but in the nature of the industry: How many firms? Are the products differentiated? Are there barriers to entry and exit? What is the shape of the demand curve? We now examine each of the three noncompetitive market models, beginning with monopolies.

Pure Monopoly

The precise opposite of competition is monopoly. The pure monopoly market model describes a situation in which

1. There is a single seller of a product;
2. The product has no close substitutes and could be considered totally differentiated;
3. There is no freedom of entry or exit; and
4. There is an unavailability of knowledge or information.

In effect, the monopoly situation does not meet any of the competitive assumptions. This leads to a big difference in management options and market behavior. If the competitor is a price taker, then the monopolist can be described as a **price searcher** (or **price maker**).

Why do economists describe the monopolist as a price searcher? Doesn't he or she always charge the highest possible price since consumers cannot get the product anywhere else? The answer to these questions lies in the structure of the monopoly market. By definition a monopolist is the industry. Thus, the monopolist will face the industry demand curve, which has the characteristics of the usual demand curves (downward sloping to the right). Figure 11.2 presents a representative case, where if the monopolist always charged the highest possible price he

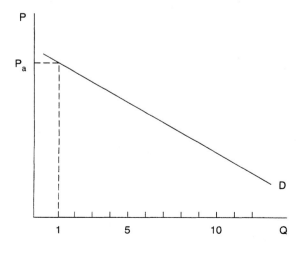

FIGURE 11.2
Monopolist demand curve.

or she would sell a very small quantity. At price P_a in Figure 11.2, the monopolist would sell only one unit of output. To increase sales, the price must be lowered. So the monopolist must search for the profit maximizing output and price.

The fact that the monopolist faces the industry demand curve is not the only difference between competitive and monopoly demand situations. Not only can the monopolist influence price, he or she can also attempt to influence the entire demand curve. The monopolist may be economically able to afford to advertise and persuade consumers to buy more of his or her product. The advertising has the effect of shifting the entire demand curve to the right as illustrated in Figure 11.3, where initially the monopolist can sell quantity Q_1 at price P_1. A big advertising campaign results in increased demand (from D_1 to D_2) for the monopolist's product, enabling the monopolist to sell additional quantity, Q_2, at the initial price P_1, or to sell the same quantity, Q_1, at a higher price, P_2. In effect, successful advertising has created a new demand relationship, D_2, for the monopolist's product.

We still lack some essential information that we must obtain before we can determine how the monopolist searches for a profit-maximizing price. We know the monopolist's demand situation and how it differs from that of the competitive firm. What about cost relationships?

Generally, the monopolist faces a cost situation similar to that in the competitive environment. Raw materials must be purchased, labor must be employed, capital equipment must be leased or purchased, and management must be trained and motivated. Productivity of resources is as important to the monopolist as it is to the competitive firm. Given this situation, we expect the cost curves under monopoly conditions to be similar in nature to what we've seen under competitive conditions. We expect to see a U-shaped average total cost curve and a steeply sloped marginal cost curve.

Determining the profit-maximizing price for the monopolist requires knowledge of the cost and demand relationships plus revenue curves. Here we find an important distinction between monopoly and competition. A competitive firm can

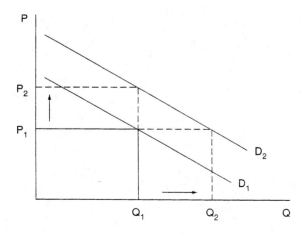

FIGURE 11.3
Effect of advertising on monopolist's demand curve.

sell all of its output at market prices and therefore P = MR and is constant. For a monopolist, price must be lowered to increase sales and so $P \neq$ MR. Table 11.1 illustrates this situation. From this table, we see that as the price per-unit is reduced, quantity sold increases as does total revenue. At a price change from $12 to $10, revenue remains constant (MR = 0), and at further reductions in price, total revenue declines (MR < 0).

Graphing the figures in Table 11.1, we can illustrate the difference between the demand and marginal revenue curves. As shown in Figure 11.4, since the monopolist must decrease price to increase sales, marginal revenue is downward sloping and more steeply sloped than the demand curve. (Since the monopolist is a profit maximizer, the negative portion of the MR is typically not shown because it represents negative returns.)

TABLE 11.1   Monopoly Price, Quantity, and Revenue Schedules

Price ($)	Quantity Sold	Total Revenue	Marginal Revenue
20	1	20	
			> 16
18	2	36	
			> 12
16	3	48	
			> 8
14	4	56	
			> 4
12	5	60	
			> 0
10	6	60	
			> −4
8	7	56	

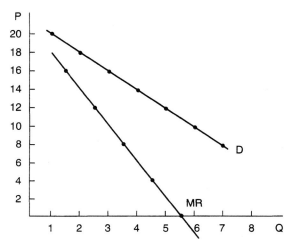

FIGURE 11.4
Monopoly price, quantity, and revenue schedules.

If we superimpose the cost curves on these demand and marginal revenue curves, we can determine the monopolist's profit-maximizing position. The rule for profit maximization in monopoly is the same as in competition: Profits will be at a maximum where MR = MC. For the competitor P = MR and the firm determines output by comparing price to marginal cost. However, $P \neq$ MR for the monopolist, but, as illustrated in Figure 11.5, the intersection of MR and MC determines the profit-maximizing output, Q_m. Why? Beyond Q_m, MC is greater than MR (MC > MR) for each unit of production. Between 0 and Q_m, MR is greater than MC (MR > MC) for each unit of production. Therefore the monopolist would find it profitable to expand output until MR = MC, at quantity Q_m and no further.

Because the monopolist has full pricing power (he or she is the only seller in the market), the monopolist prices his or her product based on what the market will bear, or based on what consumers are willing to pay for the product. Thus, the monopolist determines the product price from the demand curve. The market demand relationship indicates that for the optimal quantity, Q_m, the market-clearing price is P_m.

The profits of the monopoly firm are equal to the difference between average costs of production for each unit (from the ATC curve) and the price (from the demand curve) times the production level, area P_mABC in Figure 11.5. Because the monopolist has full pricing power, the price charged by the monopolist is higher than in a competitive market and the quantity offered is less. In Figure 11.5 the competitive price would be P_c, where the marginal cost curve intersects the demand curve (the demand curve is the marginal revenue curve for the competitive firm), with quantity Q_c placed on the market. You can see why consumers like competition rather than monopoly because monopoly decreases quantity available and increases price.

In Figure 11.5 we can see that in comparison to the competitive situation the monopolist can potentially make generous profits. There is no pressing need to

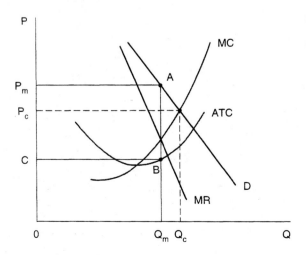

FIGURE 11.5
Profit maximizing position of the monopolist.

lower production costs since no competitors are around to provide price competition. Further, the monopolist is in a stable market condition and not at the mercy of the market conditions facing the competitor.

Monopoly and the Real World

Monopoly, as described in its "pure" form, has few examples in the real world. Again, as with competition, pure monopoly is a textbook model used in economic decision making. However, there are a few close counterparts in the real world to our monopoly model.

Public utilities. Electric, gas, water, and telephone service have all been traditionally provided by one company in a given geographic area. These were known as **natural monopolies.** They arose from the tremendous capital costs involved in supplying utility services to consumers. It was considered inefficient to duplicate distribution systems (e.g., two companies stringing electric lines in one town) to compete for service to individual customers. This is changing for telephone and electric service and the notion of a natural monopoly may soon be confined to local utility service. In electric and telephone service, new federal regulations are allowing consumers to choose their service provider in some cities, thus ending the old monopoly utility.

Large locational monopolies. A monopoly can arise from the ownership of a unique natural wonder or a particularly important piece of geography. If one company or person owned the Grand Canyon that would be a monopoly situation. Certainly no one is going to be able to duplicate the Grand Canyon and provide any competition. The Rock of Gibraltar is a unique piece of geography owned by the British. By virtue of their ownership of the rock, the British historically could exert monopoly over shipping between the Mediterranean and Atlantic oceans.

Small locational monopolies. These exist on smaller scales all across the country. A natural hot spring for bathing and swimming; the only bridge across the river; the only shoe repair shop in a small town; and a prime piece of real estate near a freeway intersection are examples of monopoly locational advantage.

Technological monopolies. The possession of unique technology or knowledge can give rise to monopoly situations. The Hughes Tool Company for many years had a patent on the most efficient drilling bit. The monopoly profits it extracted from oil and gas drilling firms built the Howard Hughes empire. In technological monopoly, it is the patent law that provides the barrier to entry and creates the monopoly situation. Now that the U.S. Supreme Court has permitted the patenting of new forms of life, it is conceivable that living inputs to agricultural production will have to be purchased from monopolies.

Statutory monopoly. Certain goods and services are provided in a monopoly manner because federal or state statutes mandate it. First-class mail service is a national example. It is illegal for anyone to provide first-class mail service in competition with the U.S. Postal Service. At the state level, an example of statutory monopoly is the sale of whiskey and other liquors through state-owned stores. That is, states such as Washington exercise a monopoly on the retail sale of distilled spirits by statute, making the state-owned store the sole source.

In the real world (outside your classroom and campus), there are few examples of either of the polar extremes of market structure models, competition and monopoly. Most businesses and industries will have characteristics of each; that is, they are somewhere in between the extremes. Economists have taken this world of in-between markets and divided them into two broad categories: monopolistic competition and oligopoly. These two market forms, along with monopoly, make up the world of imperfect competition. We continue our discussion of imperfect competition by examining monopolistic competition.

Monopolistic Competition

Monopolistic competition is a market structure that has elements of both competition and monopoly, hence, its title. Economists make the following assumptions about this market form:

1. There are a large number of firms in the market;
2. Products are differentiated;
3. There is ease of entry and exit; and
4. Knowledge and information about products and markets are readily available.

As is evident from these assumptions, monopolistic competition contains elements of both monopoly and competition. However, "large" in this case could mean 10 or 1,000 but the crucial factor is that the monopolistic competitor can expect his or her actions to go unheeded by other firms. This is a situation much like that in competition. Monopolistic competitors produce similar products (thus they can be lumped together in categories), but they try hard to differentiate their products from those of rival firms. As with the case of competition, it is assumed to be relatively easy to get into and out of this industry. There are no substantial technological, market, or legal barriers to entry.

Examples of monopolistic competition are easy to recognize. First you look for a fairly standard product available from many sources but which is heavily advertised or differentiated in other ways. Flour is an example. In 1992 the U.S. Census of Manufactures found that there were more than 350 producers of "flour and other grain mill products." To most consumers, flour is flour. But the producers of flour try hard to make consumers believe their brand is better. The differences between flour brands may be real (type of wheats used, blending, milling, additives) or perceived (created by advertising claims alone). But the important factor is that some differences exist between brands *in the minds of the consumer.*

Other examples of monopolistic competition include furniture retailing, clothing, bread, eyeglasses, and paints. In each case, the products are similar but not quite homogeneous, as we assumed under pure competition.

If a large number of firms are producing similar but differentiated products, the demand curve for a monopolistically competitive firm will take on a special shape. If the firm were a pure competitor, the demand curve would be horizontal because the products are homogeneous. Because the monopolistic competitor has a differentiated product, his or her demand curve is slightly downward sloping to the right, as illustrated in Figure 11.6.

The slope of the demand curve in Figure 11.6 is created by the brand loyalties that lie behind the differentiated products. If the firm were to raise price, say, from P_1 to P_2, not all customers would switch to other brands, even though the products are similar. Thus, the firm's efforts at product differentiation have paid dividends, giving the firm at least some influence over price. Furthermore, the other firms in the industry will not pay much attention to one individual firm's actions. If firm A raises prices, every other firm will not necessarily follow suit. The same case holds for price decreases.

There may be meaningful differences among the products in a monopolistic competitive market, but the prices of one firm cannot diverge very far from the prices of other firms in the industry or consumers will shift to another brand or product. Thus, the monopolistic competitor has limited control over price, which often results in competition being vigorous in areas other than price (i.e., customer service, warranties, etc.). Economists refer to this as **nonprice competition.**

Figure 11.7 compares a hypothetical, competitive situation with that of a similar situation in monopolistic competition where, although the cost curves are familiar (U-shaped ATC and a steeply sloping MC curve), the demand curves differ. In competition $D = MR = P$, so that the equilibrium output, Q_E, at price P_E is determined by the rule $MC = MR$, where $P_E = MR$ for the competitive firm. Because the monopolistic competitor has a differentiated product, the individual firm faces

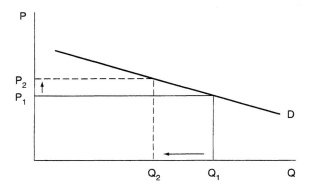

FIGURE 11.6
Demand curve for a monopolistic competitor.

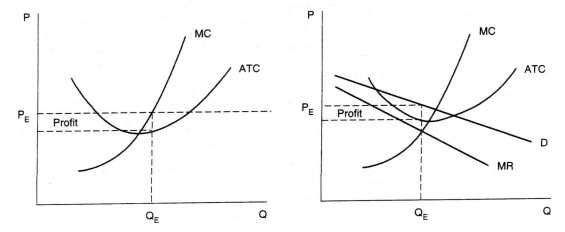

FIGURE 11.7 Comparison of a competitive firm and a monopolistic competitive firm.

a downward-sloping (but very elastic) demand curve. That is, to increase sales, the firm must decrease price. Since the firm faces a downward-sloping demand curve, the firm's marginal revenue curve will also be downward sloping, reflecting the changes in revenue associated with each set of prices and quantities. The monopolistic competitor is a profit maximizer and so determines the optimal production level using the decision criterion MR = MC. However, the firm has some control over price, so when the firm determines optimal output, Q_E, where MR = MC, the firm determines the product price, P_E (which is higher than in the competitive market), from the demand curve. (Monopolistic competition is similar to monopoly, but with much less price flexibility.)

As illustrated earlier, the manager in a monopolistic competitive firm has some options not available to his or her counterpart in pure competition. He or she exploits these options to the fullest by keying upon both the firm's demand curve (advertising and other marketing efforts) and the firm's cost curves (cost minimization).

The final market model to be examined is oligopoly. This market form lies between the extremes, but is closer to the monopoly end of the spectrum. In competition and monopolistic competition we talk of "competition among the many." In oligopoly we speak of "competition among the few."

Oligopoly

Oligopoly is a market structure where production is generated by a few large firms that dominate the industry. Specifically, economists make the following assumptions about oligopoly:

1. There are few large firms in the industry;
2. Products can be either standardized or differentiated;

3. Entry is very difficult either for market or technological reasons; and

4. Knowledge and information are imperfect and not available to all firms.

You don't have to look far to see examples of oligopolistic industries. Any industry dominated by a few large firms is a good bet. The products may be standardized, as with sugar, metal cans, light bulbs, small arms ammunition, gasoline, steel, glass, or copper. On the other hand, the products may be differentiated, as with automobiles, breakfast cereals, aspirin, cigarettes, soap, tires, typewriters, computers, or beer. In the same manner as monopolistic competition, the oligopolistic firm may differentiate its product through real quality differences or perceived differences. The best example of the latter is aspirin. Basically all aspirin contains exactly the same drug ("active ingredient"). Some firms combine this drug with additives ("buffered aspirin"), or coat the aspirin tablet ("timed release"), or claim it dissolves more quickly in the stomach ("faster acting"). Whether or not these differences have any real effect on performance has not been proven. Nonetheless, these differences are the basis for million-dollar advertising campaigns designed to create perceived differences. Don't think you are immune. If you had a bad headache right now and we placed before you brand X (650 mg) aspirin and Bufferin® (650 mg), which would you take?

Oligopoly occurs not only on a national scale, but also on a regional or local scale. Any time two or three firms dominate the market for any product in a local or regional area, it can be understood to be an oligopoly situation. For instance, if there were only two building-supply firms in a large region, we could expect the economic behavior of the firms to be oligopolistic. Local and regional examples of oligopoly arise most often when transportation costs are an important part of product costs, and it is difficult for consumers to shop around.

Economists have a rough measure for determining whether or not an industry is oligopolistic. This measure is called a **concentration ratio (CR),** and is simply *the percentage of total sales or value of industry shipment by the largest four, or sometimes eight, firms in the industry.* In a monopoly, the concentration ratio would be 100%, since all sales come from one firm. In competition, the concentration ratio would be 1 percent or less since there are so many firms in the industry. In imperfect competition, the concentration ratios vary widely and can only be used as a rough measure. Managers can use these same ratios to evaluate the competitive situation in their industry.

Table 11.2 contains some examples of concentration ratios from the U.S. Census of Manufactures (1992). For instance, the production of cigarettes is confined to four large firms (CR = 93% or 0.93). Although there are actually 160 firms in the industry, malt beverage production is dominated by four firms as revealed by the concentration ratio (CR = 90%).

Economists expect the behavior of firms in concentrated or oligopolistic industries to be different from firm behavior under conditions of competition or monopolistic competition. The crucial factor that determines this difference in behavior is the small number of firms in the market and their importance to the total market. Under conditions of oligopoly, you can expect the dominant firms to watch

TABLE 11.2 Selected Concentration Ratios for U.S. Industries

Industry	Number of Companies	Percent of Industry Shipment by	
		Four Largest Companies	Eight Largest Companies
Aluminum	30	59	82
Automobiles	398	84	91
Office machines	143	45	63
Breakfast cereals	42	85	98
Cigarettes	8	93	100
Tires	104	70	91
Glass	16	84	93
Malt beverages	160	90	98
Farm machinery	1,578	47	53

closely the price, output, and profits of their rival firms. Each firm is so large and so important to the total market that the actions of one firm will affect all firms. Thus, the oligopolistic firms are said to be *interdependent*, as opposed to independent, competitive firms.

The interdependence of oligopolistic firms leads to an unusual demand relationship facing each firm. The demand relationship is still downward sloping to the right, reflecting the inverse relationship between price and quantity. However, oligopolistic firms really have two demand curves facing them.

Situation 1. Suppose one of the dominant firms decides to raise prices in order to increase profits. Unless all firms follow suit, customers will switch to lower priced products, and the firm will face a significant reduction in sales. Thus, for the firm that increases prices while other firms do not faces an elastic (responsive) demand relationship.

Situation 2. Suppose one of the dominant firms decides to increase sales (and profits) by slashing prices in order to attract new customers. You cannot expect the other firms to stand idly by and watch their customers desert to the rival firm. Other firms will follow suit and decrease prices in order to protect their market share.

The two situations discussed can be illustrated with a special "kinked" demand curve, as presented in Figure 11.8, where the demand curve is very elastic above the existing price. If you, as manager, raise the price and your rivals do not follow suit, you will lose sales and profits decline. If you decrease the price and your rivals match each price decrease, your sales will not increase as you expected and revenues will decrease. In this "bind" created by the kinked demand curve you would expect everyone to sit tight and not "rock the boat" with regard to price. As a result, we often see stable prices in an oligopolistic market.

This "sit tight" behavior can be confirmed and explained by examining the MR curves associated with the kinked demand curve. Because the kinked demand

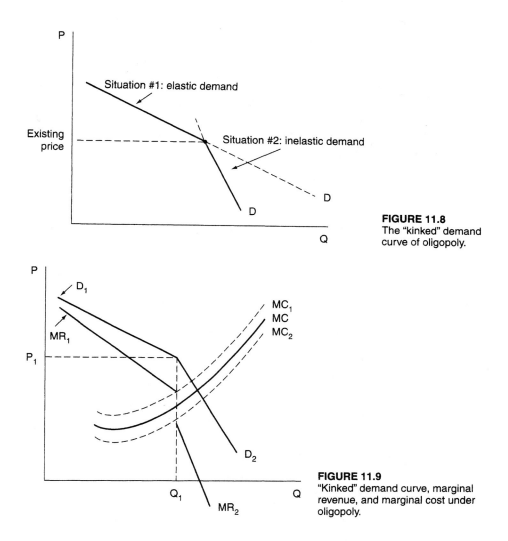

FIGURE 11.8
The "kinked" demand curve of oligopoly.

FIGURE 11.9
"Kinked" demand curve, marginal revenue, and marginal cost under oligopoly.

curve is really two demand curves (one for price increases, one for price decreases), there will also be two MR curves (MR_1 and MR_2), as illustrated in Figure 11.9. The kinked demand curve generates a break or indeterminate portion to the MR curve. If the MC curve intersects MR in this indeterminate portion or range, the old optimizing rule (MR = MC) results in quantity Q_1 produced and price P_1. And notice that even if costs change (increase to MC_1 or decrease to MC_2) as long as they still lie within the indeterminate range no changes in profit maximizing price and quantity will be seen. The firm is left in a position of depending on the actions of other firms to finally determine price and output.

Under these pricing conditions, a firm acting independently only gets itself into trouble. The solution: Act together. Thus, in oligopolistic industries economists typically observe mutual behavior. In the auto industry, vigorous nonprice competition is relieved by price competition that is limited to "rebates" or below-market interest rates on auto loans. A similar situation occurs in consumer pharmaceuticals.

However, price competition does occur in some oligopolies, which leads to unstable market conditions due to the very nature of the industry. Take airlines for instance. They are a classic oligopoly—a homogeneous product and a few large firms dominating the industry (Delta, American, United, Northwest). Every year there are "fare wars." Consumers have come to *expect* fare wars so it is almost inevitable that they will occur in the spring, prior to the peak summer travel season. Once one airline announces "Special Summer Fares!" then all the other airlines must act quickly to match these fares or risk leaving seats empty. In the end, no single airline really gains market share with a fare war. However, the unstable price competition leads to instability in the industry and has resulted in airline bankruptcies.

This mutual dependence among oligopolists means there are advantages to acting together. If the representatives of all the firms meet and decide jointly what the price will be next year for their particular product and who will sell in particular markets, this is **collusion.** In the United States, collusion is illegal and subject to criminal penalties and fines if proven in a court of law. However, outside the United States, collusion is not always illegal, thus American firms must deal with it routinely. For example, the Organization of Petroleum Exporting Countries (OPEC) is a cartel that attempts to coordinate world prices for crude oil. The DeBeers Central Selling Organization sets world gem diamond prices. These are cartels that operate freely and openly but would be illegal inside the United States.

Collusion and cartels are solutions to the pricing and output problems presented by the oligopoly market structure. A firm manager in an oligopoly market has considerable control over price and output—the envy of the competitive firm. But oligopolists are shackled by the kinked demand curve and seek to collude, formally or informally, so as not to "rock the boat" and to keep prices and outputs stable, and even achieve the results of a monopolist.

Why Are There Differences in Market Structures?

The different market structures we have discussed fit into broad categories and the exact border between each category is not clear and distinct. In addition, the categories, such as the number of firms in an industry or whether or not the product is differentiated, are not as important as is the behavior of the firms in the industry. The crucial factor is that the economic behavior is different for each category.

Why do differences in market structure arise? That is, why is soap produced by an oligopolistic firm, furniture by monopolistic competitors, and wheat by competitors? Two primary factors lie behind the existence of imperfect competition: barriers to entry and economies of scale.

Barriers to Entry

Monopoly and oligopoly often exist because there are substantial barriers to entry. Patents and copyrights present legal barriers. For many years Polaroid had a monopoly in the instant-picture business because patents protected its exclusive process. In certain industries capital costs are an effective barrier to entry by new firms. The staggering capital cost of starting a new steel mill, automobile company, or pharmaceutical drug production represents a serious obstacle to entry of new firms.

Economies of Scale

In some industries, the size of the plants and equipment necessary to achieve the lower production costs of mass production are extremely large. A new firm could not survive long enough to reach these economies of scale and establish a market share. Economies of scale motivate big firms to get bigger and keep new firms out of the market.

Why Economists Love Competition, Again

In Chapter 10 we discussed the reasons economists love the competitive market model. One result of a market organized as perfectly competitive is efficient production of products. Since efficiency governs the operation of the competitive market products are supplied to society at the least possible cost. This point is illustrated in Figure 11.10, which shows how the price and quantity decision differs between a firm operating in a competitively organized market and one operating under imperfect competition.

In Figure 11.10 the market demand curve reflects the inverse relationship between price and quantity purchased by consumers and the supply curve reflects the positive relationship between price and quantity supplied by producers. In both competitive and noncompetitive market situations, the supply curve is derived from the marginal cost curve of the individual firms. Since firms operating in a noncompetitive market have some price control, or power, over setting prices, the firm's marginal revenue curve is downward sloping. Under perfect competition an equilibrium price and quantity are determined by the intersection of the industry supply curve and the market demand curve. In Figure 11.10 the competitive price would be P_C, where the marginal cost curve (S) intersects the marginal revenue curve (D), with quantity Q_C traded. Firms operating in a noncompetitive market have some influence over price. Although these firms determine optimal output where MR = MC, they price their product from the demand curve. In Figure 11.10 the noncompetitive firm would produce quantity Q_N, where marginal revenue (MR) intersects marginal cost (S), which would sell for price P_N. As

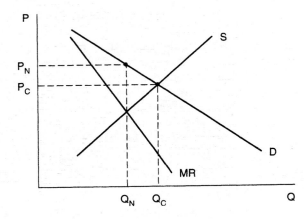

FIGURE 11.10
Market price and quantity
determination.

illustrated in Figure 11.10, a noncompetitive market structure results in higher prices and lower quantities available in the market, both of which negatively impact the welfare of consumers.

As we have discussed, differences in firm behavior arise if the market environment is not competitive. The manager of a firm has more control over pricing decisions and hence production decisions. How does a manager successfully manage a firm when operating within a noncompetitive market? We discuss some management strategies in the next section.

MANAGING IN A NONCOMPETITIVE ENVIRONMENT

Before we discuss various management strategies a couple of generalizations should be made. The environment of imperfect competition can be generally characterized by an emphasis on advertising to create product differentiation or public relations, and an absence of open price competition.

Of course, the extent of advertising is understandable. Firms advertise in order to influence their demand curve. But why, if in the imperfectly competitive world the manager has some control over price, do firms generally avoid price competition?

The answer lies in the nature of imperfectly competitive markets. In monopoly, price competition is nonexistent since one firm constitutes the market. For the many firms in monopolistic competition, price cuts may attract some new customers, but the new profits may attract new firms into the market (since ease of entry is there) and the advantage quickly disappears. In the case of oligopoly, the kinked demand curve ("If I lower price they all will follow; if I raise price they won't.") indicates that price competition is suicide for the individual firm.

Thus, in the environment of imperfect competition the management strategies generally emphasize advertising and avoid price competition. To simplify matters, we've divided management strategies into two categories: defensive and offensive strategies.

Defensive Strategies

Defensive management strategies are employed to protect a firm's position in the market. Managers "on the defensive" are worried about self-preservation and stability. The best way to protect one's position is to recognize that a "gentleman's agreement" not to compete with other firms will reduce the uncertainty about price and output and at the same time stabilize the market. But this is collusion.

A famous example of collusion on a national scale was the agreement among major manufacturers of electrical equipment to set prices and divide the market during the 1950s. Several firms, including General Electric and Westinghouse, were involved in the elaborate collusion that involved all types of electrical parts and equipment.

However, collusion does not have to be so direct and conspiratorial. On other levels it can involve an unspoken agreement not to discount prices or not to sell in certain areas because "that area belongs to company X."

In the oligopoly industries of steel, tires, glass, and other products, illegal collusion has been replaced by price leadership. One dominant firm may consistently be the first to raise prices and all other firms swiftly and quietly follow suit. Everyone acts mutually as a defensive strategy. Price changes by the leader firm often occur following a new union contract or an increase in raw material costs so the new prices can be "justified" and all firms have a reason to follow suit.

Another type of collusion is the quasi-agreement common in professional associations. The best example of this is the real estate industry where local associations of agents have an agreement among all members on what percentage the real estate commission will be, how commissions will be split among cooperating agents, and how services and information will be shared. Agent-members of the association must adhere to the agreement or else be considered "unethical" and dropped from membership. Similar agreements are used in such other professional associations, such as medicine and law. These professionals also agree not to advertise services or publish fee schedules. Several states have now struck down the binding nature of these agreements in order to encourage competition.

Another defensive strategy that avoids price competition is standardized pricing. Big corporations and small businesses have the same problems forecasting sales and determining per-unit costs of production. Therefore, many managers look for some general guidelines for determining pricing policy. One such rule is full-cost pricing or standard costing. To determine a full-cost price, the manager estimates the costs of production if the firm were operating at full capacity and then uses this cost estimate to determine price. This is not the MR = MC rule advocated by economists, and it may lead to less-than-maximum profits for the firm. However, many managers feel a rule-of-thumb is necessary if business is to get done, and it avoids price competition.

A close cousin of standard costing is cost-plus pricing. This is a pricing method based on the standard costing method of estimating per-unit production costs and then adding some allowance—10, 15, or 20 percent—for profit. In grocery stores, some wholesalers of products such as fresh vegetables even supply the manager with a preprinted sheet showing what the final price must be for each product (based on the current wholesale price) in order to generate a particular profit.

One of the most important defensive strategies utilized in imperfect competition is to create barriers to entry for new firms. Some barriers to entry exist in most industries. No one is going to challenge the Big Three in domestic automobiles or start a new steel firm. Large capital costs are the effective barrier in that case. On a smaller scale, barriers to entry are also important protection to the profitability of the firm and the stability of the market. There are many examples of barriers to entry, including these:

> You can't practice law unless you are a member of the bar association. Similarly, you can't practice medicine until you've been certified by the state licensing board.

> You can't grow and market tobacco without substantial financial penalties unless you own or lease a tobacco quota.

> You can't start a new bank or get a liquor license, in most states, unless it is "justified" by growth in a community and permission is granted by a state agency.

> You can't manufacture a product patented or copyrighted by someone else. These barriers are intended to "reward" the owner and inventor for a specific length of time.

Offensive Strategies

The nature of an imperfectly competitive market dictates that firms acting together (collusion) can maximize profits and maintain stability in the market. Not all firms desire this cooperation and stability. There are always new, hot-shot innovative managers who think they can bring new growth to a company and increase the company's power in the market. Dominant firms such as ABC-TV, General Motors, Campbell Soup, IBM, Coca Cola, or Ralston Purina are constantly challenged by smaller firms. The objective of the challenge is market power and dominance.

To build or maintain a large share of the market in an imperfectly competitive environment often means the managers must take the offensive. Defensive strategies protect market shares and maintain stability. Offensive strategies build (or try to build) market shares and disrupt stability.

A firm can take the offensive directly by declaring price warfare. Price cutting is the last thing an older established firm wants to see, but any challenger can use it to build sales and market shares. Price warfare can take many forms, open or hidden. Among wholesale firms, price cutting can come in the form of hidden discounts or other inducements to attract sales. One flour-milling firm has its salesmen offer grocery store managers selections from a catalog the salesmen carry if the grocer will place the firm's flour on the grocery shelves.

Another example of discount price warfare is that of the new automobile market of the late 1970s and early 1980s. Slumping sales in this period severely hurt American Motors. It took the offensive by offering rebates, or manufacturers discounts, on new cars. Soon all of the car manufacturers were advertising rebates on new cars, a practice that continues today.

Price warfare can be open and direct and disastrous to firms. The airline industry is an example. The now-defunct Laker Airways was the first to offer low-cost, no-frills flights to London. Laker's prices were rock-bottom ($125 initially), and he attracted customers and established a market share. Within the United States, World Airways challenged the major airlines ("David versus Goliath" says the World Airways' ad) by offering everyday, low-priced fares on cross-country routes. World Airways was attempting to move from the charter airline business to a regularly scheduled air carrier through price warfare.

Price warfare usually doesn't work. If a smaller firm cuts prices and attracts a few customers, the dominant firms will follow suit in an attempt to win back their customers. If the smaller firm has the liquidity to hang on through price cutting, it may be successful. More often, the dominant firms are stronger financially and force the smaller price-cutter into financial ruin. On an industry-wide basis with an inelastic market demand, price cutting doesn't usually attract enough new sales to compensate for reductions in per-unit revenues. Thus, when everyone cuts prices, everyone loses.

The more common offensive strategy is **nonprice competition,** a phrase used by economists to cover all kinds of advertising. Basically, the economic effect of advertising is to change the shape and position of the demand curve, that is, to make it more inelastic and increase sales at the given market price. The psychological effect of advertising is to create in the minds of consumers product differentiation (making the identical seem different) or product identification (making the new seem familiar).

In the world of business, the term *product differentiation* is not used. Instead such terms as marketing, sales promotion, or public relations are common. No matter what it is called, advertising is a multibillion dollar industry aimed at influencing the psychology of consumer well-being and happiness. Advertising intends to make the customer dependent on a product or service for his or her happiness. If advertising is successful it creates brand loyalty, or in economic terms, more inelastic demand conditions. Advertising efforts have created a brand name that became so popular it is used to describe all similar products (e.g., Xerox or Kleenex). This is a tribute to the power of mass communication and is the envy of every manager of a competitive firm.

Finally, if price warfare or nonprice competition is not used, then large and more powerful firms can take the offensive through acquisition. When one firm purchases another similar firm (or merges) the result on market power is simple and direct. Purchases or mergers between similar firms are called **horizontal integration** because it is the integration into one entity of separate firms manufacturing the same or similar products at the same level in the marketing chain (i.e., large meat packing plants, like Monfort's or IBP, purchasing smaller regional plants). This activity is closely monitored by the federal government because of its potential effect on market power and reductions in competition. On the other hand, **vertical integration** is the integration of firms at different levels of the marketing chain and is driven by firms desiring more control over their product from production to consumer (i.e., Campbell Soup grows many of its own vegetables; Tyson Chicken, which produces broilers, also manufactures and markets processed chicken products).

SUMMARY

The most common market situation in which managers find themselves is usually described as noncompetitive. This certainly does not suggest that no competition exists but rather that all the competitive assumptions (many buyers and sellers, homogeneous products, freedom of entry and exit, or perfect knowledge) are not met. The three market models of monopoly, monopolistic competition, and oligopoly have been used to characterize these situations, and each model uses the competitive model as a starting point. All of these firms have a profit-making motive, but the nature and character of the firm and industry vary.

The managers of firms operating in pure competition are price takers and face a horizontal demand curve. These managers look to cost minimization as the primary method of increasing profits. Group action appears to be the only effective method for firms in competitive industries to have any influence on product price.

Pure monopoly is the opposite of competition. In this situation, the manager is a price searcher (or maker) because he or she is the only seller of a product that has no close substitutes and entry into the market is blocked. The monopolist searches for the price that will maximize profit by identifying the output level where MR = MC and then sets the market-clearing price based on the demand curve. Examples of close counterparts to monopoly are public utilities, locational monopolies, technological monopolies, and statutory monopolies.

Monopolistic competition has elements of both competition and monopoly. The existence of both competition and slight product differentiation causes the manager in this situation to try to affect his demand curve, through advertising and other marketing efforts, and his or her cost curves through cost minimization. No mutual interdependence exists (e.g., firms don't react to each other's price changes).

Oligopoly, often called competition among the few, is a situation where a few large firms dominate an industry. Concentration ratios are often used to determine whether or not an industry is oligopolistic. These firms are usually interdependent (they react to each other's price changes) and, as a result, little price competition occurs. Advertising and cost minimizing are common tools used by managers. Collusion or price fixing, if legally allowed, is the expected price alternative to open competition.

Differences in market structure arise from legal barriers to entry, such as patents and copyrights, or from large capital costs, which block new firms' attempts to enter the market. Economies of scale in existing firms may serve to make new entry by small firms risky.

Managers in imperfect competition have both defensive and offensive management strategies available to them. The decision of which to use will reflect the manager's perception of the market structure and the characteristics under which his or her firm operates.

QUESTIONS

1. Explain the major differences between pure competition and pure monopoly. How do these differences affect the market behavior of each type?
2. How does the demand curve for a monopolistic firm differ from that of a competitive firm? Why is the monopolist's marginal revenue curve sloped the way it is?
3. True or false?: Prices may be higher under monopoly than competition, but output is greater and hence overall welfare is improved. Explain.
4. "Advertising is bad for consumers because it only adds to the price of an otherwise unimproved product." Comment. (Keep in mind radio, magazines, television, and their sources of income, and the occasional existence of new and better products.)
5. Distinguish between pure competition and monopolistic competition. Give a real-world example of the latter.
6. How does oligopoly differ from monopolistic competition? from pure competition? from monopoly?
7. Why do oligopolies occur, and how do we determine whether or not an industry is oligopolistic?
8. Describe the situation in which using MR = MC as a decision criterion is not the obvious choice for a firm operating under oligopoly.
9. What factors of imperfect competition exist to keep you from entering into the dairy processing industry?
10. Why do managers collude either overtly or tacitly? What is the difference between collusion and price leadership?

CHAPTER 12

Time, Risk, and Uncertainty in Management Decisions

INTRODUCTION

In the preceding chapters all of the economic analyses and decision making have been conducted using two major assumptions: that decisions are made and production occurred in one time period; and that the manager knows all things with certainty. In real life this is obviously not always the case.

In agriculture, as in many other industries, the production process takes time. The decisions of how, what, and how much to produce are made (and costs incurred) in a different time period than that of the resultant production (and hence revenue). In some cases the time period separating the production decision and the resultant output and even later sale is longer than one year. For example, winter wheat is planted in the fall and harvested the summer of the following year; an orchard is established in one year but takes three to seven years before producing a full crop. This time lag between when decisions are made and output is obtained raises two important issues to be considered by managers. First, since the real costs and returns are impacted by time, the time value of money needs to be considered in economic analysis. Second, some assets used in production have a useful life longer than one period, making it necessary to consider the value of these assets over time when making investment decisions or determining costs.

The time lag between decisions and the resultant output, as well as weather and other variables, also introduces an element of risk and uncertainty. Prices can change between the time production decisions are made and when the realized production is sold. Markets for commodities change over time as consumers' tastes and preferences change. Thus, the manager must also address the issue of risk and uncertainty in decision making.

In This Chapter

In this chapter we begin to conduct more realistic economic analyses as we relax the assumption of perfect certainty and introduce time into the decision-making process. We begin by examining the problem of time. We discuss and illustrate some of the difficult time problems, the general concepts of time and money, and discuss methods of handling durable assets, which are assets that last beyond the initial decision time period. In the last part of this chapter we relax the condition of perfect certainty and introduce two new concepts: risk and uncertainty. We define these terms, discuss their impact on decision making, and finally review some management strategies that can help minimize the problems of risk and uncertainty.

THE PROBLEM OF TIME

Up to this point our examination of economic analysis and decision making has been conducted without reference to time. To simplify the presentation of economic concepts, we assumed that the decisions about how much to produce, how to produce, what to produce, and market price determination occurred in one time period. In this section we relax that assumption in order to come one step closer to the real world of management and decision making. See, economics can be realistic!

Why Is Time a Problem?

The decisions made by managers everywhere do not occur in a timeless, static world. All decisions, trivial or important, happen at a point in time and in many cases the timing of the decision is of more value than the decision itself. One illustration is planting dates for agricultural crops. If all crops were grown in greenhouses under controlled climatic conditions, the timing of planting dates would be secondary to other considerations. In the real world of U.S. agriculture, proper timing of planting dates may well mean the difference between success and failure.

A more fundamental time problem for managers in agricultural enterprises is the impact of time on costs and revenues. At the most simple level, a producer may borrow money at the beginning of a production season and repay the loans out of revenues received at the end of a production season. At a more complex level, an apple grower may borrow thousands of dollars per acre to establish an orchard and receive no revenues until the trees reach bearing age in three to six years. In each of these cases, how do we compare the dollars expended (costs) now to the dollars received (returns) in the future?

Another fundamental time problem in agriculture concerns the nature of the production process. The growing of crops is really the transformation of many inputs (tractors, land, seed, fertilizer, chemicals) into one or more outputs (rice, apples, catfish). Some inputs are completely consumed in one season or time period, for example, seed, chemicals, and fertilizer. Other inputs are durable and contribute not only to the current season's production but to later time periods also, for example, tractors, trees, and buildings. How does the manager accurately determine profitability and performance when durable assets are used over several time periods?

Time and Money

A commonsense conclusion about people and money is that, if given a choice, nearly everyone would rather have a $1 today versus a $1 tomorrow, next week, or next year. Why? Because we have control over it, can earn interest on it, and so on. This is a time preference for money or wealth, and it is so nearly universal in its observation that it is taken as a given in economic analysis.

Because people have a positive time preference for money, financial markets put a price on money: interest. This price "bids" money away from present consumption and toward future consumption. There are several different rates of interest depending on the nature and conditions of the money market. For example, savings accounts have, in the mid-1990s, paid 3 to 6 percent with only minor variations in the interest rate. Thus, the savings account offers the consumer $3 for the use of $100 for one year with guaranteed safety of the principal sum and the interest earned.

From the borrower's point of view, interest is the price paid to use someone else's money for a specified time period. In essence, the borrower is persuading the money-holder to postpone current expenditures in order to earn a premium or interest return. From the lender's point of view it is what is returned for giving temporary control to the borrower.

Some of the possible interest rates available to savers and borrowers are presented in Table 12.1. Some rates fluctuate while others remain stable. Whether interest rates are high or low depends on overall economic conditions, the risk involved in the transaction, and the period of time involved.

The function of an interest rate is to act as a "price" and thus eliminate the time consideration in holding or borrowing money. The interest rate is a measure of a saver's indifference to $1 now or $1 plus interest at a future date. For the borrower, the interest rate is the price for the use of money now and a measure of indifference as to $1 spent now versus $1 plus interest to be repaid at a future date. In this way, interest rates are said to "factor out" the time element.

TABLE 12.1   Illustrative Interest Rates

	Rates (%)	Fluctuations in Rate
For savers:		
Savings account	2–6	Very little
Certificate of deposits	4–8	Considerable
Treasury (T) bills	4–9	Considerable
Savings bond	5–7	Very little
Corporate bonds	6–12	Considerable
For borrowers:		
Banks	6–12	Considerable
Credit unions	5–13	Some
Credit cards	8–21	Little
Farm credit banks	7–12	Considerable

In economic terms, interest rates are a measure of the opportunity cost of money. Money invested in a farming operation or a business instead of being held in a money market certificate has an opportunity cost of 4 to 8 percent (Table 12.1). A good money manager will carefully evaluate present returns on money invested versus the "costs" of passing up other earning opportunities.

Compounding and Discounting

Interest rates are used in a specific way to determine the changes in the value of sums of money over time. These specific methods are common financial management tools and are referred to as **compounding** and **discounting.** In both cases the interest rate is utilized to calculate the effect of time preference over several time periods.

Compounding: Future Value of a Present Sum

Compounding is a technique for using interest rates to determine the possible value of money at some predetermined, future time period. Compounding answers questions such as "What will $1,000 be worth if invested in a savings account at 6 percent for one year?" The mathematical procedure for calculating the answer can be expressed in the following formula:

$$FV = PV(1 + i)^n$$

where FV stands for the future value of the present sum (PV) of money under consideration; n equals the number of years over which compounding will take place; and i is the interest rate. Therefore, to determine the future value of $1,000 invested in a savings account for one year at 6 percent we calculate

$$FV = 1,000(1 + 0.06)^1 = 1,000(1.06) = \$1,060$$

If the same $1,000 were left in the savings account for five years, the future value would be

$$FV = 1,000(1 + 0.06)^5 = 1,000(1.338) = \$1,338$$

To simplify the compounding procedure, managers rely on financial tables that present the values of $(1 + i)^n$ for different values of i and n. Table 12.2 is an abbreviated version of a compounding table for selected values of i from 5 to 18 percent and values of n from 1 to 20.

To determine the future value of a present sum using the table, we find the factor at the intersection of the interest rate and the appropriate time period, and multiply it by the present sum. For example, to determine the future value of $1,000 invested today at 6 percent for five years we locate the factor at the intersection of 6 percent and five years. Multiplying the tabular value, 1.338, by $1,000 yields $1,338 (the future value of $1,000 invested at 6 percent for five years), the same answer we calculated earlier. In other words, $1,000 today has the same worth as $1,338 five years from now, if the interest rate is 6 percent.

TABLE 12.2   Future Value of $1: FV = PV$(1 + i)^n

Year	5%	6%	7%	8%	9%	10%	12%	14%	18%
1	1.050	1.060	1.070	1.080	1.090	1.100	1.120	1.140	1.180
2	1.103	1.124	1.145	1.166	1.188	1.210	1.254	1.300	1.392
3	1.158	1.191	1.225	1.260	1.295	1.331	1.405	1.482	1.643
4	1.216	1.262	1.311	1.360	1.412	1.464	1.574	1.689	1.939
5	1.276	1.338	1.403	1.469	1.539	1.611	1.762	1.925	2.288
6	1.340	1.419	1.501	1.587	1.677	1.772	1.974	2.195	2.700
7	1.407	1.504	1.606	1.714	1.828	1.949	2.211	2.502	3.185
8	1.477	1.594	1.718	1.851	1.993	2.144	2.476	2.853	3.759
9	1.551	1.689	1.838	1.999	2.172	2.358	2.773	3.252	4.435
10	1.629	1.791	1.967	2.159	2.367	2.594	3.106	3.707	5.234
11	1.710	1.898	2.105	2.332	2.580	2.853	3.479	4.226	6.176
12	1.796	2.012	2.252	2.518	2.813	3.138	3.896	4.818	7.288
13	1.886	2.133	2.410	2.720	3.066	3.452	4.363	5.492	8.599
14	1.980	2.261	2.579	2.937	3.342	3.797	4.887	6.261	10.147
15	2.079	2.397	2.759	3.172	3.642	4.177	5.474	7.138	11.974
16	2.183	2.540	2.952	3.426	3.970	4.595	6.130	8.137	14.129
17	2.292	2.693	3.159	3.700	4.328	5.054	6.866	9.276	16.672
18	2.407	2.854	3.380	3.996	4.717	5.560	7.690	10.575	19.673
19	2.527	3.026	3.617	4.316	5.142	6.116	8.613	12.056	23.214
20	2.653	3.207	3.870	4.661	5.604	6.727	9.646	13.743	27.393

Discounting: Present Value of a Future Sum

Discounting is a technique for using interest rates to determine today's value of sums to be received at some future date. Discounting answers questions such as "A promise of $1,000 to be paid to you at the end of three years is equivalent to receiving how much today?" From the discussion of time preference, we can surmise that the answer will be less than $1,000. The amount can be calculated using the mathematical procedure expressed by the following formula:

$$PV = \frac{FV}{(1 + i)^n}$$

The expression PV equals the present value of some future sum (FV) to be received after a certain number of time periods (n) and utilizing a discount rate (i). The formula can also be written as $PV = FV(1 + i)^{-n}$. The present value of $1,000 to be received in three years at a discount rate of 6 percent would be calculated as

$$PV = \frac{\$1,000}{(1 + 0.06)^3} = \frac{\$1,000}{1.19} = \$840$$

Therefore, at a discount rate of 6 percent a person would have no time preference over (i.e., wouldn't care about) receiving $1,000 in three years or $840 today. If she

TABLE 12.3 Present Value of $1: $PV = FV/(1+i)^n$

Year	5%	6%	7%	8%	9%	10%	12%	14%	18%
1	.952	.943	.935	.926	.917	.909	.893	.877	.847
2	.907	.890	.873	.857	.842	.826	.797	.769	.718
3	.864	.840	.816	.794	.772	.751	.712	.675	.609
4	.823	.792	.763	.735	.708	.683	.636	.592	.516
5	.784	.747	.713	.681	.650	.621	.567	.519	.437
6	.746	.705	.666	.630	.596	.564	.507	.456	.370
7	.711	.665	.623	.583	.547	.513	.452	.400	.314
8	.677	.627	.582	.540	.502	.467	.404	.351	.266
9	.645	.592	.544	.500	.460	.424	.361	.308	.225
10	.614	.558	.508	.463	.422	.386	.322	.270	.191
11	.585	.527	.475	.429	.388	.350	.287	.237	.162
12	.557	.497	.444	.397	.356	.319	.257	.208	.137
13	.530	.469	.415	.368	.326	.290	.229	.182	.116
14	.505	.442	.388	.340	.299	.263	.205	.160	.099
15	.481	.417	.362	.315	.275	.239	.183	.140	.084
16	.458	.394	.339	.292	.252	.218	.163	.123	.071
17	.436	.371	.317	.270	.231	.198	.146	.108	.060
18	.416	.350	.296	.250	.212	.180	.130	.095	.051
19	.396	.331	.277	.232	.194	.164	.116	.083	.043
20	.377	.312	.258	.215	.178	.149	.104	.073	.037

received the $840 today, she could put it in the bank at 6 percent and would have a total value of $1,000 three years from now.

How do we choose what rate to use for the discount rate? This is the decision of the manager. Opportunity cost plays a role in this decision. In discounting future sums, the intelligent manager will choose a discount rate at least equal to the market rate of interest. Thus, if certificates of deposit were paying 10% and a manager were analyzing financial decisions on sums to be received at a future date, 10% would be one appropriate rate of discount.

In discounting computations, financial tables can be used to determine the factor for $(1 + i)^{-n}$. Table 12.3 is an abbreviated version of a discounting table for selected discount rates from 5 to 18 percent and values of n from 1 to 20.

To find the present value of a future sum using the table, we multiply the future sum by the appropriate factor, found at the intersection of the desired interest rate and time period. For example, we can determine the present value of $1,000 to be received in three years at a discount rate of 6 percent by multiplying $1,000 by the factor 0.840 (intersection of 6 percent and three years). Thus, $840 is the present value of $1,000 to be received in three years (discounted at 6 percent).

Compounding and Discounting in Use

Compounding and discounting are methods to determine the future value of a present sum or the present value of a future sum of money. It is important to recognize that compounding and discounting are opposite sides of the same concept. For example, the future value of $100 invested at 9 percent for five years is $153.90 ($100 × 1.539); the present value of $153.90 to be received in five years (discounted at 9 percent) is $100.04 ($153.90 × 0.65). (Rounding error accounts for why the two are not exactly equal.) However, this example illustrates how $1 today exchanges for (1 + i) dollars one period in the future, or alternatively a $1 payment made one period in the future exchanges for (1/1 + i) dollars now. The exchange *price* between present and future dollars is given by the interest rate.

Compounding and discounting provides the manager with two important tools for evaluating economic decisions. We illustrate how to use these concepts in the following examples.

Example 1. *Compound interest versus simple interest.* To illustrate the impact of compounding on the future value of money, consider $1,000 placed in the two different savings accounts listed in Table 12.4. In account A the $1,000 earns 6 percent in each time period (year) and is paid to the account owner at the end of each period. This is referred to as *simple interest* since interest is earned only on the principal amount. In account B the $1,000 earns 6 percent in each time period, but interest earned is deposited to the account and in subsequent time periods the interest also earns interest at 6 percent. This is referred to as compound interest since the depositor is "earning interest on interest."

As shown in Table 12.5, after one year both accounts have equal sums of interest plus principal amount. But in successive years the compound interest account grows more rapidly as interest is earned on interest.

Example 2. *Discounting a future revenue.* Suppose an investment firm guaranteed that all partnership shares would be worth $250,000 at the end of five years. How much would you be willing to pay today for a partnership share?

To find the present value of the $250,000 you must discount it at some rate for the five years. What is an appropriate rate? If you put your money in a bank sav-

TABLE 12.4   Alternative Savings Accounts

	Amount of Deposit	Type of Interest
Account A	$1,000	6% simple
Account B	$1,000	6% compounded

TABLE 12.5   Simple Versus Compound Interest

	Account Value After			
	One year	*Two years*	*Five years*	*Ten years*
Account A	$1,060	$1,120	$1,300	$1,600
Account B	1,060	1,124	1,338	1,791

ings account you might earn 6 percent. For certificates of deposit, you might earn 9 percent. For a good corporate bond you might earn 10 percent. Since all of these alternatives are fairly safe, the discount rate you use in your investment decision should be at least 10 percent. If you use 10 percent, then referring to Table 12.3, the appropriate discount factor is 0.621 (intersection of 10 percent interest and five years). Multiplying 0.621 times the $250,000 equals $155,250.

Therefore, if you can purchase a partnership share for less than $155,250 you may have a good deal. At prices greater than $155,250 the partnership is not as attractive as your alternatives. At a price equal to $155,250, you don't care whether you invest or not.

A reporter from *The New York Times* utilized discounting to look at the real cost of Steve Young's first football contract with the L.A. Express, a team in the old U.S. Football League (*N.Y. Times Service*, March 7, 1984). Young had signed the largest professional sports contract of the time, $40 million. But the reporter noted that much of the money ($34 million) was to be paid over thirty-seven years in installments and Young would only receive a salary of $200,000 and a signing bonus of $2.5 million in the first year. Using present value principles, the reporter discounted all the future revenues (using a bank interest rate of 11 percent) in Young's contract and concluded that the real value of the contract was $5.5 million—not $40 million. Thus, Steve Young's first professional contract was not exactly what it seemed to be in the sports headlines of 1984!

Example 3. *Discounting costs and discounting revenues.* A more common application of compounding and discounting in agriculture concerns the comparison of costs paid now to revenues received in the future. Suppose you were considering the establishment of an apple orchard and faced the hypothetical cost and revenue streams presented in Table 12.6. To decide whether or not to invest in the orchard, you would want to compare the total costs of establishing the orchard to the expected income from the orchard over its productive life.

TABLE 12.6 Establishment Costs and Estimated Income of an Apple Orchard

Year	Establishment Costs per Acre ($)	Estimated Net Income per Acre ($)
1	700	0
2	300	0
3	200	0
4	100	0
5	0	100
6	0	200
7	0	300
8	0	300
9	0	400
10	0	500
Total	$1,300	$1,800

In nominal terms it appears to be a good decision. Total revenues over the ten-year period will be $1,800 per acre and total costs are only $1,300 per acre (for a per-acre profit of $500). But the problem here and in many applications in agriculture is that the costs must be borne now while revenues will not be received until some time in the future. That means money will have to be borrowed and interest costs incurred (or interest foregone from a savings account) for several years before any income is available to repay the establishment costs.

Using compounding and discounting techniques we can evaluate costs and revenues at *one point in time*. By doing so, a rational economic choice can be made because the time element is factored out and all sums are compared on an equal basis. Assuming costs are paid at the beginning of the year and revenues are received at the end of the year, we can evaluate this decision at the beginning of year 1. For example, if all costs are discounted to the beginning of year 1 and all revenues are discounted to the beginning of year 1, then the time values of all sums will be on an equal footing and can be compared directly. (Remember, we could have picked any year for comparison as long as all costs and revenues were at the same point in time.) For illustrative purposes, if 6 percent is used in the discounting calculations then the sums appear as shown in Table 12.7. Assuming that costs are paid at the start of each year means that over the ten-year period the costs incurred in year 1 are paid in year 1; the costs incurred in year 2 would be discounted by one year; the costs incurred in year 3 would be discounted by two years; and the costs incurred in year 4 would be discounted by three years. The discount factors (taken from Table 12.3) used in Table 12.7 reflect this.

Assuming that revenues are received at the end of each year means that over the ten-year period the revenues received in year 5 would be discounted by five years; the revenues received in year 6 would be discounted by six years; and so on to the revenues in year 10, which would be discounted by ten years. The discount factors (taken from Table 12.3) used in Table 12.7 reflect this.

TABLE 12.7 Establishment Costs and Estimated Income of an Orchard

Year	Establishment Costs ($)	Estimated Net Income ($)	Discounted Costs (at 6%) to Year One		Discounted Revenues (at 6%) to Year One	
1	700	0	(1.00 × 700)	$700		
2	300	0	(0.94 × 300)	$282		
3	200	0	(0.89 × 200)	$178		
4	100	0	(0.84 × 100)	$84		
5	0	100			(0.75 × 100)	$75
6	0	200			(0.71 × 200)	$142
7	0	300			(0.67 × 300)	$201
8	0	300			(0.63 × 300)	$189
9	0	400			(0.59 × 400)	$236
10	0	500			(0.56 × 500)	$280
	$1,300	$1,800		$1,244		$1,123

By adding the values of the discounted costs and the discounted revenues, a more rational economic choice is possible. On an equal time footing, the present value of establishment costs is $1,244 per acre, while the present value of orchard revenues is $1,123. Revenues do not exceed costs, but a $121 per-acre loss results; hence, by incorporating time into the decision, this manager is able to avert an economic disaster.

Depreciation: Handling a Special Time Problem

A special time problem arises in any production process such as agriculture and agribusiness where some inputs are used in more than one time period. Examples of this are tractors, grain-drying facilities, irrigation systems, and buildings. These are commonly called **durable assets,** or **durable inputs,** because they are *used over more than one production period.* But the very nature of this type of input presents a problem for the manager: How do you measure the economic contribution of a durable input? Or, how do you allocate the costs of the input over its useful life?

To illustrate this problem, imagine how to handle a $50,000 expenditure for a new tractor when determining the year-end performance of a farm. Do you compare the entire $50,000 against the current year's income statement, or do you charge off a certain fraction of the cost each year until the tractor is sold, traded in, or retired from use? The second answer sure seems the correct one to your authors and should to you too.

In theory, the amount charged off as a cost in the current year should match the amount of the input "used up" in the current year. After all, durable assets do not last forever. Therefore, by matching the useful life to annual profit-and-loss calculations the manager can gain a measure of the economic contribution of a durable asset over several time periods. This procedure is called **depreciation,** and it is a commonly used concept among economists, accountants, and managers.

The concept of depreciation as a management tool is straightforward: *Inputs that are used over more than one year in the production of income should have a portion of their costs deducted for each year of useful life.* In other words, the input depreciates (decreases) in value as it is used. This concept is really a recognition of three economic sources of depreciation:

1. Depreciation should be a measure of the proportion of the asset used in the current time period. This is the idea of measuring "wear and tear" on machinery, buildings, and so forth.
2. Depreciation is also used as a measure of economic obsolescence of durable inputs. This is a recognition that some inputs become obsolete before they are "worn out" simply because new, more efficient inputs are available. An example of this is the replacement of four-row planters with new eight-row equipment, even though the four-row planters may be perfectly operational. New technology often affects the rate of economic obsolescence.

3. Depreciation, in theory, should also represent the reduction in the market value of an input over time. Some types of machinery may "hold their value," while other inputs decline in value rapidly after only one or two years of use. A new automobile is a prime example of the latter.

The method of handling depreciation, in theory, should be linked closely to the type of input in question and its unique characteristics of use, market value, and obsolescence. In reality, it is not possible to reflect accurately the useful life, decline in market value, and relative efficiency of every asset in the production process. The actual method of using depreciation for income accounting and tax purposes is defined by the Congress and the Internal Revenue Service (IRS). However, in terms of handling depreciation for individual investment decisions the manager is free to choose the method of depreciation. This section discusses three traditional general methods of depreciation and then identifies the options available under recent tax laws.

Methods of Handling Depreciation

For income tax purposes, the rules used for depreciating durable assets depend on whether the property is classified as nonrecovery property or recovery property by the IRS. Nonrecovery property includes all property purchased before 1981 or purchased after 1980 under very specific conditions; recovery property includes all property purchased after 1980. For property termed nonrecovery property there are three methods for determining the rate of depreciation: straight-line method, declining balance method, and sum-of-the-year's-digits method. The rules governing the rate of depreciation for property classified as recovery property depend on when the property was purchased. If the property was purchased between 1980 and 1986, depreciation is calculated using the Accelerated Cost Recovery System. For property purchased after 1986, depreciation is calculated using the Modified Accelerated Cost Recovery System. Confusing? Maybe, but we now discuss each of these depreciation methods.

Straight-line method. The most convenient depreciation method to calculate is the straight-line method. Annual depreciation under this method is calculated by estimating the useful life of the input and any value the input may have when salvaged (sold, traded-in, or junked) at the end of its useful life, or

$$D_s = \frac{OC - SV}{L}$$

where D_s is the annual depreciation using the straight-line method, OC is the original cost (purchase price), SV is the salvage value; and L is the expected life (useful life) of the asset. The straight-line method results in equal annual amounts for depreciation, an economic assumption that the input is used up proportionally over its useful life.

To illustrate straight-line depreciation, consider a tractor that is purchased for $10,000. In its intended use the tractor would normally be used five years before being sold and a replacement purchased. An estimate of its salvage value after five years is $1,500. Straight-line-depreciation would be calculated in this manner:

$$\frac{\$10,000 - \$1,500}{5} = \frac{\$8,500}{5} = \$1,700$$

Annual depreciation for this tractor is $1,700 per year with the straight-line method.

Declining balance method. A second method of depreciation is declining balance. This method causes much higher estimates of depreciation in the early years of an input's life. Since depreciation is a deduction in calculating business expenses, this method produces a favorable tax advantage. It also reflects the economic facts of input usage where market value for a machine or other piece of equipment drops markedly after a year of use. Declining balance depreciation is calculated by multiplying (depreciation rate/useful life) by the remaining book value each year, or

$$D_d = R \times RV$$

where D_d is the annual depreciation using declining balance method, R is the depreciation rate, and RV is the undepreciated value of the asset at the start of the accounting period. In the first year, $RV = OC$. In succeeding years RV is equal to the previous year's RV less the depreciation in the previous year. Under this method, the salvage value is not deducted from the original value before computing depreciation, and the rate of depreciation (R) may be up to twice the rate of decline ($1/L$) allowed under the straight-line method. For example, a common depreciation rate used is 200 percent, which is referred to as a double declining balance, where the rate of decline is twice that of the straight-line method. Using the previous example with a depreciation rate of 200 percent and a remaining book value of $10,000, the first year's depreciation would be

$$\frac{200\%}{5} \times \$10,000 = 0.4(\$10,000) = \$4,000$$

In this case the depreciation rate is 0.4 (200%/5) which is twice the rate, 0.2 (100%/5), used in the straight-line method. Accordingly, the first year's depreciation would be $4,000. In the second year the depreciation would be 0.4(10,000 − 4,000) = $2,400 (purchase price less the depreciation in the first year times the depreciation rate); depreciation in year 3 would be 0.4($10,000 − 4,000 − 2,400) = $1,440 (purchase price less the depreciation taken in years 1 and 2 times the depreciation rate); and so forth until the remaining book value equals salvage value. Other applicable rates for use under this method include 50 and 125 percent, which would reduce the amount of annual depreciation accordingly.

Sum-of-the-year's-digits method. A third method of calculating depreciation on durable assets is the sum-of-the-year's-digits method. This procedure is based on the application of a diminishing rate to a constant value. This produces a rate of depreciation higher than would be found under the straight-line method but generally lower than that found under the double declining balance. In this method, the useful life of the asset is divided by the sum of a year's digits, which is multiplied by the asset's purchase price less salvage value to determine the depreciation taken in a given year, or

$$D_y = \frac{RY}{S}(OC - SV)$$

where D_y is annual depreciation using the sum-of-the-year's-digits method, RY is the estimated number of years of useful life remaining, S is the sum of the numbers representing years of useful life (i.e., for an asset with five years of useful life S would be $1 + 2 + 3 + 4 + 5 = 15$), OC is the original cost (purchase price), and SV is the salvage value of the asset. Using the previous example with a useful life of five years (the sum of which is used in the denominator of the formula) the first year's depreciation would be

$$\frac{5}{15} \times (10,000 - 1,500) = \$2,805$$

The first year's depreciation would be $2,805. In the second year the fraction would be 4/15, in the third year 3/15, and so forth, being multiplied times the constant factor of $8,500 (purchase price less salvage value).

The three methods of handling depreciation are illustrated in Table 12.8 using the example of the $10,000 tractor. Since the amount of depreciation affects the amount of income taxes, those depreciation schedules that "accelerate" depreciation by lumping it into the first two or three years are favored methods. The method actually chosen for a particular return depends on what the tax laws permit for certain classes of inputs, the ease of computation, and the importance of the tax advantage (or avoidance) to the taxpayer.

Prior to 1981, managers could choose among these three methods for calculating depreciation on property for tax purposes. However, comprehensive revisions in the federal tax law passed since 1981 have simplified depreciation accounting for tax purposes and left managers with little choice between depreciation methods.

TABLE 12.8 Comparative Depreciation Amounts Under Alternative Calculation Procedures[a]

Year	Straight Line	Declining Balance[b]	Sum-of-the-Year's Digits
1	$1,700	$4,000	$2,805
2	1,700	2,400	2,295
3	1,700	1,440	1,700
4	1,700	660[c]	1,105
5	1,700	0	595

[a]Assumptions: Tractor cost $10,000; salvage value of $1,500, and useful life of 5 years.
[b]Calculated using double declining balance method (or 200%).
[c]Tractor is fully depreciated at a salvage value of $1,500 thus leaving only $660 for depreciation in the fourth year and zero for the fifth.

Accelerated cost recovery system. The standard procedure for handling depreciation on property purchased after 1980, but prior to 1987, is the Accelerated Cost Recovery System (ACRS). Under the ACRS all depreciable property is classified as three-year, five-year, ten-year, or fifteen-year real property. (Certain types of property may not qualify for ACRS and are depreciated under the old methods already mentioned.) This ACRS classification system has little relationship to actual useful life of property or the year-to-year change in value. The ACRS is a tax-saving depreciation method that permits a more rapid write-off (or cost recovery) for tax accounting. In addition, calculations of depreciation no longer depend on the determination of the estimated useful life of the property, the salvage value, or the choice of depreciation method.

For managers using ACRS, the important decision is the proper classification of depreciable property. In agricultural operations, some examples of property in each classification are as follows:

Three-year property: Personal property with a short useful life, such as automobiles, pickup trucks, and tools. Race horses over two years old in stud service and breeding hogs are three-year property in most cases.

Five-year property: Most farming and office equipment, single-purpose farm buildings such as grain storage bins or greenhouses, and breeding beef and dairy cows.

Ten-year property: This class includes certain real property, such as mobile homes.

Fifteen-year property: Includes all real property, such as farm buildings and most land improvements not included in the first or second categories.

In general, ACRS calculations to determine annual depreciation involve multiplying a fixed percentage from the tax law by the original cost of the property. The

ACRS, introduced in the 1981 Economic Recovery Tax Act, removed areas of dispute farmers often had with the IRS over the useful life of an asset, the salvage value of the asset, and the method of depreciation.

New property, and most used property, purchased between 1980 and 1987 is subject to the depreciation rules set out under the ACRS. However, new and used property purchased after 1986 is subject to the depreciation rules set out under the Modified Accelerated Cost Recovery System introduced in the 1986 Tax Reform Act and further modified under the Omnibus Reconciliation Act of 1993.

Modified accelerated cost recovery system. The current standard procedure for handling depreciation on tangible property purchased after 1986 is the Modified Accelerated Cost Recovery System (MACRS). MACRS consists of two systems that determine how property is depreciated: the General Depreciation System (GDS) and the Alternative Depreciation System (ADS). The main difference between the two is that ADS generally provides for a longer recovery period and uses only the straight-line method to figure depreciation. However, unless the manager is required by law or elects to use ADS, GDS is typically used to calculate depreciation. How much of the cost is recovered in a given year is determined by the asset's cost recovery class and whether ADS or GDS is used.

To figure the MACRS deduction each year, the manager needs to know the following details about the property:

1. The basis;
2. The placed-in-service date;
3. The property class and recovery period;
4. Which convention to use; and
5. Which depreciation method to use.

The basis of an asset is essentially the purchase price, plus any sales tax, freight charges, and installation and testing fees, paid for the property.

The placed-in-service date applies to when the asset is purchased and available for a specific use. Thus, even if the property is not being used it is in "service," and hence depreciation can be taken, when it is ready and available for its specific use.

Under MACRS the property class and recovery periods are specified by the IRS. The property classes (three-year, five-year, seven-year, ten-year, and fifteen-year property) are the same as under ACRS. The recovery period for each property class depends on whether the GDS or ADS system is used and can be found in the *Farmer's Tax Guide.*

To figure the annual depreciation for both GDS and ADS, one of three conventions is used. The half-year convention is generally used on property other than nonresidential real and residential rental property. Under this convention the manager treats all property placed in service, or disposed of, during a tax year as placed in service or disposed of at the midpoint of that tax year. This means that no matter when in the year you begin or end the use of an asset, it is treated as if its use began or ended in the middle of the year.

Under certain specific circumstances a farmer might have to use either the mid-month convention or the mid-quarter convention. Under the mid-month convention all property that is placed in service or disposed of during a month is treated as though it were placed in service or disposed of in the middle of the month. The mid-quarter convention treats all property placed in service or disposed of during a tax year as though it were placed in service in the middle of the quarter. However, unless otherwise required by the IRS, the half-year convention is typically used.

In general, the MACRS calculations to determine annual depreciation involve multiplying a fixed percentage from the tax law by the original cost of the property. The most common MACRS used for the property classes is 150 percent declining balance over the recovery period specified by GDS using the half-year convention. The percentages from the *1996 Farmer's Tax Guide* are presented in Table 12.9. For example, if you purchase a tractor (seven-year property class) for $20,000, the annual depreciation for year 1 is $2,140 ($20,000 × 10.7%). For year 2 the annual depreciation is $3,820 ($20,000 × 19.1%); the annual depreciation for year 3 is $3,000 ($20,000 × 15%); and so on.

The manager can choose either the straight-line method over the ADS or GDS recovery periods or 150% declining balance over the ADS recovery period as an alternative depreciation method. However, electing to use one of these alternatives affects the depreciation taken on other assets purchased in the same tax year. Deciding to use the straight-line method for one item in a property class means the straight-line method applies to all the property in that class purchased in the same tax year, and once the choice is made, it cannot be changed. When the ADS election is taken on one asset in a property class it applies to all property in that class purchased during that tax year, and once the election is made, it cannot be changed. Thus, choosing an alternative depreciation method is very constricting for the manager and will affect other decisions concerning tax management.

TABLE 12.9 Modified Accelerated Cost Recovery System Depreciation Rates

Year	Three-Year Property	Five-Year Property	Seven-Year Property
1	25.0%	15.0%	10.7%
2	37.5	25.5	19.1
3	25.0	17.9	15.0
4	12.5	16.7	12.3
5		16.7	12.3
6		8.3	12.3
7			12.3
8			6.1

In general, MACRS is designed to permit more rapid tax write-off of depreciation. The economic concepts behind depreciation are secondary to the importance of tax accounting. The time concern for the manager is now focused on an investment decision in which profitability and tax advantage play equally important roles.

Time plays an important role in the decision-making process. However, as the time period between when the decision is made and the results of the decision are observed lengthens, uncertainty surrounding the outcome of the decision grows. Thus, understanding the concepts of risk and uncertainty, and how they affect the decision-making process, is also important for managers.

RISK AND UNCERTAINTY

In this section we take another step toward more realistic decision making. Up to this point, the economic analysis presented has assumed that the manager knows all things with certainty or "perfect knowledge" as the term is used in economics. Now we relax this condition of perfect certainty and introduce the concepts of risk and uncertainty.

Risk and Uncertainty Defined

If perfect knowledge or certainty were available to the manager, then the raison d'être for the existence of a manager disappears. With perfect knowledge, decisions become the routine calculations of formulas and comparisons of costs and returns. Of course, this condition exists only in the classroom and the simplified economic analysis of textbooks.

The uncertainties of actual decision making take many forms. For instance, in the economic analysis in this book, we have assumed we know what the product prices will be over the time period of analysis. In real-world markets, product prices change, sometimes dramatically, for reasons of inflation, business cycles, strikes, or natural disasters. Likewise, costs of production can change with the introduction of new technology (cost lowering) or increases in crude oil prices (cost increasing). Total output of almost any crop is affected by weather conditions over the growing season, diseases, pests, accidental contamination, natural disasters, and harvest conditions. The actions of the federal government often create uncertainty for managers in agriculture whenever important government policy is changed. The embargo by the U.S. government of grain sales to the Soviet Union in 1980 is an example. Market conditions can also be uncertain because U.S. exports depend on agricultural conditions in other countries, international trade policies, and world economic and political conditions. In economics, these situations of uncertainty can be divided into two types:

> *Risk:* a situation where the outcome is unknown, but the probability of alternative outcomes is known; and
>
> *Uncertainty:* a situation where the probabilities of different outcomes are unknown.

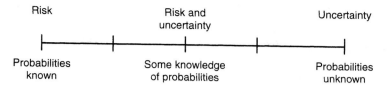

FIGURE 12.1 Risk and uncertainty as a continuum of possible situations.

In terms of these definitions, risk and uncertainty are the polar extremes.

In the real world of agriculture, though, there are many situations that have the characteristics of both risk and uncertainty. Thus, it is convenient to think of risk and uncertainty as a continuum (see Figure 12.1).

The most straightforward example of risk is a coin toss. The outcomes (head or tail) are known and so are the probabilities of each outcome (0.5 for each fair toss). This is an *a priori* risk situation because the probabilities are known *a priori*, or before the fact. If the probabilities are not known *a priori,* then the risk situation becomes statistical, and probabilities are calculated and assigned to expected outcomes based on historical records. An everyday example is the weather forecast that says, "There is a 60% chance of rain tomorrow." In risk terms, that is an expression of a six-out-of-ten probability of rain tomorrow. The weatherman has assigned the probability as a statistical risk of rain.

A common example of uncertainty is the price changes in agricultural markets. A manager has no real basis for assigning a probability to the occurrence of any price at some point in the future. Price outcomes are, generally speaking, uncertain, and influenced by conditions in world markets, government policy, monopoly power, politics, and so forth.

In many cases though, the distinction between a risk situation and an uncertainty situation is blurred. This is represented on the continuum in Figure 12.1 by the middle area where some information is known about the probability of certain outcomes. In practice, risk/uncertainty distinction is made more apparent by the strategies involved in handling the managerial problems imposed by risk and uncertainty. In short, if it is an uncertain situation, insurance is not possible and a variety of managerial strategies must be substituted.

Risk and Uncertainty in Agriculture

Agriculture is often characterized as a uniquely high-risk and high-uncertainty sector of the economy. The production of food and fiber is subject in large part to the weather conditions outside the control of the producer (an exception would be greenhouse production of vegetables and flowers). In addition, agricultural producers do not have much influence on the economic aspects of the production and marketing of their output. More detail on risk and uncertainty in agriculture can be developed by looking at two general categories: on-farm sources and off-farm sources.

On-Farm Sources of Risk and Uncertainty

Producers of crops and livestock do not know at the beginning of the production season what the total yield or output will be. The primary source of this uncertainty is the vagaries of weather—temperature, moisture, and wind. Some parts of the United States are more subject to adverse weather fluctuations than others. The Great Plains have experienced recurrent droughts, the Ohio–Mississippi River valley is subject to severe flooding, and the southern and eastern coastal states are prone to severe wind and rain episodes.

Yield uncertainty is also caused by other biological factors, such as disease and predator or insect damage. The technological advances of the past twenty to thirty years have markedly reduced but not eliminated this factor of yield uncertainty.

In addition to the yield uncertainties of weather, disease, and pests, agricultural producers face certain risks of weather-related natural disasters in which crops and livestock can be partially or totally destroyed. A partial list of these natural disaster risks would include frost, hail, floods, hurricanes, tornadoes, wind, earthquakes, volcanic eruptions, and snow.

Man-induced disasters are becoming a greater danger in agriculture. An example is toxic contamination of animal feeds or misapplication of herbicides and pesticides to field crops.

Another on-farm source of risk and uncertainty derives from the family nature of American agriculture. Despite strong incentives to incorporate the family farm operation, most farmers retain individual proprietorship. Increasing age, poor health, or death causes uncertainty about the management of farm resources.

Off-Farm Sources of Risk and Uncertainty

Agricultural producers sell products in what is probably the most competitive market in the U.S. economy. Their crops and livestock with few exceptions carry no "brand names" (SUNKIST on oranges is an exception), and individual producers cannot economically advertise products. The result of this situation is that each producer has no impact or control over the price received for products. For the producer, prices, and ultimately income, are uncertain.

Most agricultural markets are characterized by substantial short-run and long-run price variability. This price variation is caused by business cycles and inflation in the domestic and world economies, changes in the amounts produced during each season, and such other factors as population growth, personal income, and changes in food tastes and preferences.

New technological advances in agricultural machinery, chemicals, or plant varieties can disrupt traditional production and cause some confusion and considerable uncertainty. In the past few years, many crops that required a great deal of manual labor during a growing season and harvest can now be handled entirely by machines. Sugar beets in the West, tomatoes in California, potatoes in Maine, and tobacco in North Carolina are examples. New plant varieties and new herbicides and pesticides are also examples of potentially disruptive technological advances. Many producers may be uncertain about whether or not to adopt the new technology until it has been "tried and proven."

The federal government is itself responsible for uncertainty in agriculture. Farmers are subject to new regulations each year regarding the use of such inputs as pesticides and herbicides and the handling of such commodities as milk and vegetables to prevent toxic contamination. A useful pesticide can be removed from the market by the Environmental Protection Agency because suspected residues may be harmful. Livestock producers may have to adjust feed lot production facilities to meet environmental requirements for water quality and waste disposal. These regulatory rules are by no means unique to agriculture, but they do represent important sources of off-farm uncertainty.

Minimizing Risk

In the face of a risk situation, economists have in theory identified three behaviors (or attitudes) toward risk:

1. Risk seeking;
2. Risk indifference; and
3. Risk averting.

The risk seeker is the person who desires the risky situation. This is the "gambler" behavior where, given the choice between a more risky and a less risky venture, the risk seeker will choose the more risky alternative. The risk indifferent person is not affected by the risk involved but only by the expected values. Finally, the risk averter will always choose the less risky alternative.

Most managers in agriculture (and students for that matter) can be described as risk averters. Given two alternative money-making ventures with the same expected returns, the typical behavior is to select the less risky venture. Alternatively, if a manager is faced with risk, the typical behavior is to minimize the risk.

Managers face several options in minimizing risk, as shown in Figure 12.2. If the risk situation has known or estimated probabilities for economic loss, insurance may be available. If the risk situation has only subjective (personal) estimates of probabilities, most commonly found in investment decisions, then financial management strategies must be employed.

Minimizing Risk Through Insurance

The decision to buy insurance should not be automatic. People react to risk in different ways. For many, tradition governs the purchase of insurance with maxims such as these:

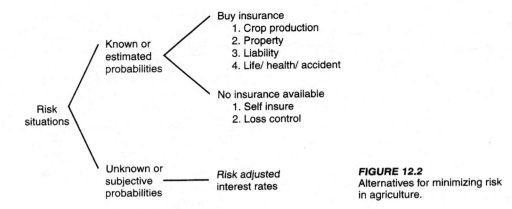

FIGURE 12.2
Alternatives for minimizing risk in agriculture.

"Insure the crop in the bin, not in the field."
"Life insurance on Dad but no one else."
"Fire insurance on the house and barn."

Whether or not to buy insurance against risk in agriculture should be an economic decision. In making that decision, two factors are critical:

1. How much loss can the manager withstand without insurance?
2. What are the trade-offs between insurance costs and potential losses?

Evaluating how much loss a manager can withstand will include a consideration of the financial health of the enterprise, the impact of loss on current operations, and the willingness of the manager to take risks. Would the loss of the entire crop due to a hailstorm bankrupt the enterprise or is there sufficient income from other crops, livestock, and other sources to maintain solvency in the face of severe hail damage? Some people avoid risky situations despite the chance to be a "winner" if events turn out satisfactorily. The less able a person or business is to withstand significant economic losses, the more likely insurance is an appropriate option.

Even if insurance is available for certain risk situations, it may not be the correct economic choice. For instance, buying collision insurance on a beat-up farm pickup with a market value of $300 is not economic. The trade-off between the insurance costs and the small potential benefit suggests the risk should be borne without insurance. This type of trade-off should be examined in each risk situation facing the manager.

In financial jargon, this evaluation of trade-offs between insurance costs and benefits can be expressed as an expected value. To calculate an expected value, you must know the alternative outcomes and the probabilities associated with each outcome. For example, in a coin toss there are two outcomes (heads, tails) and the probability is one in two (or 0.5). The expected value of a coin toss is equal to the probability of each outcome multiplied by the money involved. Thus, in a coin toss where you win $1 for heads or lose $1 for tails the expected value (EV) is

$$EV = \$1 \,(0.5) + \$1 \,(0.5) = \$1$$

The calculated expected value, $1, tells you what you already know about coin flips: You'd win about as much as you'd lose.

Expected values are the basis of the insurance industry. For an individual corn producer, one crop a season represents one coin toss. Over a lifetime the corn producer faces a relatively few number of "tosses," maybe thirty to fifty crops. But the hail insurance company may insure thousands of fields each year. The company knows from experience that a certain number of hail losses will occur each year, maybe 1 field in every 100. The company can then use its experience to calculate the expected values of hail damage claims. If it insures 1,000 fields for $25,000 each, then the expected value of losses is:

$$1,000 \times (1/100) = 10 \text{ fields at } \$25,000 \text{ each, or } \$250,000$$

Given this expected value of annual claims for hail damage, the insurance company can determine a policy premium sufficient to cover losses, operating expenses, and profits.

Between the individual producer and the insurance company, the policy becomes a bet or gamble. The corn producer pays the premium, "betting" he might lose, and the company issues the policy "betting" it won't. The producer in effect has one coin toss a year. The insurance company has hundreds or thousands.

Expected values can help analyze the trade-offs between insurance costs and benefits. The probabilities for individual losses will necessarily be subjective estimates based on personal experience.

Types of Insurance

Insurance comes in all types, terms, conditions, coverage, and costs. The unique risks of agriculture can be met with four general types of insurance: crop production, property, liability, and life/health/accident. An individual program of insurance coverage depends on personal characteristics, willingness to assume risk, and financial attractiveness of alternative policies.

Crop production insurance. The Federal Crop Insurance Corporation (FCIC) has insured against some of the risks of crop production since 1938. The FCIC began on an experimental basis in a limited number of counties covering only a few crops, mostly wheat and cotton. Over the years its insurance program has expanded to include 26 crops and 1,522 counties.

Only a fraction (14%) of eligible crop acreage is insured each year by FCIC. This is mainly due to the low level of protection—50 to 60 percent of costs of production. Legislation in 1980 thoroughly upgraded coverage and expanded eligibility to a wide variety of crops and provided for federally subsidized premiums. Recent farm legislation supplemented the crop insurance program with a new catastrophic coverage level available for a small processing fee. In the past, producers were required to have crop insurance to be eligible for farm program benefits. However, with the passage of farm legislation in 1996 this is no longer the case if producers waive all emergency crop loss assistance.

Besides FCIC, crop production can be protected against losses from hail and fire through private insurance companies. About twice as much acreage is under private hail-insurance policies as is under current FCIC insurance.

Property insurance. Basic property insurance is protection against damage, destruction, or theft of farm machinery, buildings, livestock, and building contents (tools, equipment, and so forth). Not all property need be insured. Fully depreciated or used equipment and old buildings having small market values would be a poor trade-off against policy premiums.

Liability insurance. Liability insurance provides compensation for property damage or personal injury to other parties for whom an individual may be legally liable. This insurance protects against lawsuits arising from such things as damage by livestock, accidental injury to a farm visitor, or misuse of chemical sprays. Workmen's compensation laws and certain liability insurance policies will also cover injuries to employees on farms or farm businesses.

Life/health/accident insurance. Family management and labor are an important resource in agriculture. Life insurance can compensate for some of the economic losses and disruptions caused by the death of a farm owner or operator. Health and accident policies can insure against the medical costs resulting from sickness or accidental injury.

Minimizing Risk When Insurance is Unavailable

Insurance is not available for all the risks in agriculture. Two classic examples are egg breakage and grain spillage. In many risk situations such as these the costs of insurance have been too prohibitive to be profitable for private companies. Another reason insurance may not be available for certain risks is the lack of actuarial historical (statistical) experience to form a basis for issuing policies. Damage from nuclear accidents is an example of this. Self-insurance is the only alternative available to the manager.

If no insurance is available for a particular risk, then the manager must focus on loss-control measures and other management strategies. The criterion for management decision in loss control is similar to that used in buying insurance: the costs of risk protection versus the potential benefits. In the case of egg breakage, the costs of new packaging materials or machinery must be weighed against the gains of fewer broken eggs.

Risk-Adjusted Interest Rates

Financial investments are risk situations. The probabilities of the expected returns are not known or subject to actuarial estimate, as in insurance situations. Most often, financial risk is subject to personal estimates of the possibility of a favorable outcome. Sometimes sophisticated statistical techniques are employed to estimate the chances (probabilities) for success.

In the face of risk with unknown probabilities for success, the most common management behavior is to adjust the interest (or discount) rate for the level of risk. Government bonds are considered a risk-free investment. If the return on a government bond is 7 percent (and risk-free), then the risk-averting manager will demand a higher return on other investments where there is risk involved. For instance, an investor may be indifferent between the 7 percent government bond, a 10 percent expected return on company stock, or a 25 percent expected return on an Alaskan oil well. In the case of the stock and the oil well, the investor had reacted to risk but adjusted the interest rate (or discount factor) that he or she considers necessary before he or she makes an investment.

The adjustments to the interest rate are considered risk premiums. For increasing risk, the money manager wants a premium (or return) over and above the amount that could be earned on risk-free investments.

Risk-adjusted interest rates are a routine behavior by managers in agriculture. For example, a farmer may have years of experience in growing corn with seed from company X. A salesman for company Y tries to persuade the farmer to switch to his seeds, hybrid Y. This is a situation of risk, and the farmer will want to be "rewarded" for taking a chance on hybrid Y. This reward most commonly comes in the form of the promise of increased yield (or disease or drought resistance). The farmer asks himself, "Is the chance for higher yields worth the risk of using hybrid Y, a seed totally new to me?" The consideration is the same as in financial investment decisions: Is the probable expected return sufficient to balance the risk involved?

Minimizing Uncertainty

Uncertainty describes those situations in which there is no knowledge of the probabilities of certain outcomes. For example, the dairy price-support program could be terminated by the next Congress. This is uncertain. The eruption of the volcano Mount St. Helens could affect long-term weather patterns. This is uncertain. The export market for soybeans in Europe could change. This also is uncertain.

Uncertainty is not an entirely negative condition. Economic growth and progress are also conditions of uncertainty. In situations of no change, the raison d'être for the manager disappears. Decisions can be made by adherence to routine, and historical tradition, and custom.

In the face of the opportunities and problems of uncertainty, the decision maker cannot buy insurance to guarantee an outcome or compensate for a loss.

Other strategies must be pursued in order to manage effectively an economic unit in uncertain situations.

Flexibility

To meet the challenge of uncertainty, the manager should plan for flexibility. In essence, flexibility involves modifying the most profitable business plan to avoid losses or pursue new opportunities. Flexibility implies a modification to specialization in production. Here are some examples of flexibility planning:

The use of dual-purpose or multipurpose farm structures instead of specialized buildings.

Annual crop production instead of perennial orchard or vineyard crops.

Retention instead of disposal of seemingly obsolete machinery and equipment.

Flexibility is a characteristic of management attitude. The flexible manager is willing to try out new ideas, seek new information sources, test new techniques, and experiment with new production processes. The potential for growth and profitability is the reward for a flexible management attitude.

Diversification

Some uncertainty in agriculture can be minimized by diversification in production. Simply put, in the product–product choice (Chapter 6) diversification is the selection of two or more products for production. This management strategy is aimed at reducing the production uncertainties (yield and price) of a single crop. It is often argued that proper diversification will protect against the possibility of wild swings in the economics and weather affecting a single commodity or livestock.

Again, diversification comes at a cost. The "best" plan may be a specialized production plan for corn or wheat or cow/calf production. A diversified plan may include corn/soybeans, wheat/peas, and row crops/cattle combinations. Proper diversification combines outputs that are *subject to different physical and economic conditions*. These are the two basic motivations for diversification:

1. Proper diversification will divide the physical and economic uncertainty between two or more outputs. This amounts to "not putting all the eggs in one basket."

2. Diversification can improve efficiency in utilizing the resources available. For example, machinery used in the fall for wheat can be used in the spring for peas. Land that is subject to erosion can be better utilized with multiple-cropping of row crops and forage.

The trend in American agriculture has been toward less diversification in recent years. To take advantage of some new technological advances, such as four-wheel-drive tractors, electronically controlled harvesters, and disease-controlling drugs for livestock, producers have had to specialize and expand operations. This growth has been required in order to make the new equipment economical. The result has been larger and more specialized farms where risk and uncertainty have increased.

Production Contracts

Price uncertainty can be reduced through various forms of contracting. A contract can be signed for the sale of a crop before harvest and sometimes before planting. This removes the price uncertainty from the producer, leaving him to concentrate on good production management, especially cost minimization. These production contracts are common among producers of row crops such as potatoes, onions, and beans and have been used since colonial times in the United States.

Production contracting has been used in another way in the poultry and egg industry. The economic advantages to coordination of poultry production have led to a system of contracting between the marketing and producing phases. One firm may control breeding, feeding, slaughter, processing, and distribution of chickens and turkeys. These firms are described as being vertically integrated since they control poultry production from egg to supermarket.

Poultry farmers can contract with vertically integrated poultry firms to supply the buildings, labor, and management. The firm provides the baby chicks, feed, and other supplies. At the end of the contract period the farmer receives a fixed payment and the birds proceed to the next step in the processing channel. Price uncertainty for the farmer is eliminated.

Production contracts can also be used on a collective basis with groups of producers. The contract idea is the same: fixed price and forward delivery date. The contract is signed between one or more processors/buyers and a group of producers acting through an association. The grower association guarantees delivery of a certain amount and quality of product at harvest time and may distribute the proceeds of the sale back to the association members. Group production contracts are common in fruit crops and sugar beets.

Future Contracts

Another contracting arrangement that can be used to reduce price uncertainty is the futures contract. A futures contract is a promise to buy or sell a standardized quantity of certain agricultural products at a fixed price and future date. These contracts have been traded on commodity markets such as the Chicago Mercantile Exchange and the Chicago Board of Trade since the 1800s. Not all agricultural commodities are traded on these exchanges but the ones that are include wheat, corn, soybeans, pork bellies, beef carcasses, live beef, and live hogs.

Unlike a production contract, actual delivery of the product almost never occurs when the futures contract matures. Fewer than 1 percent of futures contracts are fulfilled by actual delivery of the product. The remaining 99 percent are fulfilled by an equal but opposite contract promise to buy or sell.

The prices for futures contracts represent the free market's best estimate of what market prices will be during the delivery period. These price quotations can be obtained from newspapers or financial institutions. Of course, these prices fluctuate, sometimes wildly during given time periods, as new information is made public about conditions surrounding a particular crop or livestock (for example, drought, disease, or yield).

Individual producers can use futures contracts to reduce price uncertainty by hedging. Basically hedging amounts to protection from price changes by making a futures market transaction that will offset a cash market transaction. For example, if a corn farmer had a crop to be harvested in the fall and she planned to sell it for cash but was worried that the price might fall below current levels, that farmer could enter the futures market (through a broker) and sell a contract for corn. That action guarantees a fixed price for the futures contract. The farmer offsets the futures position by buying a contract for corn and sells her corn on the cash market. By using the futures market, the losses/gains sustained in the cash market are offset by the gains/losses in the futures market. This is the perfect hedge in simplified form.

Futures contracts can be complex; the manager who uses them to reduce uncertainty needs the advice of a reputable broker. Knowledge and experience become necessary requisites for proper use of the futures market.

Liquidity

Liquidity is a term from financial management that refers to *the ability of a business to meet its financial commitments as they come due.* It is a short-run concept. A firm may be a profitable enterprise in the long-run, but if it lacks liquidity it is bankrupt in the short-run.

Liquidity relates directly to uncertainty. A manager should handle cash and credit resources to insure liquidity for purposes of routine obligations and unanticipated future events. In its simplest form, liquidity is a small cash reserve against unanticipated needs, for example, Mom's money in the sugar bowl. In more sophisticated financial terms, liquidity involves managing ideal cash balances and proper rationing of internal and external credit sources. Regardless of the level of sophistication, the need for liquidity is clear; the firm that cannot maintain liquidity cannot meet the challenge of uncertainty.

The seasonal nature of agricultural production presents special liquidity problems in most cases. Cash may be needed in large lump sums every once in a while instead of in smaller amounts on a steady monthly or weekly basis. On a dairy farm the cash receipts are received on a more regular basis than on a wheat farm. Both have regular financial obligations (electric bills, taxes, wages, and so forth), but each faces a different problem of maintaining liquidity. The wheat farmer may have large seasonal deficits that are reduced only in the fall.

How is liquidity maintained? The most obvious source of liquidity is cash balances. The sale of products generates cash for the firm. A certain percentage must be kept in reserve to maintain liquidity. Credit borrowing is also a method of maintaining liquidity. A farmer may routinely borrow a certain sum to meet anticipated expenses and also have a "credit line" of additional credit for borrowing money needed to meet unanticipated expenses or profitable new opportunities.

Government Intervention

The unique high risk and uncertainty in agriculture have spawned many public programs designed to mitigate these problems. Many of these have been designed to reduce price and income uncertainty for farmers. National programs of price supports and production controls have been used in the past on such important commodities as corn, wheat, rice, tobacco, peanuts, and cotton. These programs have reduced uncertainty because the USDA has had programs to loan money on harvested eligible crops and thus set a floor on farm gate prices.

Government intervention to reduce uncertainty operates indirectly through the marketing agreements and orders for milk and several fruit and vegetable crops. All marketing agreements and orders are programs supervised by the USDA and are instituted only after a favorable referendum by producers. A marketing agreement is binding only on the producers who sign up. A marketing order is binding on all producers.

The purpose of the marketing order is to regulate the quantity of a commodity entering the market in order to influence and stabilize prices. Marketing orders do this by controlling the quantity and quality of a product. The most important example is fresh milk. The milk price is regulated by Congress and the USDA. The milk market order regulates how much milk can be sold as Class I for fresh consumption, receiving the highest price, and how much milk is diverted to cheese production and other products.

SUMMARY

This chapter has attempted to introduce time into the economic analysis that has been developed in the preceding chapters. Time considerations are important in economic decision making because the real world is a dynamic rather than static place where the passage of time affects the value of money and the ultimate consequences of decision. This is especially true in the broad field of agriculture where the production process is seasonal in nature and subject to the whims of Mother Nature.

We examined two important time problems in this chapter: the time value of money, and the time problem of durable inputs in the production process.

Because people have a strong, positive time preference for money, compounding and discounting are used to facilitate decisions where costs and revenues are spread out over several time periods by factoring out the time element dealing with costs and revenue at the same point in time. The performance of a durable input presents another problem in economic analysis, and this has been handled through the use of depreciation.

When we relax the assumption of perfect certainty in economic analysis, we must deal with conditions of risk and uncertainty. Theoretically risk and uncertainty are polar extremes on a continuum of situations where knowledge about the future is limited and unreliable.

Agriculture is an industry with uniquely high levels of risk and uncertainty from two sources. On-farm sources of risk include yield, weather, disease, natural disaster, age, and health. Off-farm sources of risk include price variability, technological changes, and government policy.

Managers react to risk and uncertainty in different ways but universally seek methods of reducing risks and minimizing uncertainty. For risk situations, this can be done through the purchase of insurance in some cases and through the adjustment of interest or discount factors in investment decisions. In uncertainty situations, the manager seeks minimization through flexibility, diversification, liquidity, production contracts, futures contracts, and government intervention.

QUESTIONS

1. Calculate the following present values:

    $100 received in five years, discount rate = 6 percent
    $100 received in ten years, discount rate = 9 percent
    $100 received in fifteen years, discount rate = 12 percent
    $100 received in ten years, discount rate = 6 percent
    $100 received in fifteen years, discount rate = 9 percent
    $100 received in twenty years, discount rate = 12 percent

2. Explain the differences between on-farm and off-farm sources of risk and uncertainty.
3. Define compounding and discounting. Why are they important considerations for a manager?
4. Several states now have legalized lotteries and associated games of chance. One type of game makes the winner an "Instant Millionaire." If you hold the winning ticket you win a million dollars. Your winnings do not come in one check but in twenty checks spread over twenty years ($50,000 each). Question:

If you are the winner are you really a millionaire? What is the present value of the million dollars received in twenty annual increments of $50,000? (Use a discount rate of 12 percent.)

5. What would the future value of $10,000 be if you invested it at 7 percent over a three-year period? If the same $10,000 is invested at 14 percent over a ten-year period?

6. How does compound interest differ from simple interest? If you were a bank manager, which type of interest would be more profitable to offer? Remember, you're not the only bank in town.

7. Discuss why the concept of depreciation is important to a farm manager.

8. Explain how time complicates the decision-making process of a manager.

9. What is the formal difference between risk and uncertainty? Give some examples of each as it affects agricultural production.

10. Why is agriculture considered a highly risky and uncertain sector of the economy?

11. Suppose you are a risk-averse producer of durum wheat. What is it worth to you if the expected value of durum wheat is 10 percent higher at the Chicago Board of Trade next year versus no change in price now? Explain your evaluation process.

12. Why are flexibility and diversification important considerations for an agricultural producer?

13. Give an example of how a farmer can reduce his or her price uncertainty by entering into a production contract.

PART III

THE DECISION-MAKING ENVIRONMENT

CHAPTER 13

Marketing Activities and Management Decisions

INTRODUCTION

In the beginning the producer was his or her own consumer. Only those items that were needed for immediate, personal consumption for survival were produced. However, as early man and the early settlers in the United States began to specialize into hunters, farmers, and merchants, society became more complex. With the introduction of bartering, occasioned by the development of specialized "production firms" (hunters versus farmers), the exchange activity began to spread over broader geographical areas and wider spans of time.

This separation has climaxed in the creation of a commercial agriculture industry that has totally separated the agricultural producer from the urban consumer. This commercial agricultural sector of the economy is extremely productive. However, it is also extremely dependent on the system of markets and marketing to move this production efficiently, effectively, and profitably to each consumer when, where, and in the form it is desired. The questions that an economic system must answer as discussed in Chapter 1, of what, how, and when to produce, and who should receive the income from sales have become more complicated because of this separation of producer and consumer and the advent of the marketing system.

In This Chapter

The objective of this chapter is to provide an overall picture of the marketing system operating in today's economy. We begin by examining the marketing system, the efficiency of the signals emanating from the consumer to the producer, and the importance of those signals. Then we define marketing and examine *how* products are marketed and *why*, in addition to examining the role marketing plays in the economy. The analytical framework used by economists to study marketers and their marketing activities is also discussed, and the reasons underlying international trade are examined.

MARKETING SYSTEM

As the producer began growing products desired by distant consumers, it became important that a "signal" device be developed so producers would know what was desired and under what conditions. Price serves as the signal in a market economy such as the United States, and "habit" or quotas serve to guide producers in traditional or command societies. The price signal originates with the final consumer and serves to guide resources at the national economy level and to provide needed information to managers as they face day-to-day decisions.

Efficiency of Signals

A question useful to society as a whole and to managers is how well does the market system perform this function of directing resource use? This evaluation is commonly referred to by economists as "pricing efficiency." This simply relates to how quickly and correctly the price signals are sent to the producer, through marketing channels, from the consumer. Furthermore, it describes how well the system reacts to changes in consumer demand and generates reaction and adjustments by the producer.

Efficiency of Physical Activities

Another type of efficiency desired in marketing relates to how many physical inputs and how much of each is utilized in performing the marketing functions the consumer desires. As we learned in Chapter 1, efficiency, in this case referred to as *technical efficiency*, is a ratio of valuable output to valuable input. So the best possible situation arises when our technical efficiency is such that no change will allow greater output to be produced with the same amount of input or the same amount of output to be produced with less input.

This technical efficiency is extremely important because it comprises the noticeable portion of marketing, referred to as the *marketing margin* or *marketing bill*. If the bill is too large, it can appear to both agricultural and nonagricultural producers to arise from a group of "parasites" that exist on the lifeblood of the American farmer or small businessman. This subject will receive much attention in this chapter. The area is so specialized it has created an entire subsector of the agricultural industry: agribusiness and agribusiness management.

Customer or Consumer?

The approach in this discussion of marketing will be that of examining and comparing individual firm decisions (microeconomics) to those at the society or economy level (macroeconomics). For example, to the firm selling a good or service, the person buying that good or service is a customer and the individual firm manager orients the firm's activities toward obtaining and satisfying that customer. On the more aggregate level (macroeconomics), we look to consumers as the general originators of signals and the ultimate decision makers for society's allocation of resources. Thus at the microeconomic level, we identify and examine the types of decisions and decision-making rules that will allow the firm manager to increase efficiency, revenues, and net income to the firm. At the macroeconomic level, we evaluate the activities undertaken in the market as to the resulting performance. In the latter case, performance can be judged as efficiency. This consists of maximizing the output of desired products over time while using a minimum of scarce societal resources.

What Is Marketing?

Probably because of the historical farm or production orientation of the agricultural economics discipline, marketing has often been defined as whatever is done to the product after it passes the farm gate. That is, marketing encompasses all of the business activities performed in directing the flow of goods and services from the producer to the consumer or final user. These activities are usually classified into six stages:

1. *Production:* Takes place at the farm level, output is the raw agricultural product (e.g., wheat, hogs, etc.).
2. *Assembly:* The process of getting like items together for further distribution (e.g., grain elevator, apple packing house, etc.).
3. *Processing:* In this stage product form changes (e.g., apples into apple juice, milk into ice cream, etc.).
4. *Wholesaling:* This stage is the link between the production area and consumption area for fresh/processed commodities (both domestic and international).
5. *Retailing:* Breakdown of products into smaller units for consumer (includes food service/restaurant industry).
6. *Consumption:* The product in its final form is used by the final consumer.

In agricultural marketing, the point of production (farm or ranch) is the basic source of supply. The marketing process begins at that point and continues until a consumer buys the product at the retail counter or until it is purchased as a raw material for another production phase. This traditional approach views marketing as everything that happens to a product from farm gate to final consumer. [This rather simplistic view of marketing has evolved into the concern with the creation of form, time, place, and possession utilities.]

It is this definition that we will use in this chapter to describe the molding of a raw agricultural product into the product whose characteristics the consumer desires. Since consumption is both the purpose and end result of production and marketing activities, marketers must focus their activities on satisfying consumer wants and needs. It is through the performance of these marketing activities that marketers add value to raw agricultural commodities and provide the final product that consumers want, and in the process create time, place, form, and possession utilities. Thus, it is imperative that firms, and managers of those firms, realize that marketing is consumer or customer oriented and why much marketing activity is concerned with discovering or affecting consumer desires. In this regard, our discussion of marketing will focus on those marketing activities geared toward the consumer.

Those Utilities

In terms of marketing, the concept of utility simply means usefulness or desirability of a product for a given reason. The four utilities commonly identified with marketing are time, place, form, and possession.

Without these utilities the value or desirability of a product or commodity is greatly diminished. Or, alternatively, the role of marketing is to add utility and hence value and desirability to a product. But, as we'll see, these marketing activities have costs associated with them, the summation of which is the marketing margin or marketing bill.

Time utility consists of, or is created by, getting the product to the consumer *when* it is desired. Wheat is generally produced at one time of the year in the United States, late summer to early fall; however, consumers desire it all year long. Thus, storage is the function that provides time utility to wheat by making it available during the entire year. At the firm level, the use of an inventory by a farm supply firm is an example of how firms "hold" items until the customer needs that item. Both wheat storage and inventory maintenance activities involve costs of capital, market price decline, product quality deterioration, and so forth. Someone has to pay these costs.

Place utility involves getting the product to *where* it is desired by the consumer. Wheat and cotton are produced at great distances from the urban populations of the United States and even further from the growing markets in other parts of the world. Transportation of these commodities to the location where they are desired generates place utility. For the farm supply manager, place utility generates decisions about how best to deliver fertilizer to the field or where to buy the commodities to be later resold. For many commodities, transportation costs are equal in magnitude to the raw product value on the farm.

Form utility is fairly recognizable. It is the process of putting the raw product into the *style, appearance,* or *quality* desired by the consumer. Processing is the marketing activity generally associated with the creation of form utility. Trimming of field-run celery, aging of beef, and pasteurization of milk are all examples of this type of economic activity. The firm manager is faced with how best to produce a commodity having final characteristics that meet those desired by his customer. For instance, customers may want fresh, brined, or canned sweet cherries or some combination of each. Associated with form utility are the decisions of how to package the product so as to merchandise the commodity well and who should be doing the packaging.

Possession utility relates to transferring the product's ownership to the person *who* wants it. Obviously, the corn farmer in Illinois has no use for all of the corn he or she produces annually. The marketing system must provide the means for willing sellers to meet willing buyers. This utility is created and supported by many people in the marketing channels. Brokers serve to find buyers or sellers; government provides grades and standards, so that risk and uncertainty can be decreased; and speculators help spread the ownership over a palatable time frame until the final consumer is identified.

To summarize, as products move through the marketing system, or channel, the raw product is transformed into the final product. But the separation of consumer and producer in effect separates the demand for the final product (consumer) from the supply of the raw product (producer). This separation is measured by the marketing margin and represents the summation of the costs associated with the creation of the utilities—time, place, form, and possession—within the marketing system of the economy. These utilities must be created if the raw product is to be of value to the consumer and therefore generate an economic return to the producer.

Marketing Margin

As you are well aware, very few products move directly from the farm gate to the consumer. Often products move through several stages of processing before reaching the final consumer. As the raw farm product moves through the marketing channel or system, the price of the product changes, resulting in a *difference in the price the producer receives for the raw product and the price the consumer pays for the final product*. The difference in these prices is referred to as the **marketing margin.** It represents the costs of providing marketing services and is the cost of creation of the utilities.

Demand and Marketing Margins

Basic demand, or primary demand, is observed at the retail level. Thus, the demand for any product at the wholesale, processing, or farm level exists if and only if there is a demand at the retail level. For example, wheat is the main ingredient in wheat bread; the demand for wheat by processors at the farm level is derived from the demand for wheat bread at the grocery store level. Since the demand at all previous market levels is derived from the basic retail demand, retail demand less the relevant marketing margin(s) determines the demand at any lower level in the marketing channel.

In Figure 13.1, the distance between the primary demand, D_R, at the retail level and the derived demand, D_F, at the farm level is the marketing margin. As illustrated, the difference between the retail price and the farm level price at any output level is the marketing margin. For example, the marketing margin for quantity

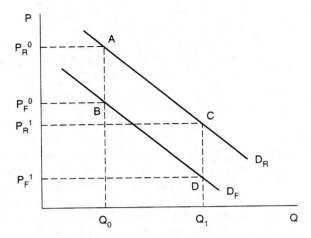

FIGURE 13.1
Demand and marketing margins.

Q_0 is the vertical distance A–B, the difference between the retail price (P_R^0) and the farm price (P_F^0). At output level Q_1, the marketing margin is the vertical distance C–D, which is the difference between the retail price (P_R^1) and the farm price (P_F^1).

Supply and Marketing Margin

Because all food production is based on farm raw products, the basic forces of supply occur at the farm level. Very few American consumers are equipped to utilize most farm products in the condition in which they are sold from the farm (i.e., the form of the product changes as it moves through the marketing channel). The supply at any subsequent market level is derived from the farm supply function. For example, the supply of pork chops in the meat department at the local grocery store is derived from the supply of hogs in the finishing barn. Thus, the derived supply curve at other market levels is the primary supply plus the marketing margin.

In Figure 13.2, the distance between the primary supply, S_F, at the farm level and the derived supply, S_R, at the retail level is the marketing margin. As illustrated, the difference between the farm level price and the retail price at a given quantity is the marketing margin. For example, the marketing margin for quantity Q_0 is the vertical distance A–B, the difference between the retail price (P_R^0) and the farm price (P_F^0). At output level Q_1 the marketing margin is the vertical distance C–D, which is the difference between the retail price (P_R^1) and the farm price (P_F^1).

Price Determination and Marketing Margins

The retail price is determined by the intersection of the derived supply curve and the primary demand curve. The farm price is determined by the intersection of the primary supply curve and the derived demand curve. The difference between the

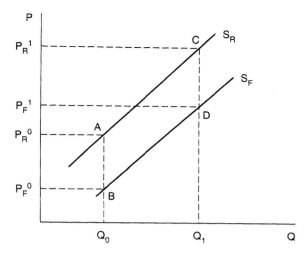

FIGURE 13.2
Supply and marketing margins.

two prices is equal to the marketing margin. This is illustrated in Figure 13.3 where quantity Q_0 trades for price P_F at the farm level and P_R at the retail level. The marketing margin (MM) is equal to the difference between the farm level price and the retail price, or MM $= P_R - P_F$.

As shown in Figure 13.3, the farm level price is lower than the retail level price. This difference in prices arises as a result of the need to pay for marketing services. The consumer is responding to the supply information only after that information has been filtered through the marketing agencies. The producer is responding to similarly filtered information. The difference between P_R and P_F represents the "price" that is paid for the filtering of the information and creating utility through the various activities (or stages) of the marketing system. It is this filtering process that determines the pricing efficiency of the marketing system.

Size of Marketing Margin

The time, place, possession, and form utilities must be created if the raw product is to be of value to the consumer and therefore generate an economic return to the producer. No hog producer likes to be told that the value of his hog enterprise, on the farm, is zero, but it is true! Until slaughtering, cutting, curing, packaging, transporting, and selling of the pork occurs, it is not ready for human consumption so it is useless to the household of the consumer. Even if the pork is intended for farm home consumption, each of these marketing activities and associated utilities must be undertaken before the farmer, as the end consumer in this case, can enjoy the product.

Each of these marketing activities has costs associated with it. When you add up these costs, at the various stages in the market process, you have the marketing margin. For example, Figure 13.4 illustrates the difference in the price the wheat farmer receives at the farm gate, $3.60/bu, and the price the importing country must pay for the same wheat dockside at the export harbor $4.80/bu. The difference

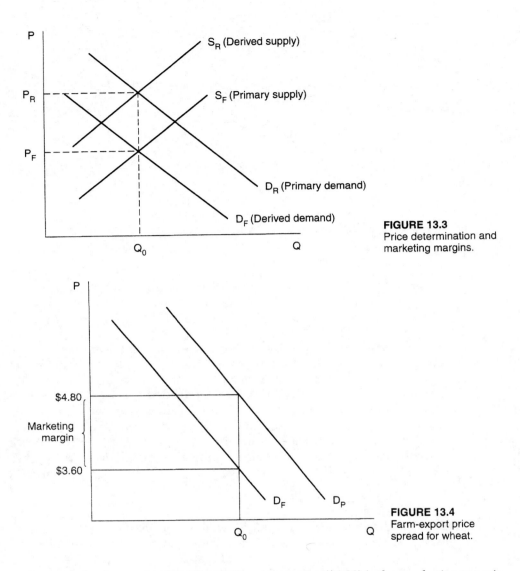

FIGURE 13.3
Price determination and
marketing margins.

FIGURE 13.4
Farm-export price
spread for wheat.

between the farm price ($3.60) and the export price ($4.80) is the marketing margin, which in this example is $1.20/bu.

This marketing margin ($1.20/bu) reflects the handling, storage, and transportation costs incurred in moving the wheat from the field to the port. All are necessary activities, yet producers continue to consider middlemen as "parasites" existing on the lifeblood of the American farmer. Obviously the amount of processing, storage, and so forth will vary considerably among products—consider the difference between turning tobacco into cigarettes, and marketing fresh, sweet cherries. Yet there is no question that these functions must be performed. We can question the size and distribution of the marketing margin on a product or the

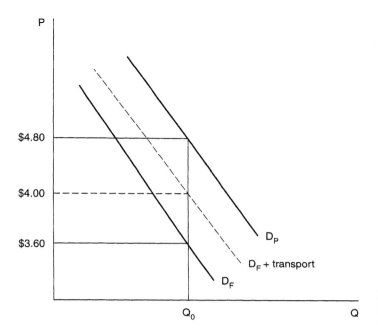

FIGURE 13.5
Internalizing the market-
ing margin.

marketing bill for all agricultural products, but we cannot nor should not question whether the functions can be eliminated.

The size of the marketing margin is affected both by the characteristics desired by the consumer and by the efficiency of the marketing system itself. There has been some movement on the part of agricultural producers, via cooperatives or vertical integration, to internalize some of these functions and move the farm demand curve to the right, thus shrinking the marketing margin from $1.20 to $0.80 (see Figure 13.5). However, even if different firms provide these marketing services, at different levels of efficiency, they are still being provided and have not been eliminated.

Marketing margins result from the cost of the marketing activities performed for the products after they leave the farm gate but before they reach the final consumer. The size of the marketing margin indicates the share of the consumer's dollars spent on food that the marketer (and the farmer) receives. In 1994, the farm value share of the food dollar spent in grocery stores was 24 percent. In other words, out of every dollar spent on food, marketers received 76 cents and farmers received 24 cents. The share of the food dollar left to farmers normally reflects the amount of processing necessary, the perishability of the product, the bulkiness of the product relative to its value, and the seasonal nature of the product. That is, the more processing the product has to go through before it reaches the consumer, the smaller the proportion of the food dollar the producer receives. Thus, the farm value share of the food dollar differs by commodities. Consider that in 1994, for every dollar consumers spent on poultry the farmer received 43 cents; for meat products, 36 cents; for dairy products, 34 cents; for fresh fruits, 18 cents; and for bakery and cereal products, a mere 8 cents.

At first glance, many people interpret the farmer's seemingly smaller share of the food dollar compared to that of marketing firms and marketers to mean that marketers are taking unfair advantage of farmers. Before statements like this can be made, the activities performed by marketing firms, and the costs associated with the various activities, must be evaluated. The fact is that consumers are the driving force behind many of the marketing functions being performed today. Increased processing (cookie and cake mixes), increased convenience (refrigerated cookie dough and microwave meals), and time-saving packaging (precut salad and the "just add meat" frozen meals) have increased the amount of marketing activities performed on many agricultural commodities. These activities must be paid for, and as long as consumers continue to demand these products, the activities will continue to be performed. As the number of activities performed on a product in the marketing system increases, so will the marketer's share of the consumer's food dollar, as suggested by the decreasing farm value share of the food dollar from 47 percent in 1950, to 37 percent in 1980, to 24 percent in 1994.

APPROACHES TO STUDYING MARKETING

To study the various roles that managers of marketing firms must play, an investigative framework is needed. Marketing economists have developed various approaches to studying marketing that can serve as that framework. The role of marketing and market firms will be examined in the following approaches:

- Functional;
- Institutional;
- Commodity;
- Systems; and
- Structural-evaluation.

Each of these approaches is quite traditional and has evolved over time under the writings of various authors. You, as the student, should evaluate each of the approaches and the items within each approach as to what type of utility is being added to the raw product and at what probable costs. Furthermore, since we are concerned with management decisions, you should be relating these utilities and costs to the decision-making rules developed in earlier chapters.

Functional Approach

Examining the functions performed in marketing offers the simplest framework of analysis. This approach receives the most attention in introductory agricultural marketing texts, as it should. It allows easy identification of the utilities being created and serves to identify the activities being examined in the other approaches. More formally, the **functional approach** is *the study of the activities performed in changing the product of the farmer into the product desired by the consumer.* Essentially, it involves the business activities performed by firms in the marketing system. Or, informally defined, it is what's done to the product.

The most common classifications of the functions performed are exchange, physical, and facilitating. Here are the specific functions in each group:

- *Exchange:* buying and selling;
- *Physical:* transportation, storage, processing; and
- *Facilitating:* grading and standardization, financing, risk-bearing, market information.

Exchange Functions

These functions relate directly to the buying and selling that must take place between producer and consumer. They create possession utility but have associated transfer costs, promotion costs, and so forth. Costs of brokerage, speculation, and so on are often considered extreme by producers because the individuals handling these functions did not do anything tangible to the product. Yet, without the service of price discovery and that of bringing the buyer and seller together, the product has little economic value.

Physical Functions

These functions deal with the tangible change in appearance, availability, and location since the three functions are processing, storage, and transportation. Transportation of commodities from production point to the dining tables of consumers creates place utility. Time utility is produced by storage because it spreads the available amount of agricultural production out over time, allowing consumers to have products when they are desired. Processing, as indicated earlier, changes the raw product into the form desired by the consumer. This function has increased greatly in level and cost of activity over time. For example, originally the consumer wanted to purchase wheat; then it was flour; then blended flour; then bread; then packaged bread; then sliced packaged bread; then sliced packaged bread of many different varieties; and so on. The "forms" the consumer desires increase the costs associated with marketing that "raw wheat."

These physical changes have been more accepted by producers and consumers although the small cost of wheat in a $2 loaf of bread is still a favorite topic of discussion in rural areas. The point that holds is that consumers encourage all of these services whenever they buy that "enriched bread" loaf so—apparently—it is economically appreciated by consumers.

Facilitating Functions

As the name suggests, these functions make the market systems operate more smoothly and with greater technical and pricing efficiency. Grading and standardization entail decreasing uncertainty in the marketplace and allow spatially separated buyers and sellers to make a purchase and sale. The use of standard measures allows all products in all markets to move as desired by consumers, with no element of risk or confusion as to what is actually being bought or sold. An associated function is the provision of market information. This activity provides buyers and sellers with precise, up-to-date information to help them evaluate market conditions and examine desires. This information is often provided by the government, regulators, and trade associations.

Two other facilitating functions, also related to each other, are financing and risk bearing. Financing relates a cost to the span of time from production to consumption. During this time, the money tied up in ownership of the product may have been borrowed and interest paid, or it could have been earning a return in some alternative activity. This opportunity cost (return) is a real cost incurred in marketing the product, and someone has to perform this expensive financing function.

Just as time creates a need for financing, it also creates the function of risk bearing. Since physical damage could occur to a product as it moves through the marketing process, some risk of loss exists, a **physical risk** that must be covered by costly insurance or a related precaution. The second type of risk is related to the possibility that the price or value of a product might decrease over time. This risk, called **market risk,** can be assumed by the owner. Futures markets, vertical integration, and contracts can be used to spread this risk among the various market participants.

In summary, the functional approach allows us to specialize as we study the marketing system. As managers operating within the system, we look at what is being done to or with the product. This approach allows both the identification of the utilities being added to the raw product, and an evaluation of the costs (marketing margin) of generating those utilities. It is important to realize that these functions cannot be eliminated. However, the functional approach should always be set in a broad understanding of the entire marketing system—or we can't see the forest for the trees. Too much specialization can develop into narrowsightedness and a shallow understanding of the overall marketing system.

Institutional Approach

The second, very common, approach to studying marketing is to emphasize who is doing the marketing functions. The institutional approach identifies the business organizations and managers that add utility to the product. These are the people often considered "parasitic middlemen" by the agricultural producer. These middlemen, often differentiated by the services that they perform, are usually classified as follows:

- *Merchant middlemen:* retailers, wholesalers;
- *Agent middlemen:* brokers, commission men;
- Speculative middlemen;

- Processors and manufacturers; and
- Facilitators.

Merchant Middlemen

Those retailers and wholesalers who actually take title to the product and own it as they move it through the market system are considered to be merchant middlemen. Retailers deal directly with the consumer, but purchase most of their product from wholesalers who deal in larger volumes of fewer commodities than the retailer. Wholesalers, often called jobbers and car-lot receivers, purchase their goods from processors or manufacturers and provide merchandising and other services to the retailer. Both the retailer and wholesaler buy and sell for their own gain, hoping to "buy low and sell high."

Agent Middlemen

This type of middleman serves as an agent or representative for buyers and sellers but does not take title to the goods, unlike the merchant middlemen. Brokers serve to find a market for the willing seller, and a supplier for the willing buyer, by providing market information and contacts. Commission men serve a similar role but have more flexibility than brokers. They usually sell at what they consider to be the best price possible, but the broker is often told at what price he or she can sell. Also, commission agencies are often involved in the physical activities of moving the product through the marketplace, but the broker is more detached and serves only as a sale facilitator.

Speculative Middlemen

Speculators buy and sell on their own account but expect profits made from price movements rather than other market functions. They take ownership of the product and assume price risk as a means of generating profits, if price movements are favorable. They do little physical handling and merchandising but rely heavily on short-term fluctuations in prices.

Processors and Manufacturers

Processors and manufacturers are principally concerned with form utility and earn a profit by transforming inputs into a more desirable product. Some processors, such as meat packers, have moved heavily into vertical integration from the production area to retail meat stores. These firms have also become more involved in retail advertising and merchandising as competition between processors and between substitute products has increased.

Facilitators

Facilitators provide the extremely necessary functions of market information, grading, standardization, and so forth. They aid the middlemen in performing marketing activities. Their organizations can be trade associations that disseminate information to specific groups, or firms that provide the physical facilities for bringing buyers and sellers together. The Chicago Board of Trade and your local livestock auction yard are both facilitators of marketing.

In summary, the institutional approach identifies who performs marketing functions. When the massive numbers and types of middlemen are identified, the questions often become "Why do we have so many?" "Are they necessary?" "Couldn't producers do their job?" The answer appears to be that producers could do most, if not all, of the activities performed by middlemen, but at what cost? There is a need for specialization in marketing activities so that cost efficiencies and scale economies can be realized. The same efficiencies that cause agricultural producers to specialize also make it appropriate to have firms and firm managers specialize when performing marketing functions, thus, it is hoped, decreasing the total physical distribution costs.

Commodity Approach

An approach to studying marketing that is receiving less emphasis in recent years is the commodity approach. This approach simply follows one product, such as cotton, and studies what is done to the commodity and who does it as it moves through the market system. This approach is quite simple and allows both the functional and institutional approach to be combined.

It is extremely useful to the person who is interested in only one product since it does allow an in-depth analysis. However, this is also a disadvantage because it ignores the interactions between products and market alternatives and also ignores multiproduct firms. Indeed, it is now rare to see a large, institutional, agricultural marketing group handling only one commodity. Also, unless the functions performed and the institutions active in the market system are studied as a category, the student, manager, or researcher may be without a conceptual framework to investigate other commodities when the need arises.

Systems Approach

A more recent approach is to emphasize the system of marketing, dwelling on the interactions of subsystems rather than on individual functions or firms. This behavioral system allows systems to be identified with the particular problem being addressed. System types include *input–output,* which identifies motives and means of affecting the input–output ratio. The *power system* predicts and evaluates firm behavior on the basis of the desires or motives of those in command of a firm. Other subsystems in this approach are *communication,* which evaluates the performance of the information system as perceived by firm management, and *adaptation to change*, which defines approaches and options as firms adapt to internal and external change.

Behavior systems can justify actions that at first seem unreasonable to the reviewer. The interactions between subsystems may seemingly change the objectives of a firm because communications can break down resulting in inefficient input–output ratios.

The obvious disadvantages of this method are its abstract nature and the reliance on intimate knowledge of individual firm characteristics and behavioral interactions. Such data and intimate knowledge are seldom available to the student, potential manager, or market researcher.

Structural-Evaluation Approach

This approach evaluates the ultimate performance of the marketing system by examining the level of competition existing in the industry. The industry structure, including the number and size of firms, is combined with firm conduct, the price behavior, advertising, and product development, to denote a performance that can be evaluated as good or bad.

This approach is used extensively by government regulatory agencies to achieve the good of competition and avoid the evil of monopoly power. However, the lack of a precise norm against which to judge performance has caused a minimal use of this approach by economists studying marketing. The development of "workable competition" as a performance norm may allow this approach to be more productively applied, and managers to better identify their appropriate role or function.

MANAGEMENT DECISIONS

The increasing size and complexity of the activities performed by the marketing sector have affected the decisions made by producers and managers. Rather than just producing a particular product because it is traditional, the producer must be sensitive to the demands of the ultimate consumer. Should the manager be a conventional producer only or should he or she respond to the demands of the marketplace as they reflect the desires of the final consumer? The answer is fairly obvious. Guided by market price, agricultural producers have changed the type of wheat used in bread, shifted to lean hog production, and, more recently, moved to grass-fed beef. The next decisions for farmers and farm organizations and cooperatives appear to be determining what combination of raw product and services should be produced by the agricultural sector. Increased marketing activities by agricultural firms move some of the costs and associated profits to the producer undertaking marketing activities.

As producers or their representatives (cooperative marketing firms) move into the agribusiness area, they are faced with different managerial options. Decisions have to be made on a marketing plan (for example, market segmentation versus market penetration versus market skimming), product decisions, pricing decisions (cost-based versus return-on-investment versus competitive pricing, and so on), and promotional decisions (sales advertising versus personal selling versus flood marketing). In truth, the evolution of marketing has made decisions more complex and confusing. However, the rewards to good decision making have become even larger as a result.

INTERNATIONAL MARKETING AND TRADE

Marketing is providing the types and quantity of goods and services desired by the consumers. As indicated in previous chapters, the more consumers that exist in the market the more quantity can be sold or the higher the equilibrium price can be in that market. In the United States one strong area of expansion of demand for agricultural production has been in the foreign markets or in international trade. Trade between countries, regions, or individuals works best when each buys some from the other, allowing specialization and efficiencies of production for each producer. Let's briefly look at the importance of international trade to U.S. agriculture and then examine the underlying principle for trade—comparative advantage.

Importance to Agriculture

The importance of international trade to American producers is startling. In the early 1980s the export market accounted for almost 30 percent of the sales for all farm commodities; in the early 1990s it was still nearly 25 percent. Crop products are especially dependent on the export market. During the 1992–1996 period, U.S. wheat producers exported an average of 51 percent of their annual production and accounted for 33 percent of total wheat traded in the world market. During this same period, U.S. coarse grain producers exported, on average, 22 percent of their annual production, which accounted for nearly 60 percent of world trade. Rice and white wheat out of the Pacific Northwest were particularly dependent on foreign consumers; about 90 percent of both commodities moved into the international market. Analysis of most of these commodities during the past ten years indicates that the trend in dependence on foreign consumers is generally increasing. With the new production levels, and the even higher prospective levels in wheat, apples, and soybeans, it appears the foreign markets will continue to be a necessary complement to our domestic demand.

The best consumers of U.S. agricultural production continue to be Asia and the countries of Western Europe. The largest individual customer is Japan, who in recent years has imported almost $11.8 billion of American agricultural exports. The value of exports to Eastern Europe and the countries of the former Soviet Union, whose purchases from the United States had steadily increased during the 1970s, declined during the 1980s but have since stabilized or slightly increased in the early 1990s.

Concept of Comparative Advantage

Comparative advantage is the underlying reason behind trade. Based on the economies of specialization, this principle states that people, areas, or countries should produce those products in which they have a comparative advantage or in which their comparative disadvantage is smallest. The advantage of one area over another arises from the production functions or cost of inputs available in each area and the resultant differing costs of production. Consider the following hypothetical example of two countries, Mexico and Japan, producing two different commodities, oil and calculators, presented in Table 13.1. This is a situation where, for a given level of 5 units of input, Mexico could produce 30 units of oil or 7 units of calculators compared to Japan who could produce 6 units of oil or 14 units of calculators.

In this example, Mexico has an absolute advantage in oil and Japan has an absolute advantage in calculators. If no trade occurred and each country had 10 units of input, the total output of oil would be 36 units (30 from Mexico; 6 from Japan) and the total output of calculators would be 21 units (7 from Mexico; 14 from Japan). However, if trade did occur and each nation specialized in the product that it produced most efficiently, then for the same 10 units of input (in each country) we would have 60 units of oil or 28 units of calculators. Thus, specialization and trade would have increased the amount of oil available by 24 units and the amount of calculators by 7 units, thus making both countries better off.

It is possible that some areas or countries would have no absolute advantage but could have a comparative advantage. Consider the hypothetical example presented in Table 13.2. In this case Japan has an absolute advantage over Mexico in both TV sets and calculators. For every TV set produced in Mexico, Japan can produce 2 (a 2-to-1 advantage); for every calculator produced in Mexico, Japan can produce 3 (a 3-to-1 advantage). Thus, Japan has a comparative advantage in calculators (3:1 is greater than 2:1) and by specializing in calculators will be able to produce 90 units of calculators using 10 units of input. Mexico, who would specialize in the product in which it has the least comparative disadvantage, TV sets (1:2 compared to 1:3), would produce 40 TV sets with 10 units of input.

This relationship can be considered similar to the product–product decisions faced by a firm (discussed in earlier chapters). The specialization by area is seen in farming (wheat in the Great Plains, tobacco in the Southern states), in individuals (lawyer versus mechanic), and in countries (Japanese high-technology industry

TABLE 13.1 Production of Oil and Calculators in Mexico and Japan

	Total Production at Five Units of Input	
	Oil	*Calculators*
Mexico	30	7
Japan	6	14

TABLE 13.2   Production of TV Sets and Calculators in Mexico and Japan

	Total Production at Five Units of Input	
	TV sets	Calculators
Mexico	20	15
Japan	40	45

versus Mexico's pottery). Comparative advantage underlies trade, but is always dependent on the production capabilities and market availability. Free trade is usually of benefit to individuals and societies.

SUMMARY

The specialization of producers over time has resulted in a commercial agriculture industry that is totally separated from the urban consumer, making the commercial agriculture sector extremely dependent on markets and marketing. This increasingly complex marketing system has made the need for good management even more critical and the rewards to good management even more pronounced.

A market system carries signals from the consumer to the producer, thus guiding the resource usage in the economy. The efficiency of this role is evaluated by pricing efficiency and technical efficiency, and the magnitude or amount of the effort results in the marketing margin or bill.

Marketing activities are the function and source of employment for many managers. Marketing, defined in many differing ways, can be thought of as "creating utilities" or usefulness/desirability in the minds of the consumer. Time utility consists of getting the product to the consumer when he or she desires. Place utility is getting the product to where the consumer desires. Form utility is putting the product into the style, appearance, or quality that the consumer desires. Possession utility is getting the ownership of the product to the person who desires it.

All of these utilities must be created if the raw product is to be of value to the consumer and is to generate an economic return to the producer. These utility-creation activities have costs associated with them. It is these costs that comprise the marketing margin. The size of the marketing margin is affected by the characteristics desired by the consumer and by the efficiency of the marketing system itself.

Various approaches to studying marketing and management functions have been developed over time: functional, institutional, commodity, systems, and structural-evaluation. Knowledge of these approaches gives a framework by which managers can examine alternative roles and functions that they or their firms could be performing.

The increasing size and level of the activities performed by the marketing sector have affected the decisions made by producers or managers. However, the rewards to good decision making have become even greater as a result.

International trade has been extremely important to American producers and is becoming even more so. Comparative advantage underlies trade and tells a manager or country whether or not to specialize in production and, if so, in what products. Free trade is usually of benefit to individuals and societies.

QUESTIONS

1. What is the importance of price as a signal in a market economy? Explain pricing efficiency.
2. The text refers to time utility and illustrates the concept by using wheat. Do the same using peaches as an example.
3. Explain the concept of place utility with reference to American demographics: Where do most people live in the United States compared with the location of agricultural production areas?
4. Distinguish between the marketing agents associated with form and possession utility and explain what their functions are.
5. "Middlemen are the main reason that farm prices are so low." Comment.
6. How might agricultural producers work to narrow marketing margins? Why would they want to do this?
7. Give a real-world example of each of the three physical functions performed in the marketing process.
8. What is the importance of market information in facilitating marketing?
9. How does the functional approach to marketing analysis differ from the institutional approach? Can you combine the two approaches? Try it.
10. How does a merchant middleman differ from a speculative middleman?

CHAPTER 14

Government in Agriculture: Impacts on Decisions

INTRODUCTION

Government has always been involved in American agriculture. In colonial times tobacco production was regulated in the 1600s by the British government. After the American Revolution, the new United States government implemented laws designed to get land into private ownership, a direct product of the Jeffersonian ideal of an agriculturally based democracy. After the Civil War, the U.S. Department of Agriculture (USDA) was formed to disseminate information systematically to farmers. In the same era, land in the new states of the West was opened under the Homestead Act (1862), and the railroads were subsidized to promote a transcontinental tie.

In this century, the Department of Agriculture has grown into a large Washington-based bureaucracy with regulatory and policy functions. This has made possible a true "agricultural policy," developed by the Congress, the president, farmers, and agribusinesses. However, only since the early 1920s has Congress ever had a "farm bill," consolidating into one piece of legislation most of the government programs affecting crops, credit, food stamps, and regulatory activities. Thus, modern agricultural policy has a relatively short—less than seventy-five years—history in the United States.

The federal government's current agricultural policies have a direct impact on manager decision making. The programs implemented by the USDA through its Washington headquarters and the hundreds of regional, state, and county offices strongly influence many of the decisions made on farms and in agribusiness. Agricultural policy has grown complex, but much of it is based largely on economic reasoning and forecasting—more so than in any other federal agency.

In This Chapter

In this chapter we explain why agricultural problems have been the focus of government policies for decades. Agriculture has been persistently plagued by problems of chronic overproduction of certain crops and livestock, resulting in low incomes for rural families on farms. Despite major adjustments in agriculture, instability in prices and incomes remains a major problem. But the nature of price and income elasticities for agricultural commodities, the new output-increasing technologies, and the overcommitment of land and labor resources to agriculture lie at the heart of agriculture's chronic economic problems. Traditional farm policy tried to address these problems with price and income supports, production controls, and credit. But the new farm policies of 1996 have deregulated agriculture on the crop side and ushered in a new era of opportunity and risk. We will examine the tools or programs used by the government and the economics of these programs.

WHY AGRICULTURE IS SOMETIMES A NATIONAL PROBLEM

The problems facing agriculture have changed over time as the agricultural sector has developed. Examining the traditional problems of agriculture will help us to understand the contemporary problems facing agriculture.

Traditional Agricultural Problems

Agricultural policy as we know it today came into being as a result of problems in the agricultural sector of the economy during the 1920s. Although the U.S. economy did not fall into the Great Depression until the 1930s, agriculture was in an economic depression in the mid-1920s. After World War I, farm prices dropped sharply as exports to European markets dropped and wartime government support for wheat and hog prices was terminated.

Also during this time, new technologies enabled farmers to steadily increase their yields per acre—and a production revolution emerged. Led by the work of the land-grant college scientists and the Cooperative Extension Service agents in every county, farmers were urged to keep detailed production records, make effective use of fertilizers, adopt new seed varieties and animal breeds, and make best use of machines and land. The more progressive farmers helped modernize farming and move it toward a more scientific production basis.

But the Great Depression coupled with improved technology created the most enduring farm problem: overproduction. Despite a growing world population and new foreign markets, it became quite clear that American farmers had the unfortunate ability to increase the supply of products more rapidly than the demand was growing, resulting in disappointingly low prices.

Improving technologies in agriculture meant the term *overproduction* also applied to the resources in agriculture: too many farmers, too much land in production. Economists and policy makers talked about the surplus of farmers and land because the returns to these resources (profits to management, wages to labor, capital gains in land) were very low during this period. Farm mortgages were foreclosed so frequently in the 1930s that the entire traditional structure of owner-operated "family farms" was threatened.

Overproduction continued to be an agricultural problem even after World War II. Technological innovations continued, until by the 1950s, the mechanization of American farms was virtually complete. Chemical and biological advances (e.g., chemical weed and pest control, growth regulators, plant breeding) continued to create technological impacts on agricultural production. One of the consequences was that after World War II, the USDA policies to support commodity prices resulted in large stocks of government-owned commodities such as butter, wheat, corn, rice, tobacco, and cotton.

The USDA could not release these surplus stocks back into domestic markets without depressing prices. So surplus commodities were distributed to low-income families in the United States and shipped overseas under foreign-assistance programs. Not until 1973 did world demand, combined with the export subsidies and devaluation of the dollar, liquidate much of the USDA's accumulated surpluses in grains and dairy products.

Contemporary Agricultural Problems

In the last twenty years the problems facing agriculture have changed dramatically. Some painful transitions and adjustments have occurred in agriculture since World War II:

> *Farms are fewer but larger.* Since 1950 farm numbers have dropped from 5.6 million to 2.07 million, while average farm size has grown from just over 200 acres per farm to around 470 acres.
>
> *Farm population is down.* The downward trend in farm population has been dramatic: 23.1 million in 1950; about 8.0 million in 1982; and less than 5.0 million today.
>
> *Government-owned surpluses are now history.* With the farm policy changes of 1996, the government no longer directly owns large stocks of surplus production.
>
> *Farm income and wealth have increased.* The families involved in agriculture have, in the aggregate, experienced a fairly steady increase in income (farm and off-farm) since 1950. In addition, the value of land and other assets in agriculture has grown significantly in the last forty years.

Despite these and other adjustments, agriculture is still faced with many problems, of which instability and risk are two. The traditional low levels of farm prices and incomes have been improved, but variability from year to year and month to month in price and income remains a major challenge to managers. Of course, variability is also induced by weather conditions from year to year. The drought in the Southern Plains and the Midwest and the eruption of Mount St. Helens, all in 1980; the Midwestern floods of 1993; and the devastation due to flooding of the Ohio River and the Red River in North Dakota during early 1997 are all recent reminders of how farming has not escaped the sometimes tortuous conditions of the biological world.

Another problem persists, and the events of the mid-1980s indicate that it will always be a threat to contemporary economic conditions: market surpluses for certain commodities. The old days of mountains of stockpiled surpluses of grain and other crops are gone. But every producer knows that one or two really good years of production with a soft domestic and international demand may have a severe effect on prices. So the old problem of surpluses remains a contemporary problem in a different form.

Economic Factors Behind Major Agricultural Problems

Is there something special about agriculture as an industry that causes it to be plagued with serious problems?

There are several answers, and most of them are a definite "yes." The weather is always cited as a problem, but in recent decades, farming has partially escaped some of its cruel realities of nature through irrigation, pesticides, herbicides, fertilizers, and certain new machinery technology. In addition, some regions have experimented with cloud-seeding to stimulate rain; other regions have diminished historical regional flooding. Things have improved, weather-wise, for agriculture in this century.

So weather alone is not a reason for agriculture's problems. Economists have long pointed to several characteristics of the agricultural industry's structure as the underlying basis for many problems.

Price and Income Elasticities

Income elasticity of demand is a measure of the percentage change in quantity demanded with a 1 percent change in consumer income. Every manager in agriculture should have an interest in this type of elasticity since it measures a response in the quantity demanded of a product to changes in consumer income. During the last two or three decades, even in the recession of 1982–1983, consumer income has been rising, although not at a steady rate, and as incomes go up food products are purchased in different proportions.

For example, potatoes are a staple in the American diet, but as family incomes increase more desirable foods are substituted for potatoes. Thus, potatoes have a low income elasticity. On the other hand, economists have documented a higher income elasticity of demand for such items as ice cream, veal, cheese, fruit, and turkey. These are more desired foods. But for almost all foods, the income elasticity is measured at less than one. This means that as consumer incomes grow, a smaller proportion of the budget is spent on food.

Even more important than income elasticity is price elasticity of demand, also discussed earlier. Price elasticity measures the percentage change in quantity demanded to a 1 percent change in price. Price elasticity coefficients are always negative. (Remember the inverse relationship between price and quantity demanded?) If demand for food items were elastic, then price decreases would stimulate a very responsive purchasing pattern and total receipts to the retailer would increase. This situation would be ideal for farms: In big output years prices would fall, but consumers would buy more than enough extra to compensate for the lower prices.

Sadly, this is not the case. Demand for agricultural products is generally considered price inelastic at the retail (grocery store) level and even more inelastic at the farm level. Price elasticities are generally estimated in the -0.20 to -0.80 range. To put this idea another way, if farmers have a good crop year, prices will usually fall, but consumers will not be rushing to buy up all the extra production at the lower prices. After all, how much does a family want to eat of any one food item? How will fresh products be stored if purchased in large quantities? Thus, it is argued that in the short-run, price elasticity of demand causes uncertainty and instability in prices and incomes.

Output-Increasing Technology

Technology is the current level of knowledge or information applied in the production of goods and services. Some obvious forms of technology are machines: computers, tractors, combines, and electric generators. Technology is embodied in many other improvements in the production of food and fiber: pesticides, hybrid corn seed, corn sweetener, machine-picked tomatoes, and so forth.

Technological changes have had, and continue to have, profound impacts on agricultural production. The change from horse- to machine-power, traditional corn seed to hybrid seeds, manure to commercial fertilizer, and use of chemical pest and disease control are some of the most important technological advances. Although some scientists believe the rate of technological improvement has slowed, agriculture is still characterized by rapid changes in production technology. An example is the recent introduction of precision farming, which uses global positioning system technology to apply inputs in the exact amounts needed to individual fields.

Any improvement in the production process changes the physical input–output ratio to generate more product or service per unit of input. Machines are more efficient per unit than man- or horse-powered harvesting. Pesticides result in more output with no change in other inputs. This is the common characteristic of technological change in agriculture: Outputs increase. It is this blessing which enables American farmers to feed 260 million Americans while exporting more than $60 billion of agricultural products overseas.

Output-increasing technology has curses in addition to blessings. As technology improves the production process, it also has an impact on prices. A new technology that stimulates large increases in supply of a crop or livestock may eventually lead to a decline in price.

Thus, a treadmill effect has occurred in agriculture. The farmers willing to test and eventually adopt new technology realize lower costs per-unit and increased production. Other producers follow the leaders in attempting to lower average costs and achieve higher outputs. As more and more producers become technologically efficient, supply increases, and (if demand has not also increased) prices fall. Producers now search for new ways to lower costs in order to earn a profit. The treadmill continues.

Resource Fixity

Another factor that some economists have identified as a possible reason for recurrent farm problems is resource or asset fixity. Despite low returns, families stay in farming and keep too much land in production in the United States, Europe, Japan, and other countries.

This leads some economists to argue that the problems of overproduction and low incomes could be solved if there were fewer farms and . . . farmers. They contend that the situation of overproduction reflects a resource disequilibrium—too many inputs. One obvious adjustment has been made—the massive migration from farm to city in the 1940–1990 period. From a farm population of 30.5 million in 1940, out-migration reduced that number to 15.6 million in 1960 and 5.0 million by the mid-1990s.

Despite this migration, people stay in agriculture in the face of unstable prices and low incomes. One reason may be that farming is considered a way of life and many people are reluctant to leave despite difficult economic times. Another reason could be that older farmers lack the educational training and/or the technical skills required for today's job market, making it difficult to secure full-time employment off the farm. It is also the case that the capital used in agriculture is very specialized (combines, balers, tomato pickers, and so forth) and has an opportunity cost or alternative return outside agriculture of near zero (that is, these machines cannot be easily reworked for new uses).

For these reasons, many resources in agriculture are considered "fixed" or immobile. If the resources are immobile, then the problem of recurrent overproduction continues despite efforts to achieve equilibrium.

PROGRAMS TO ADDRESS FARM PROBLEMS

Agricultural, or farm, policy is the set of government programs directly influencing agricultural production and marketing decisions. As the problems facing agriculture have changed over time, so has the policy implemented to address those problems.

Traditional Farm Policy

Every president and Congress have found it politically important to address the serious problems in American agriculture. These actions are embodied in policies known traditionally as "farm programs" and they were created over decades of gradual change and revision. The ideas in the first modern farm programs of the 1920s and 1930s remained important features of farm programs until 1996: price supports, direct payments, production controls, disaster assistance, credit, and research and education. Although certain commodity programs were terminated by the 1996 farm legislation, the basic structure remains in permanent law and could be restored when the current legislation expires in 2002.

Without going into great detail, we will discuss the traditional farm programs and then outline the new farm policy initiated in 1996. Three basic responses to the persistent agricultural problems were embodied in traditional farm policy: price and income support, production controls, and credit.

Price Supports and Direct Payments

The most basic farm program is the combination of **price support** and **target price,** which applies to the major food and feed grains (see Table 14.1). To the general public, price supports are expensive "subsidies" paid to producers of such crops as wheat and corn. To farmers, price supports are loans to enable crops to be marketed in an orderly and more efficient manner. Target prices are the policy mechanism used to make direct payments to farmers to support incomes. Cynics and urban congressmen of course considered target prices to be farm welfare payments.

Price supports began in the 1920s and 1930s when agriculture was in an economic depression. The idea was developed to guarantee a farmer a minimum price for selected crops that were politically important (e.g., grains, cotton, peanuts, and tobacco) by granting a nonrecourse loan once harvest was complete. The terms of

TABLE 14.1 Grains Eligible for Price and Income Support, 1995

| Commodity | Unit | Loan and Target Price Levels, 1995 | |
		Loan Rate	Target Price
Barley	Bu	$1.58	$2.36
Corn	Bu	1.94	2.75
Oats	Bu	1.00	1.45
Rice	Cwt	6.50	10.71
Sorghum	Bu	1.84	2.61
Soybeans	Bu	4.92	None
Wheat	Bu	2.28	4.00

a nonrecourse loan state that if the borrower defaults, no recourse will be made to recover any loan losses. In the case of USDA price support loans, eligible farmers borrow against crops in storage. During the loan period (usually 9 to 12 months) the farmer can decide to sell the crop and repay the loan plus interest or, if prices remain at or below the level of the loan, the farmer could default on the loan by simply allowing the government to keep the crop. No legal action (that is, no "recourse") would be taken by the USDA.

Prior to 1977 the loan rates for each crop/marketing year were written into the legislation. However, beginning with the Food and Agriculture Act of 1977, the USDA was given more flexibility to adjust loan rates to reflect changing market conditions. In 1985 the method used for calculating the basic loan rate was modified so that it would be 75 to 85 percent of the "Olympic" average of the five previous season average prices (i.e., excluding high and low prices). This was modified to 85 percent in the 1990 legislation.

Within the USDA, the little-known but well-funded Commodity Credit Corporation (CCC) is responsible for overseeing the operation, both financial and storage functions, of the nonrecourse loan program. The law requires the secretary of agriculture to extend price support loans through the CCC for grain crops at the rates listed in Table 14.1. Other crops like cotton, tobacco, and peanuts are also eligible for price support through nonrecourse loans.

Farmers who comply with program requirements can obtain loans through the USDA at the specified levels once the crop is harvested. For example, the nonrecourse loan rate for barley in 1995 was $1.58 (Table 14.1). If prices rise above $1.58, barley producers will likely sell the crop and repay the loan plus interest. But if barley prices remain below $1.58, the USDA becomes the owner of stored barley when producers default on loans.

Price supports are also available to dairy producers but are managed by an entirely different mechanism. Because milk is perishable, the price-support program operates by commodity purchases rather than nonrecourse loans. The USDA is required to support the price of milk at a level specified by Congress, $10.35 per hundredweight (cwt) in 1996. If market prices paid to dairy producers fall below the minimum, the USDA begins to buy dairy products (powdered milk, cheese, and butter) for storage. These stocks are later resold or distributed to public institutions for feeding programs such as school lunches.

Price supports increase farmers' incomes indirectly by guaranteeing minimum prices for certain products. The problem of low and unstable farm income is directly approached by a program of direct payments under the mechanism of a **target price.** The most well known of these payments has been the deficiency payments, introduced in 1973. Other direct payment programs include supplementary, diversion, and disaster payments.

The target price/deficiency payment program was designed to support farm income without influencing market price. Under the program, the Congress sets the price for the producer—the target price—but the market determines the price to the consumer. The legislated target prices effective in 1995 are listed in Table 14.1. For crop producers with a target price program, USDA paid a **deficiency payment** equal to *the difference between the target price and the market price.* Thus, if the market price for wheat falls below $4/bu, producers become eligible for a deficiency payment based on the difference between the average market price and the target price of $4.

Farmers who participated in the price-support and target price programs were also limited to production on their farm's historical **base acreage.** That is, producers with a guaranteed minimum price (the price-support level) and a possible direct payment outside the market (the target price) could not just plant unlimited acres, with the government shouldering much of the market risk. So wheat and feed grain producers were limited to production on a *crop acreage base* equal to the average of acres planted during the previous five years. But since participation in these farm programs has been voluntary, the target price/deficiency payment program also required cropland set-asides in years when overproduction of the program crops was probable. In effect, the set-aside programs required farmers to "set aside" or divert cropland from production of such crops as wheat, corn, or barley in exchange for a deficiency payment based on normal crop acreage and production. The diverted land had to be placed in an approved conservation use.

To control the size of deficiency payments and reduce budgetary outlays, producers were required to comply with acreage restrictions to be eligible for deficiency payments, and payments were limited to $50,000 per person. Still, the overall cost of farm programs was very expensive in the 1980s, the modern depression era for American agriculture. As shown in Figure 14.1, USDA farm program costs reached an all-time high in 1986, almost $26 billion, but more recently costs have been less than $10 billion.

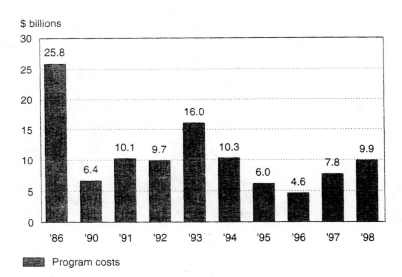

$ billions

FIGURE 14.1   Cost of USDA farm programs, 1986, 1990–1996, and estimated 1997–1998. (*Source:* United States Department of Agriculture.)

Production Controls

Price supports and direct payments inherently reduce risk and induce farmers to "farm the farm programs," meaning you produce the crops required on your base acres if you are going to be a participant in the program. This stimulated overproduction of these crops because farmers responded to government program incentives rather than market prices. The result was the need for government control of production. USDA and the Congress attempt to "balance" supply and demand such that market prices are acceptable (that is, not too low, not too high) through various programs to control the amount of land in crop production or the amount of one crop that could be legally marketed.

Major production controls began in the 1920s with voluntary agreements between the USDA and farmers to reduce acreages of the so-called "basic" crops—corn, cotton, hogs, milk, rice, tobacco, and wheat. These early programs were unsuccessful and Congress then moved to a more direct form of production control, marketing quotas.

As designed by the USDA, the marketing quota is the most direct method to control how much of any crop is produced each year. Since they are produced in only a few Southern states which have powerful politicians in Congress, peanuts and tobacco are the only two crops that still have a quota system. The Secretary of Agriculture is authorized to proclaim a national production quota for them. This national quota is broken down into state and farm quotas for all eligible participants. The law provides serious penalties for growing a quota crop without being properly eligible.

The most popular production control program has been land retirement. In the 1950s, Congress established the Soil Bank, a large-scale program to take land out of production. Under the Soil Bank, farmers were paid to reduce plantings of certain crops or to "retire" land from crop production into such soil-conserving uses as pasture. The Soil Bank consisted of two major components: an acreage reserve and a conservation reserve. The acreage reserve represented short-term land retirement as producers were paid to divert land from current use. The conservation reserve was aimed at long-term retirement of all or part of a producer's land. During this period, the land retirement program became known as "paying farmers for growing nothing."

Longer term land retirement was initiated in 1985 with creation of a Conservation Reserve Program (CRP) to help encourage the removal of highly erodible land from crop production. The idea was to pay landowners to take this land out of production and thereby reduce soil erosion, reduce surplus grain production, and expand wildlife habitat. Eligible land included crop land that had been in production for at least two of the five years between 1981 and 1985. Farmers contracted to idle highly erodible land under ten-year contracts. CRP land had to be planted in grasses, trees, or other vegetative cover, and could not be hayed or grazed except in emergency situations as determined by the secretary of agriculture. In return for putting this highly erodible cropland into "reserve," the landowner receives an annual rental payment, about $49 per-acre per-year on a national average.

At its high point, the CRP had more than 36 million acres enrolled in ten-year contracts. The 1996 farm legislation extended this program but reduced the funding so that enrollment is not likely to exceed 28 million acres.

Credit

Other than price supports, perhaps no farm program is more important to American agriculture than the federal credit programs. Beginning in the early 1900s, Congress authorized special credit sources to meet the high-risk and seasonal nature of agriculture financial requirements. The Federal Land Bank, later to become part of the Farm Credit System of banks, began offering farm real-estate loans in 1916. In 1930, the forerunner of the Farmers Home Administration (FmHA) was established to deal with the problems of poverty and credit in rural areas. Today FmHA is the principal public-credit source for both farmers and rural communities.

The USDA and the independent Farm Credit System did a good job of supplying the enormous credit demands of an agricultural system that shifted from horse to machine power. The financial capital now required by today's mechanized agriculture is staggering. In 1950, the total farm debt was $12 billion. By 1980 farm debt had grown to $158 billion. In 1993 farm debt actually decreased slightly to $142 billion.

American agriculture depends on outside investment capital to finance land and other inputs necessary for efficient production. General credit institutions (banks and insurance companies) and specialized credit institutions (Federal Land Bank, Production Credit Associations) can supply the bulk of these credit needs. The public sources of credit, primarily FmHA, remain an essential supply of capital, especially for limited-resource farms and those affected by natural disasters.

Deregulation of Farm Programs in 1996

After the longest and most contentious farm bill debate in history, Congress finally voted in the spring of 1996 to deregulate much of American agriculture by removing the link between income support payments and crop production. This legislation, the Federal Agricultural Improvement and Reform (FAIR) Act, represents a dramatic change in the way government is involved in American agriculture. But the "old" farm programs are still there and could be reinstated in 2002, when FAIR expires.

The mechanism in FAIR that decoupled the cropping decisions from farm programs was the **production flexibility contract.** The target price/deficiency payment program was suspended until at least 2002 and replaced with "transition payments" to the farmers who had been participating in the old target price programs. These payments were meant to ease the transition to a more market-oriented crop agriculture over a seven-year period, from 1996 to 2002. Under FAIR the market will play a much larger role in production decisions because the old farm program rules on acreage and crops planted are reduced or eliminated. However, because payments are fixed and no longer related to the level of market prices, producers will bear greater income risk.

For individual farmers to be eligible for the payments under FAIR, they must have participated in a program in any of the crop years 1991 through 1995. To receive benefits on program commodities, producers must enter into a production flexibility contract for the period 1996 through 2002. The contracts require participating producers to comply with the existing conservation plans for their farm, the wetland provisions, and the planting provisions. Contract payments are limited to $40,000 per person, a $10,000 reduction from the current payment limit.

The market will be more important to individual decision making as planting flexibility to participating producers is increased. Under FAIR, participating producers are allowed to plant their total contract acreage plus additional acreage in any crop (with limitations on fruits and vegetables) with no loss in program benefits. The participating producer has greater planting flexibility under the 1996 legislation primarily due to the elimination of annual acreage restrictions and the option to remove land from the Conservation Reserve Program before contract maturity.

FAIR maintained a "safety net" for the major crop producers by retaining a modified version of the price support loan. Now known as *marketing assistance loans*, the loan rates continue to be based on 85 percent of the five-year Olympic average (i.e., excluding the high and low years) for wheat and corn. Loan rates for grain, sorghum, barley, and oats are set at levels considered "fair and equitable" relative to the feed value of corn.

BASIC ECONOMICS OF FARM PROGRAMS

The goals of farm policy focused primarily on raising farm incomes and prices for agricultural products. In this section we examine the economics of these major farm policy tools to understand how they work to achieve their goals.

Economics of Price Supports

Price supports amount to minimum price guarantees and short-term credit during marketing periods. One general effect of price supports is to reduce risk for producers. With a minimum price guarantee and a nonrecourse loan during the marketing period, a manager experiences considerably less risk than those producers not participating in the program.

Price supports have a second important economic effect besides risk reduction. If the price-support level set by the USDA is higher than market equilibrium levels, the quantity of product supplied during a production period will exceed the quantity demanded. In Figure 14.2, if P_0 is the market-clearing price for wheat, a price support of P_1 will result in surplus production (that is, $Q_S > Q_D$). If P_0 remains below P_1, wheat

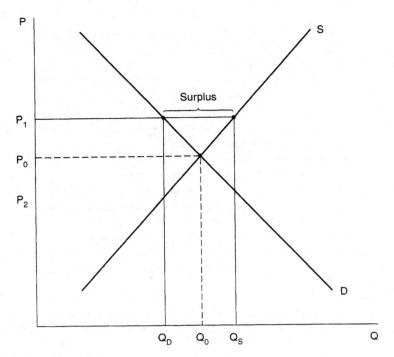

FIGURE 14.2 Economic effect of price-support levels above equilibrium market prices.

producers will default on price-support loans and the government will accumulate stocks equal to $Q_S - Q_D$ and incur costs. At price support levels below P_0, for example P_2, P_0 would hold in the market as a market-determined equilibrium price. Thus, the USDA would incur a minimal program cost, but not end up with surplus wheat stocks.

The job for economists and USDA program managers is to carefully set price-support levels such that:

They are high enough to provide short-term credit needs and reduce risk; but

They are not consistently above equilibrium levels such that large surplus government stocks are accumulated.

Price supports through nonrecourse loans in effect create a price floor for the commodity. Since the loan places a floor under the price a producer receives for a product, it in effect places a price floor under the market price. Thus, to avoid accumulating large volumes of stocks, the government must be very careful about where the loan rate is set. If continually set above the equilibrium market price, as the United States did in the early 1980s, government stocks will continue to grow as farmers forfeit grains to receive the higher price. Because P_0, the market-clearing

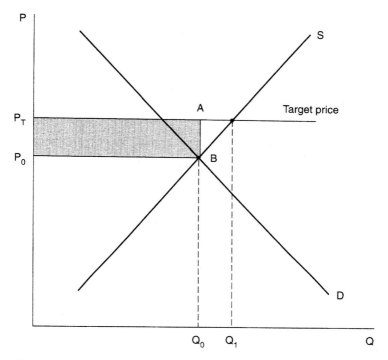

FIGURE 14.3   Economic effect of a target price-direct payment program.

price, fluctuates significantly for most agricultural commodities, the job of managing the price support program is a difficult one.

Economics of Direct Payments

The inherent possibility of a price-support program to generate surplus production is substantially reduced with the target price/direct payment approach to guarantee farmers a minimum price for certain crops. In the 1970s and 1980s, the target price deficiency payment program was the largest direct payment program operated by USDA. Under this program, target prices were announced for eligible crops in advance of planting. If final market prices fell below the announced target, the USDA paid producers the difference, the "deficiency" between target and actual prices. Producers even received an *advance deficiency payment* in the spring, based on market price projections, which helped pay seed and fertilizer costs.

The economics of a target price program is illustrated in Figure 14.3. The congressionally mandated target price is shown at P_T. If P_T were to be a price support, farmers could plant enough acreage to produce Q_1. However, if the market operates freely, the crop at harvest is Q_0 and equilibrium price settles at P_0. In this case the individual eligible producer would receive a payment equal to AB (the difference between the target price and the equilibrium market price). The entire shaded area, $P_T A B P_0$, represents the amount paid to all producers, which is also the cost of the program to the taxpayer.

Because a direct payment program is geared to production, a target price of P_T would stimulate increased production as long as P_T remains above P_0. To limit surpluses of crops and reduce budgetary outlays, producers are required to comply with acreage restrictions and production controls to be eligible for deficiency payments, and payment size is limited. Therefore, the incentive to increase production if P_T exceeds P_0 is substantially reduced and market forces should govern supply and demand.

With a direct payment program the government determines the price the producer receives while market operation determines the price to the consumer. Thus, the market price can be well below the target price, benefitting consumers, while producers' incomes are still protected. However, instead of dealing with acquisitions of surplus production the government is faced with the costs of deficiency payment expenditures, which are highly visible to nonagricultural taxpayers.

Economics of Production Controls

The most direct production control program is the quota system. Only two crops, peanuts and tobacco, remain under a production quota system which limit the amount of a crop which a farmer can grow and still qualify for government benefits. For each of these crops, the USDA determines a national production quota which will produce a "fair" price, always a price below parity. This price is linked to production costs and is guaranteed to eligible producers who participate in the price support program. The result is a substantial reduction in market risk, leaving the producer with mainly production risk.

Quota systems must be tightly controlled because the benefits can be very attractive—low risk and guaranteed price. For tobacco the right to produce a quota crop is reserved to holders of production allotments. These allotments were given to tobacco producers in the 1930s when the production controls were first initiated. Allotments are tied to specific farms although they can be leased, that is, the "right" to grow tobacco can be purchased or leased from an allotment owner.

Because the tobacco quota program has been successful in controlling supply and guaranteeing prices, the allotments have become valuable. Farms with allotment rights are now worth more than identical farms without these rights. Thus, the value of the allotment right has been capitalized into the value of the land.

Other production control programs operate indirectly on supply by attempting to reduce cropland devoted to certain crops. The intended effect of these acreage diversion and cropland set-aside programs is illustrated in Figure 14.4. If sufficient cropland is retired from production or diverted to other uses, the supply curve should shift to the left. This is exactly what happened when the CRP took 36 million acres of cropland out of production beginning in 1985–1986.

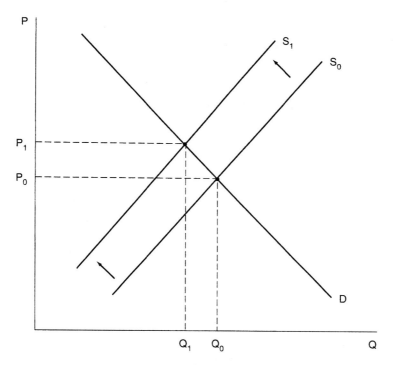

FIGURE 14.4 Economic effect of acreage diversion or cropland set-aside programs.

In Figure 14.4, if P_0 were an equilibrium price but unacceptably low, an effective diversion of cropland should have the effect of reducing supply (from S_0 to S_1) and raising market prices to P_1. This is the theory behind land-based production controls. However, because these programs operate indirectly in controlling supply, the exact impact of the shift in supply is difficult to estimate.

Crop production control programs operate on the premise that market prices can be increased through the removal of productive resources, namely, land. The less land in production, the less is produced of a certain commodity. However, growers have succeeded in increasing the productivity of the remaining land, by new technology or by use of nonland inputs; thus, these programs are seldom successful in achieving the desired decrease in supply.

Management Challenges Under the New Agricultural Policy

The significant deregulation of agriculture under FAIR has had several important management consequences for farmers and agribusinessmen. Beginning in 1996, farmers and other managers in agriculture had to operate in an environment with more freedom to make on-farm decisions but also with considerably more risk.

These new risk factors come from several sources: income variability, market risk, farmland values, and increased planting flexibility.

Income Variability

Because the new production flexibility payments are not tied to market prices (i.e., payments are fixed even when crop prices go down), farm income is potentially more variable from year to year. This was experienced immediately by crop producers when prices for wheat, corn, and soybeans plunged from almost record-high levels in 1996 to more "normal" levels in 1997. The new marketing assistance loans continue to provide some income protection, but this "safety net" is at a relatively low level compared with the market prices of 1996.

Market Risk

In the past, market price risk was partly shouldered by the government through the old deficiency payment scheme—deficiency payments went up as market prices went down—and through the land-based supply management programs (e.g., cropland set-asides, diversion, CRP). In addition, the USDA was a major holder of grain and dairy surplus stocks and financed the Farmer Owned Reserve. Only the CRP remains. Thus, with suspension of most of the government programs designed to temper the impact of low prices in large crop years, market price risk is shifted to the individual producer. This implies that producers should expect more price variation from season to season and greater income volatility.

Farmland Values

The major asset in American agriculture is farmland. Under FAIR, farmland values are projected to increase since the value of the annual payments from production flexibility contracts is capitalized back into the value of cropland historically devoted to wheat, corn, other feed grains, rice, and cotton. But the USDA has already projected an 8 to 10 percent drop in land values if Congress does not continue the production flexibility contracts after 2002. Why? Per-acre income will decline without the annual USDA payment. Thus, if payments are continued after 2002, farm income and farmland values will continue to increase—assuming no other major economic trend is occurring to drive down land values (e.g., deflation similar to the 1980s).

Freedom to Farm Changes Decisions

In Washington, D.C., FAIR was commonly known as the "Freedom to Farm" bill since it decoupled payments from cropping decisions. This new planting flexibility under FAIR is permitting farmers to alter their production practices. For example, in 1997 farmers increased planted acreage of major field crops by 16 million acres over 1996. The major shift was toward soybeans and away from wheat, corn, and sorghum. Increased acreage combined with favorable growing conditions resulted in a bumper crop of soybeans and a drop in prices from $7.38 in 1996 to around $6 in 1997. We can expect more acreage shifts like this in the future. Of course, this will make crop forecasts and financial planning more difficult since Freedom to Farm may lead to more uncertainty about the aggregate potential supplies of major crops, resulting in less accurate price forecasts a year or more in advance of planting decisions.

The redesigning of government support and involvement in agriculture also means that managers must find ways to manage the increased risk effectively. We discussed several ways to reduce risk and uncertainty in Chapter 12. Under the new policy environment, crop producers will need to consider new planning and management strategies to offset some of the new risks. Each manager will have to develop a strategy suited to his or her own farm or agribusiness. Some of the possibilities for managing risk may include the following:

1. Diversify the farm or business with new or alternate enterprises.
2. Produce commodities with less variable yields and price trends.
3. Use futures markets or forward-contracting for grain and livestock marketing.
4. Reduce debt load and increase equity to minimize cash flow problems in low-income years.
5. Go to contract production in poultry, hogs, or specialty crops like popcorn or white corn.
6. Purchase crop insurance and revenue insurance.
7. Expand the use of information and analysis in managerial decision making.

Careful planning will be necessary for every manager to capitalize on the new choices and opportunities in farming and agribusiness while minimizing the risks brought on by the shifting of risk from government to farmers under FAIR. Many managers may need to rely more on the assistance available from the Cooperative Extension Service at the land-grant college in their state or from commercial financial planning services. The new economic environment will reward those managers most able to address risk factors and develop strategies to improve production efficiency, marketing, and financial performance.

SUMMARY

Government has had a long history of involvement in agriculture, and its impact on managerial decision making is changing. Traditional agricultural problems were ones of overproduction relative to demand, with resulting surplus production and general resource disequilibrium. Contemporary agricultural problems now involve more market volatility since the government is unwilling to hold large grain and dairy stocks; the aging of the family farm owners and the difficulties with entering farming; and continuing declines in farm numbers and farm population. The causes of the agricultural problem are related to the structure of the agricultural industry and markets. Price and income elasticities remain very low, thus large crop yields result in even larger price drops and income increases do not appear to be a solution to the farmers' problems. Output-increasing technology has a decreasing effect on per-unit costs but may eventually lead to a decline in product price. Resource fixity causes resources, both people and inputs, to remain in agriculture even as incomes and returns decrease, thus further complicating the problem of recurrent overproduction.

Traditional government programs to deal with agricultural problems included price supports, direct payments, production controls, and credit. Price supports can set prices above equilibrium prices in a market, resulting in program costs to the USDA and surplus commodity stocks. Direct payments allow the market-clearing equilibrium price to be established, with the difference between this price and a "target" price being paid to the producer. This is a production incentive unless controls are instituted. Production controls, such as quotas or acreage diversion, seek to estimate a market price that will generate a reasonable return on investment and then constrain supply so that price will hold in the market. The effects of these problems are to control supply, reduce risk to producers, and capitalize program value into cropland.

The Federal Agricultural Improvement and Reform Act of 1996 made substantial changes in farm policy. Farmers have new freedom to respond to market signals with planting decisions and marketing strategies. However, this new legislation shifts much of the market risk from government to producers. This is requiring managers to investigate new risk management options in order to survive profitably in the new era of reduced government involvement in agriculture.

QUESTIONS

1. Why is the problem of overproduction not encountered by durable-goods manufacturers to the same extent as it is by wheat farmers?
2. What major changes have taken place in the economic nature of American agriculture during the past five decades?
3. "The family farm is the backbone of a strong America; without it the nation will suffer." Is this statement as true now as it might have been in 1800? 1900? Explain.
4. Explain how the technology treadmill affects farmers.
5. Remembering that one characteristic of pure competition is no barriers to entry or exit, what can you say about farmers' ability to leave farming for more productive pursuits? What can you say about agriculture's overall compliance with this assumption of pure competition?

6. Explain how marketing quotas and acreage diversion programs control supply and help stabilize commodity prices.
7. "I think we ought to quit spending so much money on all these university research projects. All they do is increase our yields and drive down the market price from oversupply." Attack or defend this position.
8. Graphically illustrate the effect of a price-support level that is higher, and then lower, than that suggested by prevailing market conditions.
9. How did the government try to control supply for commodities with a target price/deficiency payment program? Was the government successful?
10. What is the nature of the new risks in agriculture under the FAIR Act?
11. How would risk management strategies differ between the wheat production regions of the northern plains (e.g., North Dakota) and the Midwestern states (e.g., Ohio)?

CHAPTER 15

Macroeconomic Policy Linkages to Agriculture and Agribusiness

INTRODUCTION

In addition to the direct government policies that affect agriculture and agribusiness, it is important to recognize the ways in which general economic policy and the overall economy are linked to agriculture. Economic costs and returns to farms and agribusinesses (part of the "microeconomy") are not only influenced by market conditions but also by general economic policies implemented to promote economic growth and temper business cycles. These policies are known as **macroeconomic policies** and include *monetary and fiscal policy actions taken by the federal government and the Federal Reserve Bank.*

Although the primary aim of macroeconomic policy is to temper business cycles—the tendency for the economy to grow and contract—these policies often have a profound impact on agriculture and agribusinesses. The most significant example of this impact was the monetary policies implemented in late 1979 to stop double-digit inflation. This tight monetary policy in conjunction with depressed commodity prices caused an economic depression in agriculture and other natural resource-based industries. Cash flow problems became so severe that thousands of farms, 300 banks, 2,200 agricultural service firms, and hundreds of smaller farm-related businesses went bankrupt in the early 1980s. This period helped dramatize the linkages between agriculture, agribusiness, and the general economy.

In This Chapter

In this chapter we provide a broad overview of macroeconomic policy and how agriculture is linked to the general economy. We begin with a brief overview of monetary and fiscal policy and how these policies affect costs and returns in agriculture. Then we discuss the linkages between macroeconomic policy and agriculture and how these policies impact farms and agribusinesses.

WHAT IS MACROECONOMIC POLICY AND WHY IS IT IMPORTANT?

The massive and complex U.S. economy is really the amalgamation of households, businesses, and the government operating in hundreds of markets—transportation, food and fiber, energy, entertainment, etc. During any year, the estimated total value of goods and services produced in all of these markets in the United States is known as **gross domestic product (GDP).** For 1996 this added up to $7.6 *trillion,* or about $29,000 per capita.

Farm organizations like to remind consumers how essential agriculture is to America, but as a share of the overall economy, farming is a declining portion of GDP. In the last fifty years, farming has dropped from 7 percent of GDP to only 1 percent. But agriculture includes much more than just what happens on the farm. By looking at the entire food and fiber system (i.e., farming + food service + wholesale/retail trade + transportation + manufacturing + inputs), you account for 16 percent of GDP, which makes agriculture and agribusiness a much more important segment of the economy. In numbers of employees, food manufacturing and distribution companies hire about 16 million workers or nearly 13 percent of the U.S. workforce. In terms of GDP and labor, agriculture and agribusiness is an important part of the overall economy.

Monitoring the change in GDP is a measure of the relative success of our nation's economic growth. Changes in GDP are reported quarterly by the Department of Commerce and have become one of the important indicators for possible macroeconomic policy changes. When the change in economic growth is negative for several months, this period is considered a *recession.* The United States has experienced nine of these economic downturns since World War II. Although differing in length and severity, these contractions have lasted an average of four months and the GDP loss has been about −3.1 percent. On the other hand, when GDP change is positive, the economy is expanding. On average the economic growth cycles last about eighteen quarters. But in 1991 the economy launched into a record growth period that was continuing well into 1997—the longest economic expansion since World War II.

The goal of macroeconomic policy is to help the nation pursue goals of full employment, low inflation, and economic growth. There are three broad dimensions to macroeconomic policy: monetary policy, fiscal policy, and trade policy. Each of these affects agriculture and agribusiness in different ways.

Monetary policy is implemented by the Federal Reserve Bank, (or "Fed"), which is really a system of twelve regional banks established in 1913 to centralize monetary policy. The Fed pursues monetary policy through its Open Market Operations Committee, which has representatives from all the regional banks. The decisions of this committee are closely monitored by financial markets and the national media since their actions determine key interest rates, the supply of money in the economy, the value of the dollar for international financial transactions, and the overall integrity of the nation's banking system.

The Fed implements monetary policy by announcing a key interest rate, the discount rate. This is the interest rate charged by Federal Reserve Banks for loans to commercial banks. The interest rates on loans to farmers and agribusinesses are affected by changes in the discount rate. In addition, the Fed can use other open market operations to affect the money supply (e.g., buying and selling bonds, changing the reserve requirement) and achieve monetary policy goals.

To boost the economy out of a recessionary period, the Fed would pursue an **expansionary monetary policy.** Through bond purchases and lowering of the interest rate, the Fed would strive to increase the money supply in the economy and increase the level of credit available to businesses and individuals. This was the case during the last recessionary period of 1990–1991 when the Fed progressively lowered the discount rate from 7 percent in 1989 to 3 percent in early 1992. These moves helped the economy start into a long period of economic expansion. Farmers and agribusinesses benefit from expansionary monetary policy since it makes land and equipment purchases less costly and tends to enhance the overall market environment.

Implementation of a **restrictive monetary policy** means the Fed is trying to restrain economic growth by changing the key monetary policy tools: selling bonds and increasing the discount rate in order to decrease the supply of money and credit. The last period of very restrictive monetary policy came in late 1979 and on into the early 1980s. During this period the discount rate was increased to 14 percent in 1981, resulting in consumer interest rates of greater than 21 percent for car loans, real estate mortgages, and other credit purchases. The result was lower inflation, higher unemployment, a stronger dollar, and a major recession. In agriculture, land values plunged and when farmers also faced low commodity prices the result was a severe cash flow crunch which led to high default rates on loans and 200,000 farm bankruptcies.

Fiscal policy involves government expenditures and tax rates. Congress appropriates money for government expenditures and sets tax rates, so it is ultimately Congress that influences fiscal policy. Our modern concept of fiscal policy began when President Roosevelt pressured Congress into ambitious public works programs designed to create subsidized jobs and put unemployed Americans back to work in the 1930s. Congress has been involved in active fiscal policy ever since in order to pursue two of the same economic policy goals as the Fed: full employment and economic growth.

While government expenditures act indirectly to pursue these goals, tax rates changes are more direct and powerful. Lowering income taxes or capital gains tax rates puts more money in consumers' pockets and tends to spur economic activity. Congress reduced income and capital gains taxes in 1981 during the long recession and again in 1997, even though the economy was healthy and growing at the time.

Expansionary fiscal policy involves increased government spending and/or decreased taxes, resulting in a boost in economic activity. After President Reagan was elected in 1980, the government pursued a vigorous expansionary fiscal policy. A major tax cut reduced corporate and personal income taxes by $276 billion over four years and federal spending was sharply increased for military and other functions. This new fiscal policy resulted in a huge federal budget deficit and doubled the national debt in five years. However, the economy reacted with more than eight years of continuous economic growth.

Restrictive fiscal policy involves reduced government spending and/or increased taxes. These fiscal actions tend to reduce economic activity and are used to slow down an economy that is growing too fast or is inflationary. In the 1990s Congress tried to balance the budget by reducing federal spending over a seven-year period. This is a restrictive fiscal policy since it reduces the growth in federal spending. In 1997 President Clinton and Congress reached an historic agreement to balance the budget by the year 2002. However, this agreement did not *reduce* federal spending, it merely slowed its growth!

Trade policy is another dimension to macroeconomic policy that has important linkages for agriculture. The United States is the world's largest exporter of agricultural products and services. In the last three decades, the United States has had a 17 percent share of the total world market of agricultural trade. In 1996 U.S. agricultural exports totaled about $60 billion, an all-time record.

Thus, trade policy affects the overall profitability of U.S. farms and agribusinesses as well as the competitiveness of the United States in international trade. Global economic forces will therefore dictate some of the important trends for the future economic environment in farming and agribusiness.

Both Congress and the president directly influence trade policy as one aspect of macroeconomic policy and international politics. As early as 1789 Congress was imposing tariffs on imported goods. The president has the constitutional power to negotiate trade agreements, which Congress then ratifies or disapproves. Congress sets the tariff rates on imported goods, determines if quotas will be imposed on certain countries or commodities, and decides which industries will receive trade subsidies or incentives.

Trade policy is always controversial in its impact on the overall economy. Advocates of free trade believe all nations benefit from the reduction or elimination of trade barriers (i.e., tariffs, quotas, subsidies). Critics of free trade like to note the dominant position of Japan in manufactured goods and point out that Japanese markets are severely restricted for many foreign products, especially agricultural products from the United States such as rice.

However, the United States has been slowly moving toward a freer trade policy ever since the end of World War II. Recently, new trade agreements have dramatically changed barriers to trade among the major trading nations and for the countries of the North American continent. In 1995 the latest version of the General Agreement on Trade and Tariffs (GATT) was implemented and provided for substantial reductions in trade-distorting policies such as tariffs, quotas, price-support subsidies for crops, and export subsidies. Mexico, Canada, and the United States joined in the North American Free Trade Agreement (NAFTA) beginning in 1994. This agreement completely phases out agricultural tariffs for the three countries, thus promoting a more open market for most commodities. The inevitable changes have been very controversial as some American farmers have "lost" their traditional markets to competitors from across the border.

MACROECONOMIC POLICY LINKAGES TO AGRICULTURE

American agriculture and agribusiness is land and capital intensive and dependent on foreign markets for a substantial part of its income. Therefore, changes in monetary, fiscal, and trade policies will affect the profitability and competitiveness of farmers and agribusinesses through linkages to the costs of inputs, output prices, the value of land, and exchange rates. Although these linkages are complex and dynamic, they can be summarized as is done in Figure 15.1.

Figure 15.1 shows that trade policy has historically been dominated by tariffs, quotas, and other restrictions on free trade. In the 1990s, the GATT and NAFTA trade agreements dramatically changed the economic environment for

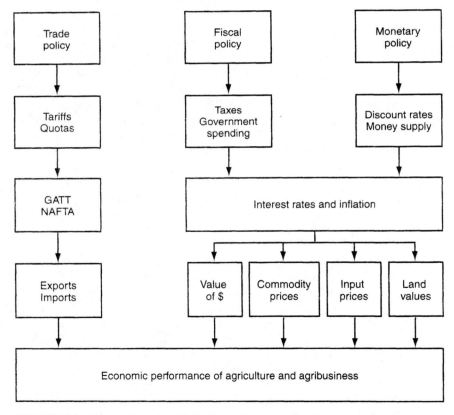

FIGURE 15.1 Macroeconomic policy linkages for agriculture and agribusiness.

agricultural trade and thus directly affected exports and imports. But the value of the U.S. dollar in international currency markets also affects trade. A strong dollar will discourage exports and encourage imports, leading to a trade deficit.

As shown in Figure 15.1, both fiscal policy and monetary policy affect interest rates and inflation. These are links to the important economic variables (commodity prices, input prices, etc.) that will affect the economic performance of agriculture and agribusiness. Macroeconomic policy is implemented to serve the overall goals of low inflation, full employment, and economic growth. Although agriculture and agribusiness are important components in the overall economy, macroeconomic policy is not made to promote the profitability of farms and agribusinesses. Thus, managers must be aware of the linkages and impacts of macroeconomic policy in order to understand how macroeconomic policy may change the economic outlook and how policy changes will impact their individual farms and businesses.

Monetary Policy Linkages and Impacts

Whenever monetary policies are changed to influence economic growth and inflation in the overall economy, the costs of production, commodity prices, the value of the dollar in international trade, and land values will be affected. These linkages can be illustrated by looking at the impacts of an expansionary monetary policy, illustrated in Figure 15.2. Recall that in pursuing expansionary monetary policy the Fed may buy government securities, lower the discount rate, and/or lower the reserve requirement. Whenever the Fed uses monetary policy to boost economic growth, it will increase the money supply and lower interest rates. These actions will have a ripple effect on agriculture and agribusiness.

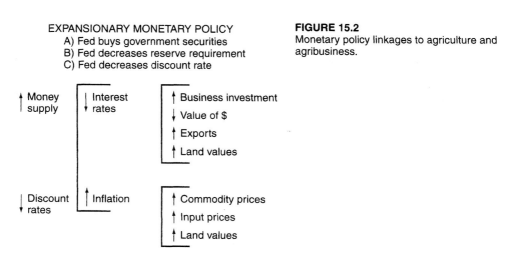

EXPANSIONARY MONETARY POLICY
  A) Fed buys government securities
  B) Fed decreases reserve requirement
  C) Fed decreases discount rate

FIGURE 15.2
Monetary policy linkages to agriculture and agribusiness.

Costs of Production

Lower interest rates reduce a farmer's cost of borrowing money for short-term operating expenses (e.g., seed, feed, fertilizer, fuel) and long-term capital investments (e.g., machinery, buildings, land). In 1996, total farm production expenses were $183 billion of which $13.3 billion was interest expense on short- and long-term debt—about 7 percent of total expenses. It is generally estimated that every one-point change in nominal interest rates will change interest expenses in agriculture by $4 billion.

The relatively low interest rates of the 1990s have made interest expense a smaller proportion of farm production expenses and stimulated new investment in land and machinery. But during the 1980s when interest rates reached 21 percent on some loans, interest expense became a much larger expense and contributed to severe cash flow problems for many farms and firms. The resulting economic depression in agriculture forced farm credit institutions to "write-off" $19 billion in bad loans—10 percent of all agricultural loans—during 1984–1988.

Commodity Prices

Monetary policy can also affect the general level of prices in the economy. This is known as the **inflation rate** and is measured at the retail level with the Consumer Price Index and at the producer level with the Producer Price Index. Expansionary monetary policy generally causes farm commodity prices to rise more rapidly than input and nonfarm prices. This happens whenever there is "demand–pull" inflation, which occurs during periods when the economy is operating at nearly full capacity, and demand for certain goods or services starts to exceed available supplies. This is especially important in agriculture since it is a more competitive market that is sensitive to changes in demand and supply conditions.

During the 1980s the economy experienced double-digit inflation—reaching almost 14 percent in 1979. This allowed farmers with large real estate debt to repay loans with "cheaper" dollars, as inflation continued to increase commodity prices and boost land values. But restrictive monetary policies in the early 1980s caused **deflation** as the inflation rate fell from double digits to 4 percent by 1983. This contributed to the farm financial depression of the 1980s. In the 1990s, inflation has remained around 3 percent despite a record-breaking cycle of economic growth. So inflation has not had any significant role in commodity price changes in the last decade.

Land Values

Monetary policy's most direct linkage to agriculture and agribusiness is through land values. Agriculture and forestry are the most land-intensive sectors of the economy, so any factors that affect land values will have an enormous impact on economic performance.

Changes in interest rates will influence the price of farmland and agricultural wealth through this linkage:

$$V = \frac{R}{i}$$

where V is the current value of a unit of land, R is the expected financial returns to that unit of land in each future time period (i.e., rent or income), and i is the interest rate. This expression calculates the present value of land (V) by discounting expected returns (R) into current dollars. It is a simplification of the basic discounting concept discussed in Chapter 12. In this case, the discount rate should reflect the opportunity cost of money for the buyers—what an alternative investment would earn (e.g., money market account, bonds, etc.). The interest rate on long-term bonds is a common interest rate used in these calculations since it represents some risk over a long period of time. This makes it an appropriate interest rate when looking at land values since real estate purchases are usually long-term investments.

Thus, if a farmer estimates that an acre of prime farmland could earn a net income of $100 per-year and the interest rate on long-term bonds is 6 percent, then the "value" of that land is $1,667 (100 ÷ 0.06 = 1,667). If the interest rate falls to 4 percent, then the value of that land with the $100 annual income is $2,500. Thus, you can see that expansionary monetary policy boosts the financial position of agriculture because of its large land base.

The monetary policies initiated in 1979 to correct high inflation in the economy had a devastating impact on agriculture. As inflation stopped and commodity prices sagged in response to contracting international markets, land values changed dramatically. In the Great Plains, farmland was selling for $750 per-acre in 1980 but after interest rates were boosted into double-digit ranges, land values collapsed to only $400 per-acre. Monetary policy increased the i and lower commodity prices reduced the R, resulting in an unavoidable collapse in land values—the major asset in agriculture.

Exchange Rates

Monetary policy, through its impact on foreign exchange rates for the U.S. dollar, plays a major role in determining the competitiveness of U.S. products in international markets. **Exchange rates** represent the price of one country's currency in terms of the currency of another country. These exchange rates are a direct reflection of the supply and demand conditions in currency markets and are available from a variety of sources including *The Wall Street Journal*.

Interest rates are the link between the value of the dollar and monetary policy. When U.S. interest rates are higher than in other countries, investors will buy U.S. bonds and other financial instruments in order to earn higher incomes. This increases the value of the dollar.

Expansionary monetary policy attempts to spur growth by expanding markets for U.S. goods and services. With lower interest rates, the value of the dollar will weaken and make U.S. products more price competitive. For example, assume it costs 600 German marks to purchase a ton of American wheat valued at $200—an exchange rate of 3:1. If monetary policy drives the value of the dollar down to an exchange rate of 2:1 (now it will only take two German marks to buy one dollar), the ton of American wheat now costs the German buyer only 400 marks. With the lower price and other factors held constant, we would expect the Germans to import more American wheat. This is why American farmers and agribusinesses tend to benefit from a weak dollar.

When the Fed pursued a tight money policy in 1979 and drove the discount rate to 14 percent, the value of the dollar rose sharply. Why? With high interest rates in America, foreign investors wanted to buy dollar-denominated investments, like bonds, to get those higher interest earnings. While the American consumer benefitted from a strong dollar (i.e., foreign products were cheaper), U.S. agricultural exports became uncompetitive in world markets. The results were dramatic. Total agricultural exports plunged from $44 billion in 1981 to only $26 billion in 1986. This was the depth of the economic depression in agriculture—primarily a result of changing monetary policies.

Fiscal Policy Linkages and Impacts

The federal government's fiscal policies, known as "tax and spend policies" to some, are also directly linked to agriculture and agribusiness. This linkage, illustrated in Figure 15.3, is through interest rates and inflation but with a less direct and slower impact than is the case with monetary policy.

Expansionary fiscal policy—increased government spending and lower taxes—tends to increase the demand for money and lead to higher interest rates. This is especially the case when fiscal policy involves **deficit spending,** a situation that has plagued Congress since the Vietnam era. In the early 1990s the budget deficit was nearly $300 billion but then began to decline to under $100 billion by

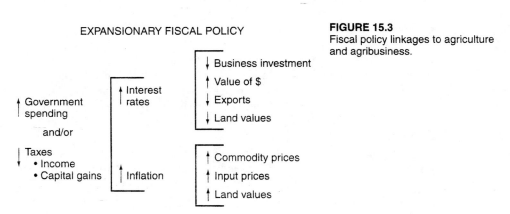

EXPANSIONARY FISCAL POLICY

FIGURE 15.3
Fiscal policy linkages to agriculture and agribusiness.

the end of 1997. However, the budget deficit is projected to rise modestly over the next ten years due to entitlement spending on Social Security, Medicare, and other programs whose beneficiaries are older Americans.

Whenever government expenditures exceed tax revenue and other income, the government must borrow money from domestic and international sources to finance the budget deficit. This federal borrowing reduces funds available for private investment and puts upward pressure on interest rates, since the government is competing with private borrowers in credit markets. These rising interest rates have a negative impact on agriculture and agribusiness since they tend to have these impacts:

1. Discourage investment in new machinery, equipment, buildings, and land.
2. Strengthen the dollar in foreign exchange markets, which reduces the competitiveness of agricultural exports.
3. Cause declines in land values.

Expansionary fiscal policy also can fuel inflation by generating excess demand for goods and services. Fears about the inflationary impacts of government spending have persisted in policy circles ever since the double-digit inflation of the 1970s. However, the fears of inflation have not stopped Congress from pursuing expansionary fiscal policy to promote growth and full employment. Rising incomes and more jobs expand the demand for farm products, agricultural inputs, and related services.

Historically, American farmers have benefitted from inflationary periods, which have raised land values and commodity prices. Since farmers are generally in high-debt positions relative to other consumers and businesses, inflation allows them to finance more debt (through higher land values) and repay debt with "cheaper" dollars in future years. However, inflation is the economic enemy of those Americans on fixed incomes and those who do not own real estate and other assets. Therefore, the Fed is determined to maintain a low inflation economy despite Congress's fiscal policies, which may tend to promote inflation.

Trade Policy Linkages and Impacts

American agriculture is trade dependent and becoming more so in the 1990s. Production of one-third of all harvested acreage goes into international markets—about 20 percent of the value of U.S. agricultural output. Approximately 57 percent of the U.S. wheat crop, 30 percent of the corn crop, and 50 percent of all tobacco moves into world markets. For soybeans, the United States held 73 percent of the world market in 1996.

Being a trade-dependent industry, agriculture's profitability is affected by trade policies. Export volume and price directly impact farm incomes and indirectly affect the agribusinesses that supply inputs, process products, and distribute farm products. It has been estimated that a one dollar increase in agricultural exports generates a second dollar's worth of related economic activity. Also, about each additional $50,000 in exports creates one new job in the United States. This is the nature of the linkage between trade and the economic performance of agriculture and agribusiness.

Free trade requires *access* to international markets. Agricultural interests in the United States have fought hard over trade barriers erected by other countries that limit access of American products and services to these markets. Common barriers imposed by other countries include (1) tariffs, which are import taxes or custom duties on goods, and (2) nontariff barriers like quotas, health and safety regulations, import licenses, and port limitations. Quotas merely act to limit the amount of a product that can enter a country in any one year. The most famous example is the quota on imported rice in Japan. The Japanese farm interests are so strong that they have been able to effectively prevent any significant amount of U.S. rice from entering their country. After intense trade negotiations and political pressure, the Japanese allowed some imports in 1997 but the United States discovered they were using all or most of this U.S. rice as donated commodities to Korea and other coun-

---

BOX 15.1 *Wheat Trade Impacts*

Both the United States and Canada are major wheat exporters. During the 1990s production of soft red winter and durum wheat declined in the United States, although demand was rising with increased consumption of spaghetti and other pasta products. The USDA estimated in 1997 that in fourteen of the past fifteen years, total consumption of durum wheat exceeded production on U.S. farms.

But when wheat prices slumped in 1997, North Dakota farmers got very upset with all the durum imports coming across the border from Canada. These imports were perfectly permissible under NAFTA, but the North Dakota farmers still claimed Canada was "dumping" durum wheat on U.S. markets. They called for the U.S. Trade Representative to negotiate limits on Canadian imports. Predictably, the Canadians have resisted limits.

Without an agreement to limit durum imports, U.S. farmers called for Congress to pay durum producers export subsidies to allow them to more "fairly compete" with the Canadian producers. However, with durum output projected to be 81 million bushels in 1997 and total use about 116 million bushels, paying export subsidies to domestic producers would raise the cost of durum wheat to all U.S. pasta mills. Consumers would be the losers.

NAFTA expands free trade and makes consumers better off in most cases. But there are always some producers and consumers who will have to adjust to the impacts of free trade.

---

tries with rice shortages. So, even though the quota had allowed U.S. access to the Japanese market, the government was still preventing the Japanese consumer from getting U.S. rice.

The United States has also used quotas to prevent the importation of foreign products. Sugar is a good example. Sugar-producing countries are allocated a limited amount of bulk sugar that can legally enter the United States. Since these quotas limit the amount of foreign sugar that enters the United States, they in effect raise the price of sugar to American consumers. In addition, sugar quotas have resulted in high prices for U.S. sugar cane and beet sugar producers. These sugar quotas are not permitted under GATT and the United States is now in a transition period to *tariff rate quotas,* a combination of tariffs and quotas that should gradually reduce barriers to sugar imports during the next several years.

Whereas GATT has reduced tariffs and is gradually eliminating most trade quotas, NAFTA has created the world's largest free trade zone. The result has been increased trade among Canada, Mexico, and the United States (see Box 15.1). For U.S. producers of livestock, meats, feed grains, dairy products, cotton, and processed foods, NAFTA has boosted trade and increased employment and profitability. However, labor-intensive producers of fruits and vegetables are seeing their markets shrink under low-price competition from Mexico.

SUMMARY

Macroeconomic policies and international markets now dictate much of the economic environment in which farmers and agribusiness managers must operate. Monetary and fiscal policies are implemented to serve broad national goals of low inflation, economic growth, and full employment. However, these policies have direct and indirect impacts on agriculture and agribusiness since there is a linkage between the policies and production costs, land values, input prices, and exchange rates.

With the growing integration of world markets through GATT and NAFTA, future policy changes may play an even larger role in determining the financial performance of agriculture and agribusiness. Thus, it is even more important that managers understand the linkages between macroeconomic policies and sound business decision making on both farms and businesses.

QUESTIONS

1. If agriculture accounts for only 1 percent of GDP, then why is macroeconomic policy important to farmers and agribusinessmen?
2. What are the main elements of monetary policy and who implements changes in monetary policy?
3. Explain how fiscal policy and monetary policy are both linked to inflation rates and interest rates.

4. If the Fed implemented a restrictive monetary policy and boosted nominal interest rates from 6 to 10 percent for real estate loans in agriculture, what would that do to overall land values? Why?

5. Explain how the expansionary fiscal policy in 1997 of decreased taxes might affect inflation and interest rates and how this might impact agriculture and agribusiness.

6. What did GATT and NAFTA do for trade policy and why is this so important to American agriculture?

7. Why were some farmers "winners" under NAFTA and other farmers "losers"? Were American consumers "winners" or "losers" under NAFTA? Why?

CHAPTER 16

International Trade and Agriculture

INTRODUCTION

U.S. agriculture and agribusiness firms have become dependent on world markets not only as an outlet for their goods but also as a market for purchasing the commodities desired by American consumers. The importance of international trade to American producers is startling. Production of one-third of all harvested acreage goes into international markets—about 20 percent of the value of agricultural output. During 1990–1995 the value of agricultural commodities exported by the United States averaged $44.6 million, and in 1996 the value reached $60 million. American consumers enjoy a tremendous selection of goods produced in other countries—coffee from Brazil, confections from Switzerland, wine from France. During 1990–1995 the value of agricultural commodities imported by the United States averaged $25.5 million, and in 1996 the value of agricultural imports equaled $33.6 million. It is clear from these numbers that **international trade,** *the buying and selling of goods on world markets,* is important to the profitability of agribusinesses and agricultural producers.

The growing importance of international trade to the fortunes of agricultural producers and agribusinesses means that the far reaches of our globe have become as important to most agricultural or agribusiness managers as the local market in the United States. In recent years, consumers have taken for granted the availability of a tremendous selection of goods and services, many of which were produced in distant fields or processing plants. U.S. producers have also begun to recognize the desirability of access to the many international consumers who have the ability (remember effective demand?) to purchase their products.

The result or cause of this expectation of available products and markets on the part of consumers and producers is the growth in international trade of agricultural commodities. Exporting of agricultural commodities allows increased market expansion, while offering rewards for growth in productivity and efficiency in basic production processes. (The more consumers that exist in the market, the more quantity can be sold or the higher the equilibrium price can be in that market.) Importing of agricultural commodities allows consumers access to goods that are produced, typically at a lower cost, in other countries.

However, international trade results in benefits and losses to individual countries. These gains and losses result from differences in costs of production. That is, open markets in all countries will mean that, in some countries, their domestic industries cannot compete against lower cost foreign products or services.

The success of exporters can also directly affect the **balance of trade,** *a measure of whether a country is spending more on foreign products (imports) than it is receiving from selling in those markets (exports).* For the United States, the recent shift during the 1980s and 1990s to a negative trade balance would have been more dramatic if it had not been for the positive balance of trade in agricultural commodities (the United States exports more agricultural products than it imports). Historically, increases and surpluses in agricultural trade have been indicators of prosperity and economic health for the agricultural sector, health that has, over time, stabilized some of the overall economic fluctuations in the U.S. economy.

In This Chapter

In this chapter we take a closer look at international trade and its importance to agriculture. We begin by reviewing the basis of trade and comparative advantage and then illustrate how trade can serve to equate the excess supply/excess demand of individual countries to determine world equilibrium price and quantity traded. Countries that engage in trade will have both gains and losses. We examine some of these gains and losses to illustrate why some firms, industries, and nations oppose free trade while others strongly support free trade. Then we discuss some of the alternative trade policies that affect movement of commodities and the international agencies and institutions that influence trade.

RATIONALE FOR TRADE

In the most direct sense, trade has the effect of shifting outward and to the right the demand curve faced by a firm or a manager. Earlier in this text, we talked about population and income being two of the determinants of demand, or demand shifters. In the United States one strong area of expansion of demand for agricultural production has been in foreign markets, or in international trade. For example, the availability of Japanese millers who desire to buy and are able to pay for U.S. wheat has greatly increased the quantity demanded for U.S. wheat, especially the soft white wheat of the Pacific Northwest. This increased demand has increased the price received by farmers. For commodities traded in world markets, the world price for a commodity equals the domestic price, minus, of course, the transportation differential between the producer and consumers.

From a broader perspective than that of the one firm and producer just mentioned, international trade offers the ability of firms, regions, or nations to specialize in the production of those products for which they have the greatest comparative advantage and to trade for commodities that are better produced in other countries. Recall that **comparative advantage,** the underlying reason behind trade, is based on the economies of specialization. This principle states that people, areas, or countries should produce those products in which they have a comparative advantage or in which their comparative disadvantage is smallest. The advantage of one area over another arises from differences in the production functions or the availability and costs of inputs in each area, resulting in differing costs of production. For example, the United States trades wheat to a Latin American country for bananas. The U.S. trades wheat for bananas not because the United States can't grow bananas but because the United States can grow wheat more efficiently (at less cost) than it can bananas. Conversely, the Latin American country can grow bananas more efficiently (at less cost) than it can wheat. Thus, by growing the commodities they can produce most efficiently, both nations are economically better off as a result of this trading and specialization. It is this comparative advantage that is the basis of international trade because every country will have a comparative advantage in at least one commodity and a comparative disadvantage in one or more of the alternative commodities. This principle is also why more and more countries and regions are moving away from total self-sufficiency in domestic production—the loss in efficiency and output is too great.

Using a simple two-country trade model we can illustrate the concept of comparative advantage and examine the effects of international trade. These are the assumptions we will be using:

1. Two countries are involved: country A and country B.
2. One product is produced, consumed, and traded: barley.
3. There are no transportation costs and everything is measured in the same currency.

In this simplified example, both country A and country B produce and consume barley. Figure 16.1 illustrates the market for barley in both countries. In the absence of trade, the equilibrium price and quantity in each country will be determined by the interaction of the supply and demand for barley in that country. Thus, country A has the supply and demand curves S and D resulting in a domestic equilibrium price P_e and quantity Q_e; and country B has the demand and supply curves D and S resulting in a domestic equilibrium price P_i and quantity Q_i.

Within any market, the resulting equilibrium price (where supply and demand intersect) provides an indication as to how efficient that country is in terms of producing a given commodity. Since the equilibrium price in country A is lower than that of country B, we can say that country A is a more efficient producer of barley than country B (or country A has a comparative advantage in producing barley). Thus, trade would benefit both countries.

If trade is allowed to take place, how much will each country export/import at each price? We introduce the concepts of excess supply and excess demand to represent the price and quantity relationships for exports and imports.

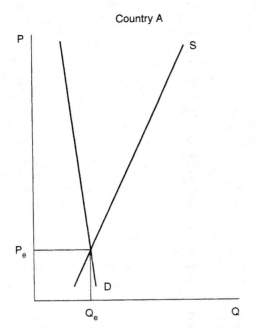

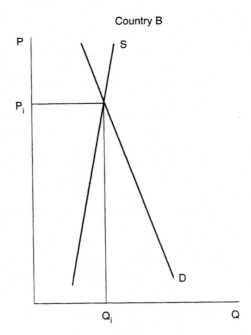

FIGURE 16.1 Market equilibrium in the absence of trade.

Excess Supply

From our previous discussion we know that the barley market in country A is in equilibrium at price P_e. If prices were above P_e, we would have a surplus of barley (i.e., the quantity demanded in a market is less than the quantity supplied). This surplus represents an *excess supply* of barley in country A, or the quantity of barley country A could export to the world market. Using this information allows us to determine an excess supply function to represent the relationship between prices and the quantities a country is able to export to world markets. **Excess supply (ES)** is defined as *the quantity made available for export at alternative price levels and is equal to the quantity supplied less the quantity demanded at each price level.*

Figure 16.2 illustrates the derivation of country A's excess supply curve for barley. We know from our discussion of market equilibrium that at prices less than P_e shortages will occur, and at prices greater than P_e surpluses will occur. This means that country A will only have barley available for export at prices higher than P_e (domestic equilibrium) when surpluses occur in the market. At a price equal to P_e country A has zero quantities to export ($Q_s = Q_d$); the intercept term for country A's excess supply curve is P_e. At prices above P_e country A will have positive quantities available for export ($Q_s > Q_d$) with the exact amount available for export equal to $Q_s - Q_d$ at that price level. For example, at price P_1 the quantity supplied equals Q_2 and the quantity demanded equals Q_1, a surplus equal to $Q_2 - Q_1$. This surplus equals the quantity placed on the world market, Q_x, at that price.

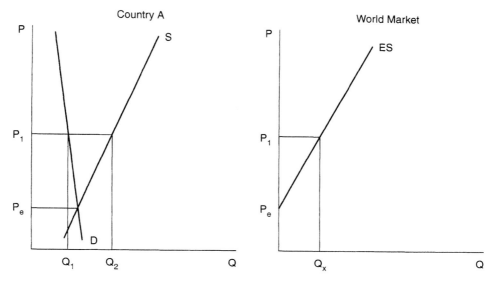

FIGURE 16.2 Excess supply curve for country A.

In essence, the excess supply curve represents the quantity of a product that a country is willing to make available after domestic consumption has been met. The relationship between the price of a good and the quantity of that good offered for trade is represented by the ES curve.

Excess Demand

In our example, country B has a comparative disadvantage in producing barley (i.e., is a high-cost producer). This indicates that purchasing barley in the world market (importing) would be cheaper than producing it domestically. From our previous discussion we know that at price P_i the barley market in country B is in equilibrium. If prices were below P_i, we would have a shortage of barley (i.e., the quantity demanded in a market is greater than the quantity supplied). This shortage represents an *excess demand* of barley in country B, or the quantity of barley country B could import from the world market. Using this information allows us to determine an excess demand function to represent the relationship between prices and the quantities a country is willing to import from world markets. **Excess demand (ED)** is defined as *the quantity desired for import at alternative price levels and is equal to the quantity demanded less the quantity supplied at each price level.*

Figure 16.3 illustrates the derivation of country B's excess demand curve for barley. We know from our discussion of market equilibrium that at prices less than P_i shortages will occur, and at prices greater than P_i surpluses will occur. This means that country B's barley producers will be unable to meet domestic consumption at prices less than P_i (domestic equilibrium) when shortages occur in the market. At price equal to P_i country B will have no need to import barley ($Q_s = Q_d$);

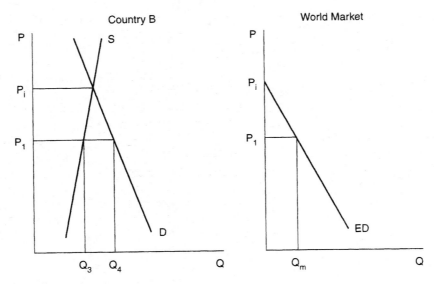

FIGURE 16.3 Excess demand curve for country B.

the intercept term for country B's excess demand curve is P_i. As prices decrease below P_i the quantity demanded from the world market will increase as shortages occur ($Q_s < Q_d$) with the exact quantity demanded for import equal to $Q_d - Q_s$ at that price level. For example, at price P_1 the quantity supplied equals Q_3 and the quantity demanded equals Q_4, a shortage equal to $Q_4 - Q_3$. This shortage equals the quantity purchased on the world market, Q_m, at that price.

In essence, the excess demand curve represents the quantity demanded of a product that a country is willing to purchase after domestic production has been exhausted. The relationship between the price of a good and the quantity demanded of that good on world markets is represented by the ED curve.

World Equilibrium

When trade between two countries occurs, assuming no transportation costs for this brief example, the excess supply and excess demand curves are brought together in what is now an international market. The interaction of excess supply and excess demand determines the price and the quantity that will be traded on the world market. This is illustrated in Figure 16.4, where for our simple two-country model, the world equilibrium would be at price P_W when quantity Q_W is traded (where $E_S = E_D$).

Responding to the world price, P_W, country A will produce quantity Q_2 and Q_1 will be consumed domestically, and the difference ($Q_2 - Q_1 = Q_W$) will be exported. In country B, Q_3 will be produced, Q_4 will be consumed domestically, and the difference ($Q_4 - Q_3 = Q_W$) will be imported. Given our simple two-country as-

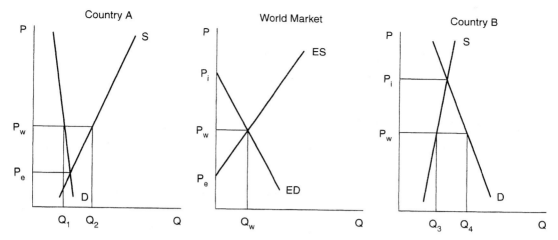

FIGURE 16.4 World market equilibrium.

sumption, the quantity traded in the world market (Q_W) is exactly equal to the surplus in country A and the shortage in country B at the world price P_W. In addition, the introduction of trade raises the price in country A (benefitting producers) and lowers the price in country B (benefitting consumers).

We could expand this simple example to include more than two countries by simply horizontally summing the excess supply and excess demand curves of each individual country to get the world excess supply and demand (remember Chapters 7 and 8?). The world equilibrium price in this multiple-country case would be determined by the interaction of the world excess demand and world excess supply.

Impacts of Trade

It is too simple and maybe even dangerous to talk about trade as if it is always desirable to everyone. Looking again at Figure 16.4, it is evident that in the country A, producers are better off because their market price increased from P_e to P_W and the amount they are marketing increased from Q_e to Q_2. However, in that same country, the domestic consumers are now paying a higher price ($P_W > P_e$) and the amount available on the domestic market has decreased from Q_e to Q_1. Conversely, the producers in country B, the importing country for this commodity, have seen their product price decrease from the domestic price of P_i to P_W and, due to their cost structure, they can now only afford to produce Q_3 rather than Q_i, the equilibrium quantity in the absence of trade. Consumers in the importing country are certainly better off because price has fallen and the amount on the market has increased from Q_i to Q_4.

In a broader sense, the reasons for a nation or region to trade include the increased employment caused by the growing export industry, the increased income realized by the region or even the firm, and the increase in product prices felt by the industry, whether agricultural or other. Underlying all of these is the potential increase in efficiencies from larger farms or plants, better technologies, and increased research and development that occur with the enlarged market of international trade.

Our discussion of the impacts of trade in Figure 16.4 has already outlined reasons why some producers would not benefit from trade. As comparative advantage works its way through international trading partners, some consumers and producing sectors will be impacted negatively. In the earlier example, domestic U.S. producers of bananas may lose as Latin American bananas are imported into the United States. It is this question of *distribution of benefits and costs* that underlies much of the negotiations, trade disagreements, or outright political fights among potential trading partners. Other reasons often cited for trying to protect the domestic industry or block trade are concerns about domestic security (is our nation too dependent on another nation's production?), cheaters in competition (will the other nation unfairly subsidize their producers?), or the infant industry argument. This latter argument, well developed in theory and political discourse, suggests that a new, emerging (infant) industry should be protected from international competition so it can grow in management skills and economies of size so as to produce even lower cost than the imports. As this lower cost goal is achieved, specialization and comparative advantage would then be allowed to achieve their acknowledged benefits.

Another potential benefit or cost of trade is the impact on the balance of payments and monetary flow of the countries' international currency, an issue discussed later in this chapter.

THE EXCHANGE RATE

Our earlier discussion on trade, its rationale, and its impacts was useful to the understanding of a manager or decision maker, but it was a bit too simple. The world price is not always stated in U.S. dollars; in fact, in most cases it is in a currency other than U.S. dollars. However, the world price is always an equivalent value, sometimes quoted in dollars, yen, pounds, francs, colon, baht, etc. The means of determining equivalent value is the exchange rate. **Exchange rates** represent *the price of one country's currency in terms of the currency of another country.* These exchange rates are a direct reflection of the supply and demand conditions in currency markets. If demand for dollars is high (i.e., the more pounds a dollar will buy), the lower in equivalent value is the price of a good purchased in England. The exchange rate is affected by economic growth of the economies (and demand for dollars for investment), the interest rate (return on investment in dollars), and the desire for investors to spread risk by holding different types of currencies.

Exchange rates affect exports by changing the price of U.S. goods. Recall our brief discussion of exchange rates in Chapter 15: Assume it costs 600 Japanese yen to purchase a ton of American wheat valued at $100—an exchange rate of 6:1. When the exchange rate changes to 2:1 (now it will only take two Japanese yen to buy one dollar), the ton of American wheat now costs the Japanese buyer only 200 yen. With the lower price and other factors held constant, we would expect the Japanese to import more American wheat. In this case the domestic price of wheat ($200 per ton) has not changed but due to changes in the exchange rate exports of wheat have increased.

In 1997, Desmond O'Rourke, the director of the IMPACT Center at Washington State University, outlined a real-world example in an IMPACT Center newsletter. In 1996, the U.S. sold apples to Thailand at an average price of $765.20 per metric ton and bought pineapples from Thailand at an average price of $1,868.14 per metric ton. The U.S. dollar was worth an average 25 baht (the Thai currency) during 1996. This meant that to buy a metric ton of U.S. apples in 1996, a Thai importer had to come up with 19,130 baht (765.2 × 25 = 19,130); and a metric ton of Thai pineapples was worth 46,703.5 baht (1,868.14 × 25 = 46,703.5).

However, in 1997, the Thai currency fell in value against the U.S. dollar to 34 baht to the dollar. Now the Thai importer would have to come up with 26,016.8 baht (765.2 × 34 = 26,016.8) to buy the same metric ton of U.S. apples. Due to the changing exchange rate, the price of U.S. apples in Thailand would have to be increased by 36 percent. Conversely, each dollar would now exchange for more baht so a U.S. importer could now buy the same metric ton of Thai pineapples for $1,373.63 (46,703.5 ÷ 34 = 1,373.63). Thai pineapples are now 26.5 percent cheaper in the U.S. market just because of the Thai currency devaluation.

In sum, when the domestic currency increases (decreases) in value relative to that of the foreign currency, it is considered to have **appreciated (depreciated)**. Appreciation helps in the sale of exports, whereas depreciation makes exports more costly and less desirable to importing countries. In the real world, the world price (as outlined in our previous example) is a dynamic floating price, largely determined by the exchange rate.

EFFECTS OF TRADE POLICIES

As discussed earlier in this chapter, there are both gainers and losers in a domestic economy from international trade. For this reason, various policy instruments have been used to block trade or enhance domestic producers' positions.

National governments throughout the world have many policy or protection mechanisms available for stopping or controlling imports while encouraging the exports of their own producers. Common mechanisms used to restrain imports are tariffs, quotas, and quarantines. Exports are often encouraged by the use of export subsidies, domestic production subsidies, and export controls such as embargoes or other diplomatic tools.

Tariffs effectively increase the importing price by imposing a tax on the product as it enters the domestic country. This is illustrated in Figure 16.5(a) where the ES curve shifts to ES'. The effect of a tariff is to decrease quantity traded (Q_w to Q'_w), to increase the price in the importing country (from P_w to P'_w), and to decrease the price in the exporting country (from P_w to P_2). The domestic price

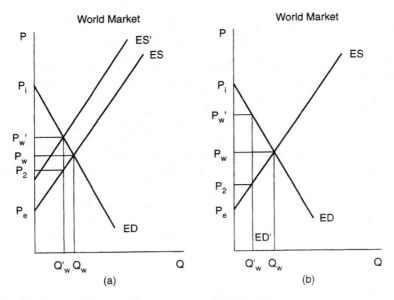

FIGURE 16.5 Trade policy impacts on world equilibrium.

rises in the importing country because fewer imports are in the market. While this is favorable to domestic producers, domestic consumers will have to pay a higher price and fewer goods will be available on the market as a result of the tariff. In the exporting country, the domestic price declines as the quantity exported to world markets declines, leaving additional quantities available for domestic consumption (which benefits consumers). Tariffs are often justified by the self-sufficiency or infant industries arguments.

Another common means of blocking or constraining trade is for the importing country to impose an **import quota** on the commodity. Quotas increase the import price by limiting the physical quantities of the good allowed in the country. This is illustrated in Figure 16.5(b) where the ED curve becomes perfectly inelastic at the quota level Q'_w (represented by ED'). The quota acts very similar to the tariff just discussed. The effect of a quota is to decrease quantity traded (Q_w to Q'_w), to increase the price in the importing country (from P_w to P'_w), and to decrease the price in the exporting country (from P_w to P_2). The quota decreases the amount of imports in the importing country, thus increasing domestic price while decreasing the quantity available to the domestic consumer. Producers of substitute goods benefit from import quotas. For example, if a quota on beef is imposed, consumers will turn to substitute goods such as chicken and pork when the price of beef increases. As a result of the import quota on beef, the producers of substitute goods such as chicken and pork will experience an increase in the demand for their products. Hence, pork and chicken producers could gain from an import quota (or other import control).

Other measures, with similar results, are used to achieve the same purpose of decreasing imports. The European Economic Community relies on variable levies to constrain imports and maintain domestic prices, whereas other countries rely on quarantines or strictly applied standards.

In striving to enhance their own exports or to slow imports to improve their balance of trade, nations often use export subsidies and/or domestic production subsidies. Export subsidies are used to lower the effective price of the good on the world market, thus stimulating the quantity demanded in the world market.

Two principal programs have been used in the United States to accomplish this effect: loans from the government to aid international marketing efforts and the Export Enhancement Program (EEP). Both programs serve to support exporters, the EEP via additional physical quantities of the exported product, and loans via a reduction in net loan payment amount.

An interesting barrier to trade, and one slightly different from those we have discussed, is the existence of export controls. Whether to avoid domestic price increases, to increase self-sufficiency, to avoid shortages, or to conserve limited resources (such as oil), export controls do interfere with free trade goals. Exports have been constrained by embargoes of wheat and soybeans, by negotiated trade agreements, and by suspension of exports for political reasons. Such activities serve to achieve short-term political goals, but do cause inefficiencies, uneven distribution of impacts, and lost opportunities in trade.

INTERNATIONAL TRADE AGREEMENTS

The international institutions, agencies, and programs constructed to aid international trade are many and they vary in scope, power, and success. The distribution of benefits and losses from trade have given rise to many special interest groups, within and between countries. As early as 1789, in the first Congress of the United States, imported goods were assigned tariffs to raise revenue for the new nation and to protect young domestic industries in the United States. Over time, the principal use of tariffs has been to achieve economic protection and development. The general position of the United States has been in favor of liberalized trade, although concerns about the balance of payments began in the 1970s and 1980s, with associated charges that our exports were being held out of our principal export customer nations by tariff and nontariff barriers.

Two approaches can be used to liberalize and balance trade conditions. One is the aggregate approach, where all relevant international parties come together to negotiate terms of trade and barriers to trade. The second approach is a local or regional attempt to bring together selected self-interested nations, to the partial exclusion of others. The United States, in its leadership role in the world economy and political arena, has and continues to use both approaches.

The aggregate approach for the United States was a precedent-setting conference in Bretton Woods, New Hampshire, in 1944. Out of this conference came the institutions that are with us today: the International Monetary Fund (IMF), the World Bank, and the General Agreement on Tariffs and Trade (GATT).

The IMF was to design a monetary system to support international trade and resultant economic development. It recognized that balance of payments was a problem, that exchange rates contributed to that problem, and that international financing was probably necessary to support economic stabilization programs, particularly of developing nations. It has been successful in providing support as well as worldwide statistical and monitoring data, allowing other nations to operate more knowledgeably.

The World Bank, really the International Bank for Reconstruction and Development, serves less developed and developing countries. Its lending has gone from World War II reconstruction, to Third World developing nations, and recently to debt restructuring in debt-laden national economies.

The third institution arising from the original conference is GATT and it is a major step away from protectionism. The overall themes of these negotiations are transparency, nondiscrimination, and reciprocity. Transparency is designed to get rid of cheaters; all trade laws and regulations are to be open and impartial. Countries are not to discriminate by giving another country or block of countries preferential trade terms. This is called Most Favored Nation (MFN) treatment. Last, reciprocity means that if one country lowers tariffs on another country's products, that nation should respond equally. And tariff barriers and levels can only be decreased, not increased, under the theme of GATT.

There have been four major negotiating rounds under GATT: the Dillon Round (1960–1962), the Kennedy Round (1963–1967), the Tokyo Round (1973–1979), and the Uruguay Round (1986–1994). Each of these difficult rounds of negotiations to lower tariffs were successful on some issues and unsuccessful in others. Agriculture was initially a major stumbling block in the Uruguay Round Agreements (URA), but after further negotiations, the URA was passed on January 1, 1995.

Major outcomes of the URA were more inclusive tariff agreements than previous and the creation of a new World Trade Organization (WTO) Committee on Trade and the Environment. WTO is to address, for 1995–2001, market access, export subsidies, internal support, sanitary and phytosanitary measures, special and differential treatment for developing countries, and a due restraint provision.

An unprecedented total of 112 nations signed to establish the WTO in 1995, including the United States' leading single customer, Japan, along with the largest consolidated export market, the European Union. The WTO will implement and monitor the international trade procedures accepted prior to 1995 and encourage further trade rounds in the future. Secondly, WTO is assigned the duty of settling disputes among its members. This aggregate international institution may be the dominant actor in near-term international trade.

Another way of improving trade is to develop regional or local entities to decrease barriers to trade among the selected members. These entities, such as the European Economic Community and North American Free Trade Agreement (NAFTA) (and its predecessor, the U.S.–Canada Free Trade Agreement), do not extend Most Favored Nation treatment to all countries. Some elements of these relationships allow subsidies within the agreement, common currency for all selected countries, and strict tariffs to outside countries.

NAFTA, unlike the Uruguay Round Agreements, completely phases out North America's regime of agricultural tariffs, with the time period for the phase-out depending on the crop or product. Nontariff barriers, such as import quotas, are to be converted to ordinary tariffs and then phased out. NAFTA is less stringent toward domestic agricultural subsidies and export subsidies, allowing their use against countries outside of the NAFTA countries.

U.S. INTERNATIONAL TRADE PERFORMANCE

In this chapter we have reviewed what trade is, the rationale for trade, the benefits and losses resulting from trade, and the institutional approaches to improving the international trade situation. Now, let's take a brief look at how the United States has performed in the past twenty-five or so years, based on statistics available from the Congressional Research Service.

During the period from 1970 to 1995, the U.S. merchandise trade balance shifted from a surplus of $3 billion to a $160 billion deficit in 1995, with expectations of the deficit increasing to around $300 billion by the year 2006. Trade with Japan and China alone accounts for two-thirds of the amount. Such trade performance reflects the ability of U.S. businesses and producers to compete with imports at home and in foreign markets.

This deficit was generated by imports growing faster than exports during the twenty-five-year period; exports grew by thirteen times, but imports increased by eighteen times. U.S. merchandise trade is normally divided into manufacture, agricultural products, mineral fuels, and other products. These products accounted for 72, 17, 4, and 7 percent, respectively, in 1970, but had shifted to 83, 10, 2, and 5 percent, respectively, in 1995. In terms of imports from 1970 to 1995, manufactures increased from 68 to 85 percent, agricultural products declined from 14 percent to 10 percent, mineral fuel remained at 8 percent, and others declined from 10 to 3 percent.

The U.S. trade balance also affects the value of the dollar. Since 1970, the value of the U.S. dollar has fallen by 29 percent overall and by 72 percent relative to the Japanese yen. Such a change has been beneficial to agricultural exports, but not for imports into the United States.

TRADE DEPENDENCY

It would be a small, but important, mistake if we did not mention one of the problems of successful international trade, namely, that the United States, or some sectors of the economy, can become too dependent on the foreign market. As overall growth exceeds the growth in domestic demand, industries rely more and more on foreign customers to buy their products, at least at a production cost recovery price. Harvested crop acreage going into the export market is now almost 40 percent of production. The state of Washington moves 80 to 90 percent of its white wheat into the international market, while its apples, cherries, and hay have found new profitable niches in overseas markets.

The result of this dependency is a sensitivity among nations and among individual firms and industries as to the trading rules, practices, and opportunities available. Agriculture continues a strong trade balance, but competition among other sectors in the international market might have political and economic impacts on this trade-dependent U.S. agricultural sector.

SUMMARY

In this chapter we have tried to introduce a complex topic, international trade, and break it down into its basic components. Producers and consumers have gained from the continual and rapid growth in world exchange of agricultural products. Exporting of agricultural commodities allows increased market expansion and also offers rewards for increases in productivity and efficiency. International trade produces gainers and losers because open markets in all countries can mean some domestic industries, in some countries, cannot compete against foreign products or services.

Comparative advantage is the basis for trade among nations. For individual firms, trade can be seen as a demand shifter, due to the increase in the population and income represented by the foreign market.

The excess demand from an importing country and the excess supply of an exporting country combine to determine world price. This price, when reflected back to each country, determines the amount to be traded.

The exchange rate, as an indicator of equivalent value, is the price of domestic currency for any country relative to the foreign currency under consideration. Exchange rates affect trade by changing the relative world price. When a domestic currency appreciates in value (buys more stuff), it helps in the sale of exports and decreases import levels.

National governments use trade policies to control their producers' and consumers' international trade opportunities. Tariffs, quotas, and quarantines are some of the tools often used to discourage imports. Export subsidies, domestic production subsidies, and export controls serve to affect exports.

International trade agreements are used to facilitate trade. This chapter discussed GATT, NAFTA, WTO, etc., relative to the world trade environment. The newer organizations seem to offer more potential in lessening trade barriers, especially from tariffs.

QUESTIONS

1. Is international trade a positive issue for a firm? Under what conditions? Could it be negative and, if so, under what conditions?

2. Develop a graphic presentation on how excess supply and demand curves can be used to determine world price. What is the amount of product traded? What happened to domestic price and quantity in each country?
3. Explain comparative advantage in words, then explain its relationship to international trade.
4. Graphically depict how a tariff affects domestic price and quantity.
5. Repeat Question 4 for import quotas.
6. How does the exchange rate affect the world price faced by producers? In your answer, define the exchange rate.
7. Explain the impact on domestic producers and consumers if a domestic currency appreciates in value.
8. Why do so many countries maintain protectionist policies in international trade?
9. Explain what GATT is and its relationship to the aggregate approach to improving trade.
10. What does NAFTA attempt to do? Why and how?

CHAPTER 17

Environmental Policy: Impacts on Decisions

INTRODUCTION

No other industry is as sensitive to environmental policy as agriculture and agribusiness. American farmers are the largest single land users, using more than one billion acres. Farms are the dominant users of water resources in the seventeen states of the western United States. Food quality and safety is a major public issue in the 1990s, raising new concerns about how production occurs on the farm and how food is handled in the processing and marketing system.

During the 1960s the term *environment* took on a new meaning for farmers, managers, the public, and economists. Farmers had long considered themselves conservationists and "stewards of the soil." But during the 1960s, the environmental movement raised questions about many agricultural practices such as heavy pesticide use (e.g., "bomb the bugs"), drug use in livestock production, damming American rivers for more irrigation and hydropower, and surface water pollution from field runoff. By the late 1960s and early 1970s, farmers and agribusiness managers were experiencing some new impacts on their decisions concerning what and how to produce.

In 1970 Congress reacted to the tremendous public outcry over environmental problems with the National Environmental Policy Act (NEPA). NEPA had a profound impact on environmental policy and resource decision making. A whole new federal bureaucracy arose, the Environmental Protection Agency (EPA), with the mandate to clean up the waterways and air across America. EPA regulations now affect every household and business in America. In addition, NEPA created the environmental impact statement (EIS), a new feature for federal regulatory decision making and project approval.

NEPA was passed at a time when economists were first beginning to specialize in scientific study of environmental subject matter. A new field emerged in the 1970s, environmental economics. Economists who specialize in environmental economics apply economic theory and methods of analysis to the problems of the use and management of ecosystems where the quality of human existence is an essential focus.

Much of what environmental economists and other scientists study relates directly to the decisions made by managers in agriculture and agribusiness. For example: "What would be the economic impact of restricting the use of growth stimulating hormones in feed lot livestock production?" "What are the economic and human safety trade-offs between pesticide use and crop-production efficiency?" or "What impact would a pollution tax on nitrogen fertilizer have on corn producers?"

In This Chapter

The tools of economic analysis play a role in the study of environmental management decisions just as they do in farm management or marketing or forestry. In this chapter we add some new concepts about environmental externalities, property rights, and environmental policy options. These ideas complement the basic tools of economic analysis used by managers. Then we review how emerging new policies to address environmental issues impacts decision making in agriculture and agribusiness.

ENVIRONMENTAL PROBLEMS

When you think of the "environment" what probably comes to mind is the air you breathe, the water you drink, and the land where you live. Thus, "environmental problems" probably mean some misuse of certain natural resources (land, air, water) or ecosystems. But when you look at the interaction of people, institutions, and the environment, it is important to think of the three major groups that make resource management decisions—how, what, and how much to produce—that have environmental implications. This is illustrated in Figure 17.1. We can simplify the economy by grouping the major actors into three categories: firms, households, and governments. Firms buy the factors of production from households and governments through various market and nonmarket transactions and they pay taxes. Households sell their labor and other resources (e.g., land, timber, water) to firms and governments and also pay taxes to all levels of government. Government is the largest single resource owner in the United States, plus it has a major role in regulating the environment since passage of much environmental legislation in the Environmental Decade of the 1970s.

If all of the positive and negative impacts of the interactions of firms, households, and governments were contained in their market transactions, then the need for environmental policy would be small. However, many market actions impose costs on third parties that are not accounted for in the buyer–seller relationship. For example, the lagoons used to handle the waste generated by large hog operations in the Midwest and South are efficient from an engineering viewpoint but create unsavory odors for neighbors. These odors may not be a health hazard and as long as the lagoon is contained there is no threat to water quality, but the firm is still creating negative impacts on third parties. It is these third-party effects that can become major concerns to managers in agriculture and agribusiness.

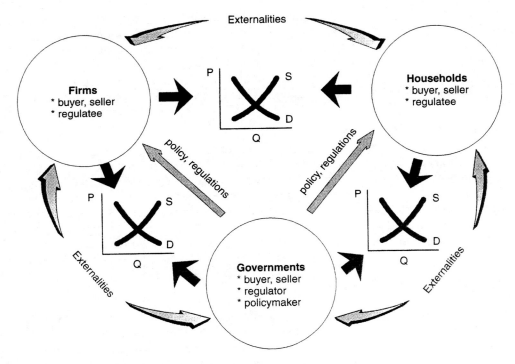

FIGURE 17.1 Roles of firms, households, and governments in resource and environmental policy.

Social Costs and Externality

Look again at the supply–demand representations in Figure 17.1. If all costs were accounted for in these transactions, then there would be no third-party effects. But everyday the newspapers and TV highlight examples of situations where some costs of economic activity fall on someone else. These are known as *social costs* and are external to the market transactions. For example, the cost of pork does not always include the cost to neighbors from unsavory smells of waste handling. The cost of the airline ticket does not include the costs of noise in neighborhoods surrounding the airport.

When a social cost exists outside of the cost computations of a firm or industry, it is called an *externality*. In most cases the term *externality* refers to some environmental costs, although positive externalities are possible. (Think about the pollination of an orchard as an externality of honey production.) The economic impact of an externality is illustrated in Figure 17.2. Remember from our earlier discussion that an industry supply curve is a horizontal summation of the MC curves for individual firms. But what about the social costs of market transactions? If externalities are accounted for by adding social costs, a new supply curve, S_1, would be created. Adding social costs would logically mean higher product prices and lower

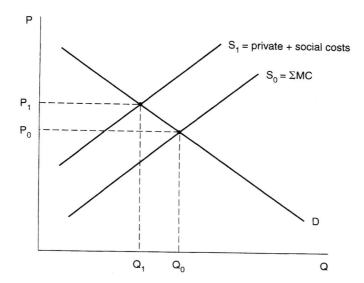

FIGURE 17.2
Accounting for social costs in price and quantity decisions.

outputs. From the environmental viewpoint, not considering social costs results in overuse of the resources at a price set too low in the market. As shown in Figure 17.2, with a new supply schedule at S_1, price rises to P_1 and output declines to Q_1.

What Causes Externalities?

This is almost the same as asking what causes environmental problems. The answer lies in the nature of the capitalistic marketplace.

Property rights govern the operation of the market system. Firms, households, and governments own resources and thus can rent, sell, save, or destroy them. The profit motive usually governs how resources are used by firms and households. But for some resources and situations, property rights are unclear or nonexistent.

Example 1. You live on a clear lake and enjoy the recreation and scenic benefits. A new coal-fired electric power plant is located near the lake and uses lake water for cooling purposes. The power plant returns clean and pure water to the lake but because the water is much warmer, algae grows more rapidly. If it continues, the fish will die and boating will no longer be pleasurable. Do you have a right to pure and cool lake water?

Example 2. You live in a western state where annual rainfall is low. Farm managers in an adjoining state begin a vigorous cloud-seeding program in an attempt to generate more rainfall. When they are successful, you calculate your rainfall has decreased. Are they "robbing" you of your rain? Do they have the right to do so?

In each of these cases the property right to cool water and rainfall is unclear. This makes legal recourse difficult, if not impossible. Thus, the opportunity arises for social costs and private costs to be different—an externality exists. Most environmental problems arise from areas in which property rights are uncertain, changing, or unenforceable.

In America, property rights are dynamic—they are modified as society's values and beliefs evolve. These changes usually come slowly but they do occur. Much of the environmental movement is an attempt to influence the nature of the property rights associated with the environment and natural resources. This can be illustrated with the changing attitudes about water use and development in America. In the early history of the United States, wetlands were a nuisance and health hazard. Congress appropriated money to drain wetlands to create farmland and home sites and even granted millions of acres of wetlands to states for drainage. By the 1960s attitudes had changed dramatically and wetlands were recognized as an important part of the ecosystem with significant water quality and wildlife benefits. Now Congress restricts wetland conversion and pays to have wetlands restored!

For managers in agriculture and agribusiness the point should be clear: Production and management practices that have been acceptable for years may become unacceptable as cultural attitudes change and property rights get modified. In considering this process, it is important to understand the environmental policy options that are possible for solving the externality issues in agriculture or related areas.

SOLVING ENVIRONMENTAL PROBLEMS

Because ecosystems and social institutions are complex, no single method is available that equitably addresses the environmental problems that impose social costs and external effects on most Americans. Over time different philosophies and techniques have developed. Their use and impact on managerial decision making have varied. These different methods for solving environmental problems can generally be divided into four categories:

1. Market solutions;
2. Common law remedies;
3. Taxes, charges, and subsidies; and
4. Standards and regulations.

Market Solutions

In some situations, problems of social costs or externalities can be solved directly by negotiations between the involved parties. For example, suppose the landowner next to you is negotiating the sale of his farm for use as a landfill by the municipal waste disposal authority. Having a garbage burial ground next to your farm is not an inviting thought—the external effects of sight and smell will impose costs on you. A market solution to this simple example would be to offer to pay your neighbor (a contract) to persuade him to sell to another buyer, rather than to the waste disposal authority. But this contract could be very expensive for you since your neighbor would probably realize substantially less money from the sale of his land to another buyer. He would want you to make up the difference. Still, expensive as it may be, the amount of the "bribe" (contract payment) could be a measure of the social costs of a landfill.

A second simple example illustrates the market solution possibility from another perspective. Sometimes external effects are beneficial and are desired by certain individuals or groups. Such is the case with the honeybee. It is in the interest of the orchard owner to make certain that plenty of bees are located near her trees to ensure pollination of their flowers. To do this, she may buy and handle beehives herself, or she may negotiate with a beekeeper. In either case, a freely negotiated contract can solve the "problem" of pollination. This is the principle behind the green-space preservation programs, which pay agricultural landowners certain benefits to maintain land in farming versus conversion to urban uses. Cities and states "buy" the development rights to farmland in order to receive the benefits of desired scenery and other environmental amenities.

Although market solutions to environmental problems are possible, they have not solved the major environmental problems of clean air and water, cropland erosion, and related issues. Two substantial obstacles block more frequent market solutions. First, many external effects are pervasive and widespread. A factory belching smoke and smells may affect hundreds of residents, but each individual may be affected only occasionally (when the wind blows a certain direction). This makes it unlikely that one person would be willing to pay the factory to reduce emissions or help organize others in order to negotiate a contract with the factory for the whole community. *The real costs of creating a market solution—time, money, and effort needed—to solve widespread environmental problems* are called **transactions costs.** When transactions costs are high, the market solution remains remote.

A second impediment to the market solution is the **free rider.** In any collective or group effort to solve an environmental problem, there will always be a few who realize they can or will benefit without any effort or cost on their part. For example, suppose the residents surrounding a small lake decide to organize, buy some chemicals, and attack the persistent mosquito problem that arises each summer. Many residents may join in the group effort. A few, however, will realize that if the others do the job, everyone will benefit even though a minority fail to participate. Thus, the free rider benefits without bearing any cost. The more free riders you have, the more remote the possibility of a market solution.

In spite of these problems, new attempts have been made to create markets that address externality issues. A prime example of this is the sulfur dioxide emissions trading program contained in Title IV of the 1990 Clean Air Act Amendments. Sulfur dioxide emissions come from coal-fired power plants and are subject to the National Primary and Secondary Ambient Air Quality Standards, which regulate

TABLE 17.1 Estimated and Actual Costs for Emissions Trading (SO$_2$), Dollars per Ton

Industry Estimates, pre-1989	$1,500
EPA 1990 estimate	$750
Early allowance trades	$250
1993 CBOT trades	$122
1995 CBOT trades	$126
1996 average trades	Low $100's
Glen Falls Middle School, 1997	$76

Source: Adapted from "Trading Emissions to Clean the Air," *RESOURCES*, Winter 1996 and National Geographic Magazine, April 1997.

concentration and timing of emissions. Under the Title IV amendments, a cap on emissions was put on the nation's 110 dirtiest power plants. Each of these plants receives an annual "allowance" for emissions, which they can transfer to other plants in their company, sell, or save for future use. This created an incentive to lower emissions with new technology and not only comply with the national standards but also be able to sell the allowances.

The electric power industry fought this new market in emissions allowances, claiming these allowances would be very expensive—possibly $1,500 or more. But EPA estimated the allowances would sell for $750. A national market was created for these allowances and is run on the Chicago Board of Trade (CBOT). The actual history of emissions trading is summarized in Table 17.1.

The new market solution to reducing sulfur dioxide emissions has been very successful. Although it is not a perfect program, it has permitted power plants to choose how to lower emissions (i.e., new technology or buying allowances from other utilities) and created a new incentive for compliance (i.e., the value of the emissions allowances). Some surprising trades have occurred in this market. For example, a middle school near Schenectady, New York, raised $24,000 to buy 313 allowances as a school project in reducing air pollution. Because they "retired" the allowances, this means 313 tons of sulfur dioxide will never be released into the atmosphere.

Common Law Remedies

Our system of justice provides for legal remedies in resolving some environmental problems. Many of these remedies are rooted in the tradition of English common law. Even though there may be no special statute on the books, a legal tradition exists that provides avenues for resolving disputes. Through the common law remedy of *nuisance*, citizens can initiate legal solutions to some environmental problems. **Nuisance** is any interference by another party with your basic right to enjoy and use your property. Air and water pollution, noise, odor, and unsafe facilities are examples of possible nuisance conditions.

The nuisance remedy is a legal action designed to show physical or financial damages and a request for an injunction to stop the nuisance and possibly for financial restitution. The courts must decide whether or not a nuisance exists and the nature of the damage. This may provide the courts an opportunity to force a firm or individual to internalize the externality or take some action to eliminate its existence.

An important point to remember is that agriculture is characterized by numerous activities that are considered nuisances by many people. Tractor noise in the early morning, misapplication of chemicals, dust from a combine, feed lot odor, and soil erosion into waterways are all examples of possible agricultural nuisances. Affected citizens have a common law remedy—nuisance suits—to bring action against farmers in these cases. In Iowa, livestock producers have been sued by neighbors for not incorporating swine waste within one day of land application. In New York, dairy producers have been sued because waste lagoons leaked and polluted groundwater. More of these nuisance suits can be expected in the future as suburban neighbors with different values and beliefs about agriculture come in direct contact with production locations (farms, feed lots, processing facilities).

Transaction costs can be high for common law actions. But it is apparent that many Americans are willing to incur these costs in order to reduce the social costs, as they perceive them, of agriculture and agribusiness. So managers of agricultural operations should be prepared to devote time, legal fees, and court costs to defend their historical practices, which may produce new conflicts.

Taxes, Charges, and Subsidies

If an environmental problem cannot be solved by a negotiated contract between parties or by recourse to legal remedies, then a market solution can be imposed by "internalizing externalities," where the firm integrates the costs of the pollution into the firm's MC. This solution can be imposed with pollution taxes, charges, or subsidies. Ideally, a system of pollution taxes, charges, or subsidies would induce or "force" all firms to account for both the private and social costs of pollution from their production processes. Since each firm's internal cost structure includes the costs of discarding or creating a pollutant, this system would operate automatically by allowing firms to adjust production levels based on their own profit-and-loss accounting with the costs of any pollution or other external effects being included in that production decision.

A simple example of how a system of taxes or charges would work involves a fertilizer manufacturer operating a factory next to a large lake. Water from the lake is used in production and for disposal of waste products. The water from the factory is discharged directly into the lake and necessitates a large expensive cleanup effort in order to preserve the fish population, sanitary swimming, and public drinking water.

The cost of the cleanup is some measure of the social costs of fertilizer production at that locale (there may be other costs). If those costs can be translated into a tax on fertilizer production or a per-unit "water user charge," then these social costs will be internalized within the firm. Figure 17.3 illustrates the impact of these

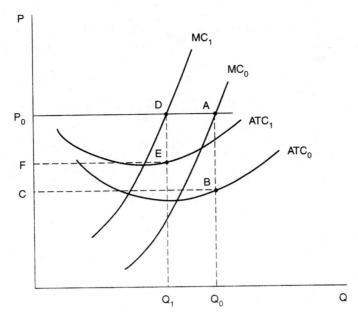

FIGURE 17.3
Impact of a pollution tax or per-unit user charge on cost structure of a competitive firm.

taxes. If the fertilizer firm were operating at the profit-maximizing level of $MC = P = MR$, then with market price P_0 output would be Q_0 and profit would equal the area P_0ABC. Imposition of a tax or per-unit charge raises the MC and ATC curves at all levels of production (from MC_0 to MC_1 and ATC_0 to ATC_1). The firm reacts by adjusting its profit-maximizing level of production. Under the new cost schedules, MC_1 and ATC_1, the level of production would be Q_1 (assuming price remained at P_0). Notice how profits have been reduced to the area P_0DEF. Further note that the manager of the fertilizer firm would utilize the same decision-making rules that we developed earlier in this text.

Actual calculations of the impact of pollution taxes are of course very complicated. Most industries will claim any tax will put them out of business, and in some cases the claim may be justified. But the important concept is that systems of taxes and user charges work by affecting the internal cost characteristics of a firm and thereby allow the firm's manager to adjust behavior in an economic manner in the face of new costs imposed by society.

If taxes and user charges are deemed inappropriate, subsidies could be used as incentives to reduce pollution. They would operate in the same fashion if offered to a polluter on a per-unit basis. That is, if a firm were paid a certain amount, say, 50 percent, to reduce polluting discharges the firm could adjust output in an attempt to "earn" the subsidy. Society may well be willing to pay subsidies in amounts up to the total costs of cleaning up or otherwise handle the pollution.

Attractive as a system of taxes, charges, and subsidies might seem, actual operation would be complex and difficult. The correct level of per-unit tax would necessarily require complicated economic analyses. Enforcement and monitoring of the system would require bureaucrats. Even a theoretically correct tax would not totally eliminate pollution, unless the MC of eliminating pollution were less than the tax at all levels of production. The cost of implementing such a system would be costly in its own right and often not justifiable when compared to the potential benefits received.

Standards and Regulations

The major public response to pollution has been complex laws that prohibit or limit pollutants. The historical root of these systems of standards and regulations was the Rivers and Harbors Act of 1899, which, among other purposes, prohibits individuals, corporations, and municipalities from discharging "refuse" into navigable streams and bodies of water. It was quite apparent from the quality of water in our rivers and lakes in the 1960s that this act had not been well enforced.

However, the basic idea in the Rivers and Harbors Act—solution by regulation—was made the central theme in the environmental legislation of the 1970s. The major new environmental laws passed during this period were the Clean Air Act Amendments (1970 and 1977), Federal Water Pollution Control Act Amendments (1972), Clean Water Act (1977), Safe Drinking Water Act (1974), Resource Conservation and Recovery Act (1976), Federal Insecticide, Fungicide, and Rodenticide Act Amendments (1975), Surface Mining Control and Reclamation Act (1977), and the Toxic Substances Control Act (1976). The general thrust of all this legislation was:

1. Outright prohibition, as with water pollution (complete elimination of the discharge of pollutants into navigable waters by 1985), or quality standards, as with air pollution (national primary and secondary standards for every air pollutant having an adverse effect on human health);
2. National and state enforcement plans; and
3. Penalties for noncompliance.

The statutes passed in the 1970s continue to be amended and modified today, constituting a regulatory framework in which firms must operate. Unfortunately, economics did not play a large role in the environmental policy making of the 1970s. In almost every case where the EPA had some discretion in treating an environmental problem, it chose regulation. Little attention was given to the economic costs and benefits of each regulation or on the level of regulation. To monitor the economic impacts of environmental programs, the Census Bureau conducts annual surveys. The survey results, which are published in the *Current Industrial Reports* series, report the expenditures by industry on pollution abatement equipment, expenditures that can be linked to environmental laws and legislation. As revealed by the numbers in Table 17.2, industrial firms in the United States devote a significant level of monetary resources to complying with pollution control legislation. Thus, the cost and burdens of environmental regulation become important public issues.

TABLE 17.2 New Pollution Abatement Capital Annual Expenditures (Millions)

	1992	1993	1994
Manufacturing:			
Air pollution	$4,403.1	$4,122.0	$4,310.6
Water pollution	2,509.8	2,294.9	2,428.9
Solid/contaminated waste	954.0	761.0	838.5
Other[a]	526.1	390.3	302.5
Total	$8,393.0	$7,568.2	$7,880.5
Mining:			
Air pollution	$105.9	$113.2	$130.6
Water pollution	144.7	161.1	190.9
Solid/contaminated waste	67.0	114.5	112.7
Other[a]	87.5	114.8	116.5
Total	$405.1	$503.6	$550.7
Petroleum and Coal Products:			
Air pollution	$1,932.9	$2,380.4	$2,627.5
Water pollution	1,380.5	1,115.1	1,113.6
Solid/contaminated waste	560.1	468.0	448.7
Other[a]	493.8	481.9	472.6
Total	$4,367.3	$4,445.4	$4,662.4
Electrical Utilities:			
Air pollution	$1,948.3	$2,929.1	$3,145.3
Water pollution	567.0	620.5	605.5
Solid/contaminated waste	37.4	482.7	428.4
Other[a]	151.2	167.7	170.2
Total	$2,703.9	$4,200.0	$4,349.4

[a]Includes expenditures for underground storage tanks, site cleanup, noise abatement, and radiation abatement.

Regulation by prohibition and standards does not permit the firm to internalize social costs by adjusting production. Instead of an economic signal regarding social costs (taxes, charges, and subsidies) the firm's manager must change production to meet a physical signal (the minimum or maximum standard). This distinction becomes especially important because economists constantly point to the advantage of marginal decision making. Under regulatory schemes of pollution standards, firms, individuals, and our society cannot compare the marginal cost of pollution abatement to the marginal benefits. Now society is pondering statements such as this: Maybe we do not want 100 percent clean water in the rivers because accepting 95 percent clean water will save billions of dollars in public and private expenditure.

IMPACTS ON DECISIONS

Because agriculture depends more on land, water, and resources than any other industry, the quality of the environment is of extreme importance. Agricultural producers have a history of concern for soil stewardship and conservation; in some senses they were the first environmentalists. But in the decade of the 1970s, the environmental movement made agriculture a focus of concern and pressed regulatory solutions to many air, water, and land problems in agriculture. The result: More than ever before, the private decisions made by agricultural managers are affected by policies and regulations announced in Washington, D.C., and state capitals.

Environmental problems and the regulations promulgated to solve them will have important impacts on several managerial decisions including output decisions, production method decisions, input decisions, and risk.

Output Decisions

The quality of air and water, especially acid-rain water, is already influencing decisions about what to produce because alternative production functions have changed. A classic example of this is reported by the National Academy of Sciences. The Zinfandel grape is being abandoned in parts of California because of damage from air pollution. Other adjustments have been reported in cigar tobacco and vegetable production decisions for air quality reasons.

Production Methods

The concern for soil erosion problems and environmental hazards of fertilizers and pesticides has created changes in agricultural production methods. Minimum tillage or "no till" for row-crop production is an adjustment in methods that is being spurred by energy costs and soil erosion concerns. "Organic farming," a cultural practice that uses no chemical fertilizers or pesticides, is the agriculture production method deemed most desirable by many environmental groups.

But these changes in production practices may not be enough to meet the emerging regulations on nonpoint source pollution coming from the reauthorization of the Clean Water Act. Nonpoint source pollution is the runoff from farms, construction sites, and urban areas. It has no pipe or discharge "point" into a waterway. Thus, regulations have been slow in coming but there is every expectation that new "best management practices" will be required by agriculture and agribusiness to protect water quality. In Kentucky, all farms larger than ten acres must have a water quality plan in place in the next five years that specifies which practices will be used in each field and around the homestead to protect both surface and groundwater. This effort is a clear attempt to be proactive on water quality before the EPA promulgates more strict regulations on farms.

A recent proposal by the EPA for more stringent regulation of airborne particulate matter is of concern to agriculture. Dust particles carried by windstorms are a leading source of particulate matter. Compliance requirements may include adoption of approved cultivation practices to minimize wind erosion of the soil.

Runoff from agricultural land is a primary source of pollution for rivers and lakes. *Precision farming*, also referred to as *site-specific farming*, is emerging as a promising method of reducing this type of pollution. Precision farming, through the use of a global positioning system (GPS), which uses satellites to determine exact locations in a field, enables the farmer to apply just the right amount of fertilizer, herbicide, pesticide, or seed to a specific area of a field. Since only the amount needed is applied, precision farming reduces the application of chemical inputs. This precision of application reduces the potential pollution problems—since less is used, less runs off—from agricultural lands, making this production method consistent with the conservation and environmental goals of reducing pollution of surface and groundwater. To be adopted, the benefits of increased yields (and returns) and decreased costs (decreased input use) from the more precise use of agricultural inputs will have to outweigh the cost of the new technology. Currently the high start-up costs have limited the adoption of the technology. However, as the technology is further refined and becomes more widely available its costs will drop, encouraging wider adoption, especially by commercial agriculture.

Input Decisions

The availability of certain agricultural inputs is not certain in the future. Chemicals such as DDT and 2,4,5-T have been removed from use. In tobacco production, the chemical MH-30 has created residue problems on leaves, and buyers have called for a reduction in use. *Integrated pest management* (IPM), a production method that utilizes all suitable techniques and methods to control pests—not total reliance on chemicals—may reduce the use of chemical inputs. The objective of IPM is to limit the growth of pest populations, using biological, cultural, and other nonchemical control methods, to levels below that which are economically damaging. Thus, a chemical application is made only when pests reach levels considered to be economically damaging, thereby reducing the producer's reliance on chemicals to control pests.

Risk

Modern agricultural techniques have reduced substantially the risk involved in agricultural production. Irrigation and chemicals have created new levels of yield and predictability. But these gains are threatened by the environmental concerns about groundwater usage and the long-term impact of chemical use. Risk could increase in agriculture if water is reduced for irrigation and certain chemicals are no longer available for farm use.

Environmental policies in certain states and at the federal level create new risks for agriculture and agribusiness. The uncertainty about what will be permissible in waste management, field practices, sanitary handling of outputs, and other issues will put more risk on the decisions made in the 1990s and beyond. So, in addition to weather, financial, market, and production risk, the good manager will have to address environmental risk management too.

SUMMARY

Environmental policy is affecting decision making by managers in agriculture and agribusiness. Once considered the stewards of the rural environment, farmers are now seen as a major source of problems in terms of many water and air quality issues. Governments at all levels are considering a variety of options to address local and national environmental issues: market solutions; common law remedies; taxes, charges, and subsidies; and new standards and regulations. Managers need to understand the consequences of these new policies and how these will impact decision making on outputs, production methods, inputs, and risk.

Different philosophies and techniques have evolved over time to address equitably the environmental problems that impose social costs and external effects on most Americans. Market solutions occur when parties to a problem engage in direct negotiations. While appealing, this solution is not widely used because of the pervasive and widespread nature of many external effects and the problem of the free rider who benefits from an action without bearing any cost. Common law remedies rely on legal tradition, such as the nuisance concept, rather than on legal statute to initiate legal solutions to environmental problems. Such activities do use the courts to decide relative positions in a complaint, but the effectiveness of this solution is again constrained by the pervasive nature of most environmental problems.

Taxes or subsidies are a means of imposing a market solution by internalizing externalities. The manager incorporates the tax or subsidy into the firm's internal cost structure, then makes appropriate profit-making decisions. In such a case the costs of discarding or creating a pollutant are being paid by the polluter. Standards and regulations are an external constraint on the activities of a firm, a constraint that modifies cost structures and could generate resource inefficiency for society.

The environmental questions are a reality affecting managerial decisions now and into the future. Output levels and by-products will be closely examined in the future. Production methods may be constrained and some technologies outlawed. Inputs that are allowed now may not be available in the future. Finally, managerial risk may increase as a result of concern about long-term use of irrigation water or chemicals.

QUESTIONS

1. How have environmental concerns tempered the traditional profit maximization approach to agricultural economics?
2. "The optimal level of water pollution is zero." Analyze this statement based on your understanding of the costs of pollution control to competitive firms.

3. "Define *externality* and give one example of an externality issue in your state. What policy options are politically feasible for addressing this externality?
4. Explain why transactions costs and the free-rider phenomenon impede market solutions to environmental problems.
5. What are the strong and weak points of the common law, taxes and subsidies, and standards and regulations approaches to solving environmental problems?
6. Why are environmental problems of special importance to people involved in agriculture, and what decisions now require reassessment that were taken for granted in times past?

GLOSSARY

absolute advantage   A country is said to have absolute advantage when it can produce a product, good, or service at less cost than another country.

Accelerated Cost Recovery System (ACRS)   Standard procedure for handling depreciation on property purchased after 1980, but prior to 1987. It permits a more rapid write-off (or cost recovery) for tax accounting.

agent middlemen   Serve as agents or representatives for buyers and sellers, but do not take title to the goods.

agricultural economics   Economics applied to agriculture and rural areas.

agricultural policy   The set of government programs that directly influences agricultural production and marketing decisions.

air/water pollution   Any undesirable or harmful substance or biological life form that accumulates in air or water.

allocation   The process of making decisions about how to use our resources or capabilities.

arc elasticity   A formula that measures responsiveness along a specific section (or arc) of a supply curve, and measures the "average" price elasticity between two points on the curve.

average fixed cost (AFC)   The average cost of the fixed inputs per-unit of output; it is calculated by dividing the total fixed costs by each output level.

average physical product (APP)   Indicates the average productivity of each unit of variable input being used.

average revenue (AR)   The average dollar amount received per-unit of output sold (produced); it is calculated by dividing the total revenue at each output level by the output level.

average revenue product (ARP)   Shows the average value of output per-unit of input at each input use level.

average total cost (ATC)   The average total cost per-unit of output; it is calculated by dividing the total cost at each output level by that output level.

average variable cost (AVC)   The average cost of the variable inputs per-unit of output; it is calculated by dividing the total variable costs by each output level.

bargaining power   Individual producers joining together and acting as a group in their marketing efforts or input purchases to influence the prices received or paid.

barriers to entry and exit   Differences in market structure that block new firms' attempts to enter or exit the market.

break-even analysis   Determines the output level, either on a total or per-unit output basis, where the costs of production exactly equal revenue.

break-even points   Points on a graph indicating at which output levels total revenue exactly equals total costs.

budget   The amount of money available for purchases in a given time period.

budget constraint   A constraint on consumption caused by price and availability of the goods in the market as well as the size of the budget.

budget line   A line indicating all the combinations of two products (goods) that can be purchased using all of the consumer's budget.

business cycles   The tendency for the economy to grow and contract.

ceteris paribus   Literally "all else constant."

change in demand/supply   Where changes in the quantity purchased/sold are a result of factors other than a change in the product's own price.

change in quantity demanded/supplied   Where the quantity of a product demanded/supplied changes in response to a change in product's own price.

close-down case   In the short-run, a firm will shut down if price (MR) is less than the minimum AVC for the firm.

collusion   Occurs when the representatives of all firms producing a certain product meet and decide jointly what the price will be next year and who will sell in particular markets.

commodity approach   An approach to marketing that simply follows one product, such as cotton, and studies what is done to the commodity and who does it as it moves through the market system.

common law remedy   Legal remedies for resolving environmental problems.

comparative advantage   Total output can be increased when each country specializes in producing the commodity for which it has the greatest advantage or the least disadvantage.

comparative static analysis   Comparison of two separate market situations (static pictures of the market) before and after a change in the marketplace.

competitive substitution   Occurs when the use/consumption of one input/good must be decreased to use/produce/consume additional units of another input/good.

complementarity   Occurs when inputs must be used in a fixed ratio; there is no choice as to what proportion of each input to use.

complementary products   Goods that are consumed/produced in combination (i.e., hot dogs/hot dog buns; wool/mutton).

compounding   A technique for using interest rates to determine the possible value of money at some predetermined, future time period.

concentration ratio (CR)   The percentage of total sales or value of industry shipment by the largest four, or sometimes eight, firms in the industry.

constant returns   For each additional unit of input used, output increases at a constant rate (the rate of change in output remains constant).

cost-minimization criterion   When the firm has more than one variable input to use in the production process, finding the least-cost combination of inputs that will produce the output level desired by the firm.

cost–price squeeze   A situation in which the costs of production are increasing and the prices received for the product remain unchanged or, in a worse situation, prices decrease simultaneously.

costs of production   The payments that a firm must make to attract inputs (resources) and keep them from being used to produce other outputs.

cropland set-aside   Withdrawal of a specific percentage of land normally devoted to a certain crop.

cross-price elasticity   Measures the responsiveness of the quality supplied/demanded of a good to changes in the price of a related good.

decision making   Evaluating alternative courses of action, with a clear idea of the objective or goal of the decision.

declining balance method   Depreciation method that is calculated by multiplying (depreciation rate/useful life) by the remaining book value each year.

decrease in demand/supply   Occurs when a factor other than the good's own price changes, resulting in less quantity demanded/supplied at the same price (inward shift of the demand/supply curve).

decreasing returns   Occur when each additional unit of input increases the production level, but with a smaller change than the previous input.

defensive management strategies   Management strategies that protect market shares and maintain stability.

deficiency payments   Direct government payments based on the difference between the target price and the lower of the average market price or the nonrecourse loan rate.

demand   Consumer's desire and willingness to pay for an item produced.

demand curve   A line connecting all combinations of price and quantities consumed for a particular good, *ceteris paribus*.

demand schedule   Information on price and quantity (consumption) combinations that give the consumer maximum utility.

demand/supply elasticity   The percentage change in the quantity demanded/supplied relative to the percentage change in price (or other economic variable) when moving from one point to another on a demand/supply curve.

depreciation   Inputs that are used over more than one year in the production of income should have a portion of their costs deducted for each year of useful life.

diminishing marginal utility   As additional units of a good are consumed, a point is always reached at which the utility derived from each successive unit declines.

direct payments   Designed to increase farmers' incomes directly during periods of low prices or natural disasters and to induce certain producers to participate in government programs of supply control.

disaster assistance   Government program provisions for assistance to farmers suffering losses from weather-related calamities, droughts, disease, and related problems.

discount rate   The interest rate charged by the Federal Reserve Banks for loans to commercial banks.

discounting   A technique for using interest rates to determine today's value of sums to be received at some future date.

disequilibrium   The price at which the quantity supplied by producers does not equal the quantity demanded.

diversification   The selection of two or more products for production.

diversion payments   Direct payments made to producers for idling land under government programs.

durable inputs   Inputs that are used over more than one production period.

economics   The allocation of scarce resources between competing ends for the maximization of those ends over time.

economies of scale   In some industries, the size of the plants and equipment necessary to achieve the lower production costs of mass production is extremely large.

effective demand   The quantity of a good consumers are willing and able to purchase at alternative price levels.

efficiency   Governs the operation of the competitive industry and enables products to be supplied to society at the least possible cost.

elastic   Refers to a change in price that brings about a relatively larger change in quantity.

elasticity   The responsiveness of quantity to changes in price or some other economic variable.

elasticity of production   The percentage change in output we receive for a 1 percent change in input.

environmental economics   Scientific study of environmental subject matter by economists.

environmental problems   The effects pollutants have on the quality of the environment in terms of health and welfare of people and how to reduce pollution to acceptable levels.

Environmental Protection Agency (EPA)   A federal agency created to carry out various regulatory functions concerning the environment.

equilibrium price/quantity   The price at which the quantity supplied by producers exactly equals the quantity demanded.

equi-marginal returns principle   A specific concept whereby returns will be maximized when scarce resources are employed so as to have equal (or as nearly equal as possible) marginal returns in each enterprise, product, or activity.

exchange functions   Relate directly to the buying and selling that must take place between producer and consumer, for example, costs of brokerage, speculation, etc.

expanding economy   Occurs when the annual change in gross domestic product is positive.

expansion path   The line connecting the combination of products that will maximize revenue for any given level of resources at the price ratio the manager considers relevant and probable.

expansionary fiscal policy   Policy involving increased government spending and/or decreased taxes, resulting in a boost in economic activity.

expansionary monetary policy   When the Federal Reserve Bank strives to increase the money supply in the economy and increase the level of credit available to businesses and individuals.

externality   Costs and benefits that exist outside of the cost computations of a firm or industry.

facilitating function   Relates to making the market systems operate more smoothly and with greater technical and pricing efficiency (e.g., grading, standardized measures, etc.).

facilitators   These people provide the necessary functions of market information, grading, standardization, and so forth.

factor demand   The firm's demand for the variable input used in production, represented by MRP for the input within Stage II of production.

factor–factor decision   The decision-making process whereby the manager must decide which combination of inputs to use in order to achieve the firm's economic objective.

factor–product decision   Utilizing knowledge of the physical production process and the costs of production, combined with the prices of products to evaluate the costs and revenue at alternative levels of production, allowing an economic decision to be made.

Farmer Owned Reserve   Government-operated stock program designed to stabilize prices and ensure dependable grain supplies by providing a means of keeping grain supplies off the market until grain prices reached a specified level.

fiscal policy   Involves government expenditures and tax rates.

fixed costs   Those costs that do not vary with the level of production; the costs associated with the fixed factors of production.

fixed factors of production Factors that have to be maintained in the short-run even if production is zero (i.e., land, buildings, factory equipment).

flexibility Involves modifying the most profitable business plan to avoid losses or pursue new opportunities.

form utility Putting the raw product into the style, appearance, or quality desired by the consumer.

free rider One who benefits from an action without bearing any cost.

functional approach The study of the activities performed in changing the product of the farmer to the product desired by the consumer.

future value Value of money at some predetermined, future time period.

futures contracts A promise to buy or sell a standardized quantity of certain agricultural products at a fixed price and future date.

graphical approach Displays relationships between prices, factors, and production levels.

gross domestic product (GDP) Over any year, the estimated total value of goods and services produced in all the markets in the United States.

homogeneous products A standardized, no "brand name" product.

immediate short-run A time span so short that no resource changes can be made.

imperfect competition Exists whenever the individual firm has some control over prices received for the goods it produces and markets.

income elasticity The percent increase/decrease in quantity demanded of a good in response to a 1 percent change in consumer's income level.

increase in demand/supply Occurs when a factor other than the good's own price change results in a greater quantity demanded/supplied at the same price (outward shift of the demand/supply curve).

increasing returns Occurs when each additional unit of input added to the production process yields more additional product than the previous unit of input.

independent commodities Commodities are considered independent if the change in the price of one good has no effect on the demand for another good.

indifference curve A line showing all the combinations of two goods (or products) that provide the same level of utility.

individual firm's short-run supply curve The MC curve above the minimum AVC curve is the firm's short-run supply curve.

industry supply curve The horizontal summing (adding up) of all the supply curves for the individual producers in the market.

inelastic Refers to a change in price bringing about a relatively smaller change in quantity.

inferior good A product whose consumption decreases as income increases.

institutional approach   Approach to marketing that emphasizes who is doing the marketing functions.

interest rates   A measure of the opportunity cost of money.

inverse price ratio (IPR)   The price of the input/product on the horizontal axis divided by the price of the input/product on the vertical axis.

invisible hand   Describes the operation of the competitive market in which, inefficient firms, those with the highest production costs, are forced out of the market.

isocline   A line connecting the least-cost combinations of inputs for all output levels at a specific price ratio.

isocost line   A line indicating all combinations of two variable inputs that can be purchased for a given, or same, level of expenditure.

isoquant   A line indicating all combinations of two variable inputs that will produce a given, or constant, level of output.

isorevenue line   A line depicting all combinations of two products that will generate a given, or the same, level of total revenue.

joint products   Occurs when the production of one product actually results in the production of another.

land retirement   Reduction of plantings of certain crops or retirement of land from crop production into soil conservation uses such as pasture land.

law of demand   The quantity of a product demanded will vary inversely with the price of that product.

law of diminishing returns   An economic concept which states that as additional units of one input are combined with a fixed amount of other inputs, a point is always reached at which the additional product received from the last unit of added input will decline.

law of supply   The quantity of goods and/or services offered to a market will vary directly with the price.

least-cost combination of inputs   The least expensive combination of inputs used to produce a given output level, determined where MRS = IPR.

liquidity   The ability of a business to meet its financial commitments as they come due.

long-run   A time span such that no inputs are fixed; that is, all inputs are variable.

loss minimization prices   Revenues are covering the variable costs of production, but not all of the fixed costs so the firm is operating at a loss.

macroeconomic policies   Policies implemented to promote economic growth and temper business cycles, and include monetary and fiscal policy actions taken by the federal government and the Federal Reserve Bank.

management   Consists of making decisions based on the ideas of choice between alternatives; and the mechanism by which choices are made is often, if not always, based on economic analysis and evaluation.

marginal cost (MC)   The increase in total cost necessary to produce one more unit of output; it is calculated as the change in total cost divided by the change in output.

marginal factor cost (MFC)   The cost of an additional unit of input and is the amount added to total cost from using one more unit of variable input.

marginal physical product (MPP)   The amount of additional (marginal) total physical product obtained from using an additional (marginal) unit of variable input.

marginal rate of product substitution (MRPS)   Describes the rate one output must be decreased as production of the other product is increased, and, by definition, is the slope of the production possibilities frontier.

marginal rate of substitution (MRS)   The rate one input can be decreased as use of the other input increases and is the slope of the isoquant.

marginal revenue (MR)   The addition to total revenue from selling (producing) one more unit of output.

marginal revenue product (MRP)   Indicates the additional (marginal) value of output obtained from each additional unit of the variable input.

market   An institution or arrangement that brings buyers and sellers together.

market demand/supply   The horizontal summing (adding up) of all the demand/supply curves for the individual consumers/firms in the market.

market equilibrium   When the quantity of a good offered by sellers at a given price equals the quantity buyers are willing and able to purchase at that same price and no conditions exist to move the market away from this point.

market models   Classification scheme used by economists to describe the characteristics and firm behavior of different market structures observed in the economy.

market price   Mutually agreeable price at which willing buyers and sellers exchange a good or product.

market solutions   Situations where problems of social costs or externalities can be solved directly by negotiations between the involved parties.

marketing   Encompasses all of the business activities performed in directing the flow of goods and services from the producer to the consumer or final user.

marketing margin/bill   The difference in the price the producer receives for the raw product and the price the consumer pays for the final product.

marketing quotas   A direct method to control how much of any crop is produced/marketed each year.

marketing system   Encompasses all of the activities performed on agricultural products as they move from the farm gate to the final consumer.

merchant middlemen   Those retailers and wholesalers who actually take title to the product and own it as they move it through the market system.

Modified Accelerated Cost Recovery System (MACRS)   Current standard procedure for handling depreciation on tangible property purchased after 1986.

monopolistic competition   A situation in which there are many sellers; when the product has many close, but differentiated substitutes; there is some freedom of entry or exit; and there is some availability of knowledge or information. Similar to monopoly, but with much less price flexibility.

monopoly   A situation in which there is a single seller of a product; when the product has no close substitutes and could be considered totally differentiated; there is no freedom of entry or exit; and there is an unavailability of knowledge or information.

negative returns   Occur when each additional unit of input added to the production process decreases the production level.

noncompetitive market   Market in which individual firms have some influence over the price they receive for their products or the price they pay for their inputs.

nonprice competition   Where competition is vigorous in areas other than price (i.e., customer service, warranties, etc.).

nonrecourse loans   Government-issued farm loans which state that if the borrower defaults, no recourse will be made to recover any loan losses.

normal (superior) good   A product whose consumption increases as income increases.

nuisance   Any interference by another party with your basic right to enjoy and use your property.

offensive management strategies   Management strategies that build, or try to build, market shares and disrupt stability.

oligopoly   A market structure in which production is generated by a few large firms that dominate the industry.

open market operations   Buying and selling bonds, changing the reserve requirement.

opportunity cost of money   Evaluation of present returns on money invested versus the "costs" of passing up other earning opportunities.

opportunity costs   The cost to an individual or firm of using a good for one purpose is equal to the value that the good could have earned in another use, its next best alternative.

optimal output   Profit maximizing output level, determined where MR = MC.

own-price elasticity   Responsiveness of a product to a change in its own price.

parity   An attempt to measure the value of farm products relative to nonfarm products to help farms achieve some economic equality.

perfect/pure competition   A market or industry with four general characteristics: large number of buyers and sellers, homogeneous products, freedom of entry and exit, and perfect information.

physical functions   Deal with the tangible change in appearance, availability, and location since the three functions are processing, storage, and transportation.

place utility   Involves getting the product to where it is desired by the consumer.

possession utility   Relates to transferring the product's ownership to the person who wants it.

present value   Today's value of sums.

price searcher   Describes the pricing behavior of the monopolist as he or she searches for the price the market will pay for his or her output.

price supports   A minimum price guaranteed to producers of a few crops.

price taker   An assumption whereby the firm can purchase a few units or many units of an input and the price of that input will remain unaffected.

price warfare   Hidden discounts or other inducements to attract sales, including rebates.

pricing efficiency   Describes how well the system reacts to changes in consumer demand and generates reaction and adjustments by the producer.

pricing power   A situation in which the individual firm has influence over the prices they receive for their product.

processors and manufacturers   They are principally concerned with form utility and earn a profit by transforming inputs into a more desirable product.

product differentiation   Making identical products seem different.

product–product decision   An approach in which the producer determines which product combination to produce that will maximize revenue with a given resource constraint.

production contracts   Contract guaranteeing delivery of a certain amount and quality of product at harvest negotiated before actual harvest.

production controls   Seek to estimate a market price that will generate a reasonable return on investment and then constrain supply so that price will hold on the market.

production function   The technical relationship between inputs and output indicating the maximum amount of output that can be produced using alternative amounts of variable inputs in combination with one or more fixed inputs under a given state of technology.

production possibilities frontier (curve)   A curve depicting all the combinations of two products that can be produced using a given level of inputs (or expenditure).

production schedule   A tabular summary of the input–output relationship.

production surface   The concept describing the "hill" of increased production obtained as we increase use of the variable inputs.

profit maximization   The economic objective of the firm.

recession   When the change in economic growth is negative for several months.

resource   An input provided by nature and modified by humans using technology to produce the goods and services that satisfy human wants and desires.

resource fixity   Occurs when the inputs in one production process have little value in alternative production processes so they remain in production.

restrictive monetary policy   When the Federal Reserve Bank restrains economic growth by selling bonds and increasing the discount rate in order to decrease the supply of money and credit.

restrictive fiscal policy   Policy involving reduced government spending and/or increased taxes.

risk   A situation in which the outcome is unknown, but the probability of alternative outcomes is known.

risk-adjusted interest rates   A method of adjusting the interest, or discount, rate to reflect the level of risk involved in a venture.

risk aversion   Attitude toward risk; someone who is risk averse avoids risk at all costs.

scarcity   Because resources are limited, the goods and services produced from those resources are also limited, which means consumers must make choices or trade-offs between different goods.

shortage   Occurs in a market when the current price is too low for producers to supply all that consumers are willing to purchase.

short-run (intermediate short-run)   A time span such that some factors are variable and some factors are fixed.

social costs   Costs imposed on certain segments of society due to the use of non-market resources.

speculative middlemen   Persons who buy and sell on their own account, but expect profits made from price movements rather than other market functions.

standards and regulations   Ways in which Congress and state governments have acted to prohibit or limit pollutants via outright bans or standards for air and water quality.

static analysis   Examination of the principles underlying market price determination at one point in time.

straight-line method   Depreciation method calculated by estimating the useful life of the input and value the input may have when salvaged (sold, traded-in, or junked) at the end of its useful life.

structural-evaluation approach   A marketing approach that evaluates the ultimate performance of the marketing system by examining the level of competition existing in the industry.

substitutability   Occurs when one unit of an input can be exchanged for another input on a consistent basis of one-to-one or some other unchanging ratio.

sum-of-the-year's-digits method   Depreciation method calculated by dividing the useful life of the asset by the sum-of-the-year's digits, which is multiplied by the asset's purchase price less salvage value to determine the depreciation taken in a given year.

supply   The amount of an item that will be available in an area or time frame at a specific price.

supply schedule/curve   The marginal cost curve above the minimum average variable cost curve.

surplus   Occurs in a market when the current price is too high for consumers to purchase all that is supplied.

systems approach   A marketing approach that emphasizes the system of marketing, dwelling on the interactions of subsystems rather than on individual functions or firms.

target price/deficiency payment   Government program in which participants receive a direct payment if the market price falls below the target price.

taxes, charges, and subsidies   May be used in an attempt to reduce pollution.

technical efficiency   Relates to how many physical inputs and how much of each is utilized in performing the marketing functions the consumer desires.

technological change   In the event a specific method of producing a product changes, the effect on the production function could be to raise the output level associated with each level of input.

time preference for money   Desire on the part of people to have money today rather than tomorrow, next week, or next year.

time utility   Consists of, or is created by, getting the product to the consumer when it is desired.

total cost (TC)   The sum of total fixed costs and total variable costs at each level of output.

total fixed costs (TFC)   Include both implicit and explicit costs of inputs that are fixed in the short-run and do not change as output level changes.

total outlay approach   Whereby the manager or producer simply calculates the total expenditure associated with each input combination and chooses the combination with the smallest total expenditure or dollar outlay in order to minimize costs.

total physical product (TPP)   The relationship that exists between output and one variable input, holding all other inputs constant. It is measured in physical terms and represents the maximum amount of output brought about by each level of input use.

total revenue (TR)   The amount of money received when the producer sells the product.

total revenue approach   Under this approach, the manager simply calculates the total revenue associated with each product combination and chooses the product combination that yields the highest revenue.

total revenue product (TRP)   Indicates the dollar value of the output produced from alternative levels of variable input.

total variable costs (TVC)   Include both implicit and explicit costs of inputs that are variable, or changeable, in the short-run and change as the level of output changes.

transactions costs   Costs in terms of time, money, and effort needed to solve widespread air or water pollution problems.

treadmill effect   New technology is constantly flowing into agriculture creating the potential for higher than normal profits for early adopters. As more producers adopt the new technology supply increases putting downward pressure on prices. Producers who have not yet incorporated the new technology have higher costs, therefore must either adopt it or become unable to compete.

uncertainty   A situation in which the probabilities of different outcomes are unknown.

unitary elastic   Whereby a change in price brings about an equivalent change in quantity.

utility   The consumer's satisfaction derived from consuming the product.

utility maximization   A decision-making process whereby the consumer must decide which products to consume, and which products not to consume, given the consumer's limited funds while striving to maximize the satisfaction from consumption.

value   The desire to obtain a good; often expressed in either monetary measure or societal worth.

variable costs   Those costs that do vary as output level changes are the costs associated with using the variable factors of production.

variable factors of production   Factors that vary as output levels change; that is, as the use of variable factors changes, output levels also change. Examples are seed, fertilizer, and number of employees.

INDEX